CALCULUS TWO: LINEAR AND NONLINEAR FUNCTIONS

CALCULUS TWO:

FRANCIS J. FLANIGAN

Department of Mathematics
University of California, San Diego

JERRY L. KAZDAN

Department of Mathematics
University of Pennsylvania

LINEAR
AND NONLINEAR
FUNCTIONS

Prentice-Hall, Inc., Englewood Cliffs, New Jersey

CALCULUS TWO: LINEAR AND NONLINEAR FUNCTIONS
Francis J. Flanigan and Jerry L. Kazdan

Current printing (last digit): 10 9 8 7 6 5 4 3

13–112334–3

Library of Congress Catalog Card Number: 70–142119

PRENTICE-HALL INTERNATIONAL, INC., *London*
PRENTICE-HALL OF AUSTRALIA, PTY. LTD., *Sydney*
PRENTICE-HALL OF CANADA, LTD., *Toronto*
PRENTICE-HALL OF INDIA PRIVATE LIMITED, *New Delhi*
PRENTICE-HALL OF JAPAN, INC., *Tokyo*

Printed in the United States of America

To our Parents

Preface

The heart of this book is the study of nonlinear functions by means of the simpler linear and affine functions. Our guide in this is the elementary calculus procedure of constructing the tangent line (a linear object) to a curve (a nonlinear object). Thus, we begin with a study of linear objects.

Linear algebra is just this study; it should be viewed as analytic geometry in modern dress. Without it, geometric ideas are lost in a fog of equations. We have developed the linear algebra by first studying the algebra of $\mathbb{R}^n$ (ordinary n dimensional space), especially $\mathbb{R}^2$ and $\mathbb{R}^3$, in Chapter 1. Geometry of $\mathbb{R}^n$, the inner product, is the subject of Chapter 2. Then we introduce the concept of a linear map in Chapter 3. And only after some familiarity with linear maps is acquired do we represent them as matrices and interpret the algebraic operations on maps in terms of the corresponding operations on matrices. This concludes the "linear" part of the book. Chapter 4 serves as a bridge from linear to nonlinear.

The differential calculus of several variables is taken up in Chapters 5, 6, 7, and 8. One measure of the success of approaching calculus of several variables using linear algebra is that students *anticipate* ideas and theorems. The criterion $f'(X) = 0$ for a maximum or minimum is but one example. This notation vastly simplifies formulas, enabling the concepts to emerge clearly.

Chapter 9 discusses multiple integrals. Instead of pushing the difficult ideas of existence of the integral, we have chosen to show how to estimate numerically the value of the integral. In this way students grasp the concepts and understand that a proof of the existence of the integral consists precisely in showing that the elementary estimation procedure can be made to yield as close an approximation as they desire. Some teachers may wish to use computers at this point of the course.

Line integrals and Stokes' Theorem are the subject of Chapter 10. We keep to $\mathbb{R}^2$, since higher dimensions force well-known complications we wish to avoid. One temptation was to introduce differential forms and exhibit the elegant modern notation for this theorem—at least in $\mathbb{R}^2$. Unfortunately, the formalism seems to swamp all but the very best students. Thus, we have taken the classical approach. However, we have tried to give some nontrivial applications. All too many students know the modern version with differential forms yet they know no genuine use of the result and have no idea why it is important (except that their teachers say so).

Scattered throughout the book are a number of "blurbs." These are very short sections that either illuminate some idea in the text, present an application, or else introduce a more advanced topic. We encourage the student to read these on his own.

This book is intended for the—possibly mythical—"average" student. Two and three dimensional cases are emphasized. We have chosen to write in an informal lecture note style with questions directed to the reader. Hopefully we have not insulted the reader's intelligence and "driven in the hammer along with the nail." But the reader should feel free to skip. This skipping is important, both over material that seems too difficult and that which seems dull and repetitive. Some teachers (including us) will often go much faster than the text, leaving the gap as reading material. Our goal throughout has been to give a feeling for the ideas involved, not just computational facility—although that is certainly not underplayed. We also hope this book is interesting. In order to keep the level of difficulty even, we have deleted most of the technical proofs, such as equality of mixed partial derivatives.

We place a great importance on the organization of this book. The logical development should be clear from the Table of Contents. This is actually an outline of the subject and should itself be used for review purposes.

Sins of omission: We do not discuss change of basis, eigenvalues, determinants (except the 2×2 case), function spaces (except for a blurb), technicalities about the real numbers, inverse or implicit function theorems, change of variables in multiple integrals (except for polar coordinates), or give the proofs for the harder theorems. These topics are more difficult and are left for more advanced courses in algebra or analysis.

This book originated as a more difficult book* which turned out to be unrealistic for many second year students, mainly due to the presence of function spaces. A thorough rethinking has led to the present volume. Many colleagues and students have generously assisted us. We give special thanks to Lawrence Corwin, Jack Gomberg, and David Ragozin. Robert Martin and James Walsh of Prentice-Hall have also been a pleasure to know and work with.

FRANCIS J. FLANIGAN
JERRY L. KAZDAN

*Kazdan, J., *Lectures on Analysis with Linear Algebra*, Prentice-Hall (to be published).

SUGGESTIONS FOR USING THIS BOOK

There is a great amount of flexibility in this book. However, it would be unwise to interchange the order of the first six chapters. Our courses from the book take roughly twenty weeks, about two thirds of an academic year (the remaining third of the year could be spent on differential equations or probability or statistics). However, there is a sufficient amount of material for a full year course. We imagine spending 7 to 10 weeks on linear algebra and 8 to 12 weeks on calculus. A variety of courses are possible, depending on the background and needs of the students. Unless stated, the outlines below take two thirds of an academic year. Of course, all "blurbs" and appendices are supplementary.

Science and Engineering Students: delete 1.4, 2.3, 2.4 (but suggest they read 2.4), 3.4i,j, 6.4, 7.2, 7.3, 9.3, all of 10; cover Chapter 4 and Chapter 9.1 quickly. A shorter course could also delete 6.1, 6.2, 8.1, 8.2, 9.1, and 9.2c,d.

Social Science Students: delete 1.4, 2.3, 2.4, all of 7, 8.3, 9.3, all of 10; cover Chapter 4 and Chapter 9.1 quickly. A shorter course could also delete all of Chapters 8 and 9.

Students knowing linear algebra: Chapters 1, 2, and 3 should be reviewed quickly to fix the notation. Then follow one of the courses outlined above.

REMARK: Answers for roughly half of the Exercises are included at the end of the book. A pamphlet containing most of the remaining answers is available for teachers by writing to the publisher.

Contents

0. Remembrance of Things Past

This chapter is included for reference. In it, we shall discuss some material on sets and functions that you may have seen already. You may discover that it is not necessary to read the chapter through but merely to refer to it occasionally, when you encounter an unfamiliar term or notation.

0.1 SETS

0.1a Sets

We describe a "set" as follows: A **set** is any collection of objects, together with a criterion for deciding whether a given object is in the set. The objects in the set are called the **elements** or **members** of the set.

For example, (1) the set of all girls with brown eyes and red hair and (2) the less picturesque set of all positive even integers.

A set may sometimes be specified by actually listing all its elements. Thus, the set of all students in some class is specified by the list of names in the roll book. The **empty set** is the set that has no elements at all.

There is a standard notation we sometimes use to describe sets. Here are some examples (note the names $\mathscr{A}, \mathscr{S}, \mathscr{B}$):

1. $\mathscr{A} = \{x \mid x \text{ is an odd integer}\}$ is the set of all odd integers.
2. $\mathscr{S} = \{(x, y) \mid x^2 + y^2 = 1\}$ is the set of all points (x, y) on the unit circle $x^2 + y^2 = 1$; the vertical bar may be read "such that."
3. $\mathscr{B} = \{-3, 1, 2, 7\}$ is the set whose elements are the integers -3, 1, 2, and 7.

Note that we use capital (often script) letters to denote sets.

Here is some further notation. If $\mathscr{S}$ is any set (not necessarily the unit circle above), we write:

1. $x \in \mathscr{S}$ if x is an element of $\mathscr{S}$
2. $x \notin \mathscr{S}$ if x is not an element of $\mathscr{S}$

For example, if $\mathscr{A}$ is the set of odd integers mentioned above, then

$$3 \in \mathscr{A}, \quad -11 \in \mathscr{A}, \quad 4 \notin \mathscr{A}, \quad \frac{2}{3} \notin \mathscr{A}.$$

0.1b Subsets

The term "subset" is a way of referring to a portion of a given set. Formally, $\mathscr{A}$ is a **subset** of $\mathscr{S}$, or $\mathscr{A}$ is **contained** in $\mathscr{S}$, written

$$\mathscr{A} \subset \mathscr{S},$$

if and only if every element of $\mathscr{A}$ is also an element of $\mathscr{S}$. Note that, according to this definition, the set $\mathscr{S}$ is a subset of itself, $\mathscr{S} \subset \mathscr{S}$, and also the empty set is a subset of every set $\mathscr{S}$. If $\mathscr{A}$ is a subset of $\mathscr{S}$ that is different both from $\mathscr{S}$ and from the empty set, we say that $\mathscr{A}$ is a **proper subset** of $\mathscr{S}$.

For example, if $\mathscr{S} = \{1, 2, 3, 4, 5\}$, then some proper subsets of $\mathscr{S}$ are

$$\{1, 2, 3\}, \qquad \{1, 4\}, \qquad \{3\}, \qquad \{3, 4\}.$$

There are others. Can you find some?

0.1c The set of real numbers; the sets $\mathbb{R}^n$

We assume that you have some familiarity with the very important set of real numbers. We designate this set throughout this book by $\mathbb{R}$. You may think of $\mathbb{R}$ as the set of all decimals (positive and negative) such as

$$-2 = -2.000 \ldots, \qquad \frac{1}{3} = 0.333 \ldots, \qquad \pi = 3.14159 \ldots.$$

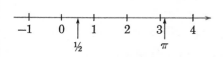

The real line $\mathbb{R}$

FIGURE 0.1

We will often be regarding $\mathbb{R}$, however, as the set of points on a line or axis, the so-called real line or x-axis from ordinary calculus. See Fig. 0.1. To every point on this line corresponds a unique real number, and vice versa. This correspondence between numbers and points on a line is a source of geometric insight we intend to exploit.

We define the **cartesian product** (after Descartes) of $\mathbb{R}$ with itself to be the set, denoted $\mathbb{R} \times \mathbb{R}$, of all ordered pairs (x, y), where $x \in \mathbb{R}$, $y \in \mathbb{R}$. We remark that the order in (x, y) is important. Thus $(1, 0) \neq (0, 1)$ and, in general, $(x, y) = (a, b)$ if and only if $x = a$, $y = b$. The elements of $\mathbb{R} \times \mathbb{R}$ are often called **ordered pairs**.

Another notation for $\mathbb{R} \times \mathbb{R}$ is $\mathbb{R}^2$. This is the one we will use.

Note that we need not stop at the product of $\mathbb{R}$ with itself. We may form cartesian products with more copies of $\mathbb{R}$:

$$\mathbb{R}^3 = \mathbb{R} \times \mathbb{R} \times \mathbb{R} = \{(x, y, z) \,|\, x, y, z \in \mathbb{R}\},$$

$$\mathbb{R}^4 = \mathbb{R} \times \mathbb{R} \times \mathbb{R} \times \mathbb{R} = \{(x_1, x_2, x_3, x_4) \,|\, x_1, x_2, x_3, x_4 \in \mathbb{R}\},$$

and so on. Thus, if n is any positive integer, we define

$$\mathbb{R}^n = \{(x_1, \ldots, x_n) \,|\, x_1, \ldots, x_n \in \mathbb{R}\}.$$

Some elements of $\mathbb{R}^3$ are $(1, 0, 0)$, $(0, 0, 0)$, $(4, -1, \frac{1}{2})$, for example, and $(1, -1, 2, 0, 5, \frac{1}{3}) \in \mathbb{R}^6$.

In Chap. 1, we will call the elements of $\mathbb{R}^n$ **points** or **vectors** and learn a way to add them.

0.1d Some notation from logic

The symbol $\Rightarrow$ is to be read "implies." For example, "x is an even integer between 1 and 3 $\Rightarrow x = 2$" or "y is an integer of the form $2x + 1$, where x is an integer $\Rightarrow y$ is odd."

We also use the symbol $\Leftrightarrow$, which is a quick way of writing "if and only if." Thus "a real number y is nonnegative $\Leftrightarrow$ it is of the form $y = x^2$ for some real number x." You might regard the symbol $\Leftrightarrow$ as being composed of $\Rightarrow$ and $\Leftarrow$. Thus "$P \Leftrightarrow Q$" means "P implies Q and also Q implies P." In other words, the two statements or sentences P and Q are **equivalent**; one of them is true if and only if the other is true.

0.2 FUNCTIONS

0.2a Functions

Let $\mathscr{A}$ and $\mathscr{B}$ be sets. A **function** f from $\mathscr{A}$ to $\mathscr{B}$, written

$$f : \mathscr{A} \longrightarrow \mathscr{B} \qquad \text{or} \qquad \mathscr{A} \xrightarrow{f} \mathscr{B}$$

is a rule that assigns to each $x \in \mathscr{A}$ one and only one element $y = f(x)$ $\in \mathscr{B}$. Some synonyms for "function" are "map," "mapping," "transformation," "operator." The set $\mathscr{A}$ is called the **domain** of the function f, and $\mathscr{B}$ the **target** of f. The subset of $\mathscr{B}$ given by

$$\{y \in \mathscr{B} \mid y = f(x) \quad \text{for some} \quad x \in \mathscr{A}\}$$

is the **image** of f. A good notation for this set is $f(\mathscr{A})$.

For example, let $f : \mathbb{R} \longrightarrow \mathbb{R}$ be given by the rule $f(x) = x^2$. Thus f assigns to each real number x its square. The domain of f is $\mathbb{R}$, the target is $\mathbb{R}$, and the image is the set of all nonnegative real numbers $y \geq 0$, since $x^2 \geq 0$. This example is familiar from ordinary calculus.

Here is another example, this time of a function whose target is $\mathbb{R}^2$. Let $g : \mathbb{R} \longrightarrow \mathbb{R}^2$ be given by $g(x) = (x + 1, x - 1)$. What is the domain? the image?

0.2b The composition of functions

Suppose that $f : \mathscr{A} \longrightarrow \mathscr{B}$ and also $g : \mathscr{B} \longrightarrow \mathscr{C}$ are two functions. We write $\mathscr{A} \xrightarrow{f} \mathscr{B} \xrightarrow{g} \mathscr{C}$. Since, for each $x \in \mathscr{A}$, the value $f(x)$ is an element of

$\mathscr{B}$, the domain of the function g, we may apply g to $f(x)$ to get $g(f(x)) \in \mathscr{C}$. The function that assigns to each $x \in \mathscr{A}$ the value $g(f(x)) \in \mathscr{C}$ is the **composition** of f and g. Note how important it was in defining the composition that the image of f was contained in the domain of g.

For example, let $f(x) = x^2$ and $g(x) = (x + 1, x - 1)$ be the functions given above. Then we have $\mathbb{R} \xrightarrow{f} \mathbb{R} \xrightarrow{g} \mathbb{R}^2$ and $g(f(x)) = (x^2 + 1, x^2 - 1)$ is a formula for the composition of f and g. Thus, if $x = 2$, $g(f(2)) = (2^2 + 1, 2^2 - 1) = (5, 3)$. Another cleaner notation for $g(f(x))$ is $g \circ f$.

1. The Algebra of $\mathbb{R}^n$

1.0 INTRODUCTION

Let us begin with an overview. The study of the differential calculus of $y = f(x)$, which you have already completed, may be divided into four parts:

1. Basic arithmetic and algebra. You learned the usual operations with numbers and equations.

2. Geometry and distance. The distance between two points on the x-axis and in the xy-plane was defined. This was essential to the notion of limit in calculus.

3. Linear functions. You studied straight lines in the xy-plane, relatively simple functions of the form $y = mx + b$.

4. Nonlinear functions. Finally, you studied more general functions— $(y = f(x) = x^2$, for example)—by constructing the tangent line (a linear function) to the graph of $y = f(x)$ above a point $x = x_0$. See Fig. 1.1.

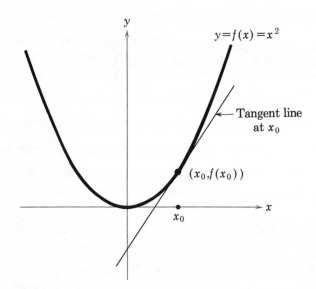

FIGURE 1.1

The slope m of this tangent line was given by the first derivative, $m = f'(x_0)$. Having "approximated" the curved graph $y = f(x)$ by the tangent line above x_0, you were able to deal successfully with several natural problems: maximum and minimum, rate of change of $f(x)$, and approximate value of $f(x)$ for x near x_0.

In the chapters that follow, we repeat this four-stage process. Now, however, we wish to study functions of *several* variables. A useful example is

$$f(x_1, x_2) = x_1^2 + x_2^2,$$

which is, as we shall see, a real-valued function of the two real variables x_1 and x_2. It is customary to think of the pair (x_1, x_2) as a "point" or "vector," denoted X, in the "vector space" consisting of all such pairs. We thereby have $f(x_1, x_2)$, also written $f(X)$, which should remind us of the ordinary (one-variable) calculus. Now, however, our variable is a vector $X = (x_1, x_2)$. The four-stage process is as follows:

1. Vector algebra. You will study addition, subtraction, and the solution of equations in vectors (Chap. 1).

2. Vector geometry. A natural concept of distance between vectors will be defined. This will provide us later with the notion of "limit," essential to calculus, and give meaning to such phrases as "let the vector X approach the vector X_0" (Chap. 2).

3. Linear functions. You will study these, and the slightly more general "affine" functions, in Chap. 3. Roughly speaking, the graph of an affine function is a plane, just as the graph of $y = mx + b$ in the one-variable case is a straight line.

4. Nonlinear functions. Finally, we will have developed enough machinery for the differential calculus of functions like $f(X) = f(x_1, x_2) = x_1^2 + x_2^2$, mentioned above. Just as in one-variable calculus, in which we constructed a tangent line to the parabola $y = x^2$ above a typical point $x = x_0$, now we will construct a tangent plane (the "best affine approximation") to the graph (a surface) of $y = f(X)$ above a typical point X_0. See Fig. 1.2.

Once we can construct tangent planes, we will be equipped to deal with the standard questions about $f(X)$: maxima and minima, rates of change, and approximate values (Chap. 4 and following).

Some words of encouragement: The brief comments given here outline an ambitious program and an important one. The concepts of "vector," "vector space," "linear function," and "affine approximation" are of supreme importance in modern mathematics and are essential to modern physics, engineering, and economics as well. You should not be disheartened if certain statements in the outline seem bizarre or incomprehensible. As we pass through the four stages, the outline will be repeated and amplified.

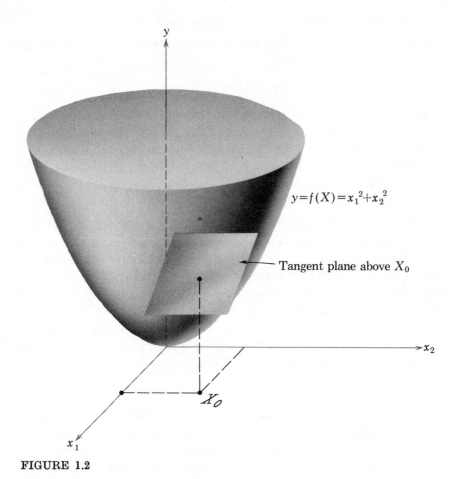

FIGURE 1.2

1.1 THE SPACE $\mathbb{R}^2$

1.1a Introduction

Let us examine informally the space $\mathbb{R}^2$ of two dimensions (whatever this means). This will serve to motivate the "higher-dimensional" spaces in the following section. We construct $\mathbb{R}^2$ as follows. Let $\mathbb{R}$, as usual, denote the set of real numbers. Then $\mathbb{R}^2$ is the so-called cartesian product, $\mathbb{R} \times \mathbb{R}$, of the set $\mathbb{R}$ with itself,

$$\mathbb{R}^2 = \{(x_1, x_2) \,|\, x_1, x_2 \in \mathbb{R}\}.$$

Thus an element X of $\mathbb{R}^2$ is an ordered pair, $X = (x_1, x_2)$. These elements are the **points** or **vectors** of $\mathbb{R}^2$. Some examples of vectors are $(3, 0)$, $(-4, \sqrt{13})$, $(\frac{1}{2}, -10)$, $(1, 1)$.

If $X = (x_1, x_2)$, the real numbers x_1 and x_2 are the **coordinates** of the vector X. For example, if $X = (-4, \sqrt{13})$, then its first coordinate is -4 and its second coordinate is $\sqrt{13}$. Note that if $X = (x_1, x_2)$ and $Y = (y_1, y_2)$ are vectors in $\mathbb{R}^2$, then $X = Y$ if and only if $x_1 = y_1$ and $x_2 = y_2$.

You are probably familiar with the pictorial representation of $\mathbb{R}^2$ as a plane, with straight-line axes intersecting at right angles at the point $(0, 0)$. It is customary to use the horizontal axis as the axis of first coordinates. See Fig. 1.3.

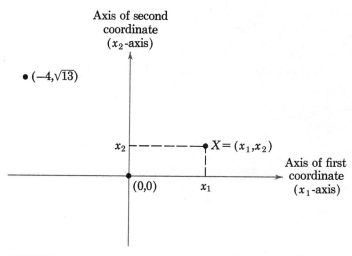

FIGURE 1.3

1.1b Algebra in $\mathbb{R}^2$

So far $\mathbb{R}^2$ is merely a set. Let us propose a reasonable algebraic structure for $\mathbb{R}^2$. If $X = (x_1, x_2)$ and $Y = (y_1, y_2)$ are any two vectors and if α is any real number, or, as we shall say from now on, a **scalar**, we define the addition of vectors as

$$X + Y = (x_1 + y_1, x_2 + y_2)$$

and the multiplication by scalars as

$$\alpha X = (\alpha x_1, \alpha x_2).$$

Note how these two operations involving vectors take advantage of the fact that we already know how to add and multiply real numbers.

EXAMPLES: Let $X = (1, 5)$, $Y = (2, -1)$, $\alpha = 7$. Then $X + Y = (1 + 2, 5 - 1) = (3, 4)$, $\alpha X = (7, 35)$, $\alpha Y = (14, -7)$, $X + \alpha Y = (15, -2)$, and so on.

Some immediate consequences of these definitions are, for all vectors X, Y, Z in $\mathbb{R}^2$:

1. Addition is **associative**: $(X + Y) + Z = X + (Y + Z)$.
2. Addition is **commutative**: $X + Y = Y + X$.
3. There is an **additive identity** or **zero element**: a vector O that satisfies $X + O = X$ for all X. In $\mathbb{R}^2$, $O = (0, 0)$.
4. Each vector X has an **additive inverse** $-X$ ("minus X") satisfying $X + (-X) = O$. If $X = (x_1, x_2)$, then $-X = (-x_1, -x_2)$. Thus the additive inverse of $(1, -3)$ is $(-1, 3)$.

You should convince yourself that these properties do hold in $\mathbb{R}^2$. In each case they follow from the corresponding property of the real numbers $\mathbb{R}$.

The following properties are true for multiplication by scalars α and β:

5. $\alpha(\beta X) = (\alpha\beta)X$.
6. $1X = X$.

Finally we state the **distributive laws**, which describe the relationships between the two different operations:

7. $(\alpha + \beta)X = \alpha X + \beta X$.
8. $\alpha(X + Y) = \alpha X + \alpha Y$.

To ensure that you too feel that these properties are obvious, let us prove one, say 7. We have

$$
\begin{aligned}
(\alpha + \beta)X &= (\alpha + \beta)(x_1, x_2) = ((\alpha + \beta)x_1, (\alpha + \beta)x_2) \\
&= (\alpha x_1 + \beta x_1, \alpha x_2 + \beta x_2) \\
&= (\alpha x_1, \alpha x_2) + (\beta x_1, \beta x_2) \\
&= \alpha(x_1, x_2) + \beta(x_1, x_2) \\
&= \alpha X + \beta X,
\end{aligned}
$$

as desired. Note that the proof consisted of replacing X by its coordinate representation (x_1, x_2) and then using the definition of multiplication by scalars and the well-known distributive law for real numbers in each coordinate slot. Properties 1 to 8, the **vector space axioms**, are not spectacular but are absolutely essential. We will meet them throughout this chapter.

1.1c Vectors as arrows

Instead of thinking of the elements (x_1, x_2) in $\mathbb{R}^2$ as points, it is sometimes useful to think of them as directed line segments or arrows with the tail at the origin $(0, 0)$ and the head at the point (x_1, x_2). The student should verify that $X + Y$ is the vector that is the diagonal of the parallelogram determined by the arrows to X and Y. See Fig. 1.4.

Multiplication of a vector X by a scalar α may be visualized as a stretching or shrinking of the arrow representing X. Thus $3X$ may be thought of as an arrow starting at $(0, 0)$ and pointing in the same direction as X but three times as long (admittedly, we do not yet have a rigorous definition of the length of a vector, but this should cause no trouble). And if $\alpha < 0$, then αX should point in the opposite direction to X. See Fig. 1.5.

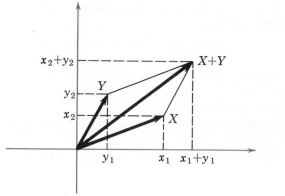

FIGURE 1.4

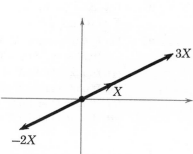

FIGURE 1.5

Exercises

These give you practice with some basic computations in $\mathbb{R}^2$.

1.   Let $X = (1, 2)$, $Y = (-1, 3)$, $Z = (0, 4)$. Compute:

(a) $X + Y$,   (e) $-3Y$,

(b) $X - Y$,   (f) $4X - 3Y$,

(c) $X + Y + Z$,   (g) $X + 2Y - Z$,

(d) $4X$,   (h) $4(X + Y) - 5Z$.

2.   Given $Y = (2, 3)$, $Z = (1, -4)$. In each equation below, compute a vector X that satisfies it:

(a) $3X + 2Y - Z = O$,   (c) $Z - 2X = 5Y$,

(b) $Z - X = Y$,   (d) $Z - 2X - 5Y = O$.

3.   Let $X = (1, 2)$, $Y = (-1, 3)$, as in Exercise 1. Locate them as points in

the x_1x_2-plane (sketch). Then locate $X + Y$, $X - Y$, $2X - 2Y$. Finally, prove that the points O, X, Y, $X + Y$ are the corners of a parallelogram in $\mathbb{R}^2$.

4. A preview. How long is the vector $Z = (1, 2)$; that is, what is the distance from $(0, 0)$ to $(1, 2)$? Can you think of a formula for the length of a typical vector $X = (x_1, x_2)$?

5. A preview of higher-dimensional spaces. How should we define $\mathbb{R}^3$; $\mathbb{R}^4$? What should a typical X in $\mathbb{R}^3$ be? How should we define the sum $X + Y$ if X and Y are vectors in $\mathbb{R}^3$?

6. Let $e_1 = (1, 0)$, $e_2 = (0, 1)$. Show that for every $X \in \mathbb{R}^2$ you can find (unique) scalars $\alpha_1, \alpha_2 \in \mathbb{R}$ such that $X = \alpha_1 e_1 + \alpha_2 e_2$. This property of e_1, e_2 will be used frequently.

7. (a) Let $Y_1 = (3, 0)$, $Y_2 = (2, 0)$. Is it possible, given any $X \in \mathbb{R}^2$, to find α_1, α_2 such that $X = \alpha_1 Y_1 + \alpha_2 Y_2$ (compare Exercise 6)? Give your reason (to begin with, what if $X = (1, 1)$?).
    (b) Same question for $Z_1 = (1, 2)$, $Z_2 = (-4, -8)$ instead of Y_1, Y_2. A sketch might clarify the issue here.

8. Let $X = (3, 2)$, $X_1 = (1, 2)$, $X_2 = (-2, -3)$. Find scalars α, β so that $X = \alpha X_1 + \beta X_2$.

9. Let a, b be scalars, X, Y vectors. Use only Properties 1 to 8 for the following (that way, they will be true for any vector space):
    (a) If $a \neq 0$ and $aX = O$, then $X = O$.
    (b) If $a \neq 0$ and $aX = aY$, then $X = Y$.
    (c) If $X \neq O$ and $aX = O$, then $a = 0$.
    (d) If $X \neq O$ and $aX = bX$, then $a = b$.

1.2 THE SPACE $\mathbb{R}^n$

1.2a Introduction

This will be a straightforward generalization of $\mathbb{R}^2$. From now on, the letter n stands for a positive integer. We define $\mathbb{R}^n$ to be the cartesian product of the real numbers $\mathbb{R}$ with itself n times: $\mathbb{R}^n = \mathbb{R} \times \mathbb{R} \times \cdots \times \mathbb{R}$ (n factors). Thus a typical element X of $\mathbb{R}^n$ is an "ordered n-tuple" $X = (x_1, x_2, \ldots, x_n)$ with $x_j \in \mathbb{R}$ for $j = 1, 2, \ldots, n$. As in the case of $\mathbb{R}^2$, X is called a **point** or **vector**. If $X = (x_1, x_2, \ldots, x_n)$ and $Y = (y_1, y_2, \ldots, y_n)$, then $X = Y$ if and only if the corresponding **coordinates** x_j and y_j are equal: $x_j = y_j$ for each index subscript $j = 1, 2, \ldots, n$.

1.2b Algebraic operations in $\mathbb{R}^n$

This parallels our work with $\mathbb{R}^2$. If $X = (x_1, x_2, \ldots, x_n)$ and $Y = (y_1, y_2, \ldots, y_n)$ are vectors in $\mathbb{R}^n$ and α is any scalar, we define addition by

$$X + Y = (x_1 + y_1, x_2 + y_2, \ldots, x_n + y_n)$$

and multiplication by scalars thus:

$$\alpha X = (\alpha x_1, \alpha x_2, \ldots, \alpha x_n).$$

For example, let $n = 3$. Then $X = (1, 0, \frac{1}{2})$ and $Y = (-1, 4, 4)$ are vectors in $\mathbb{R}^3$. We have $X + Y = (0, 4, \frac{9}{2})$, and if $\alpha = 7$, $\alpha X = (7, 0, \frac{7}{2})$, $\alpha Y = (-7, 28, 28)$, $\alpha X + Y = (6, 4, \frac{15}{2})$, and so on.

Important. We do not add vectors from different spaces $\mathbb{R}^n$ and $\mathbb{R}^q$ with $n \neq q$. Thus $(1, 0, \frac{1}{2}) \in \mathbb{R}^3$ and $(0, 6, 1, -1) \in \mathbb{R}^4$ cannot be added.

1.2c The vector space properties

Note that Properties 1 to 8 listed for $\mathbb{R}^2$ remain valid in $\mathbb{R}^n$ and with the proofs essentially unchanged (just add dots . . . inside the parentheses). This is because we have defined $X + Y$ and αX in coordinates. For instance, $(x_1, x_2, x_3) + (y_1, y_2, y_3) = (x_1 + y_1, x_2 + y_2, x_3 + y_3)$. Therefore all the proofs can be reduced to proofs within a single coordinate slot, that is, within the real numbers.

In the preceding section, we referred to Properties 1 to 8 as the "vector space axioms." What is a vector space? Roughly speaking, a vector space is any set in which we can add elements and multiply by real scalars, just as we do in $\mathbb{R}^2$ or $\mathbb{R}^n$. To be precise, a set $\mathscr{S}$ is a **vector space** provided that:

I. Addition is defined: given $X, Y \in \mathscr{S}$, an element $X + Y$ is uniquely specified in $\mathscr{S}$.

II. Multiplication by scalars is defined: given $X \in \mathscr{S}$ and $\alpha \in \mathbb{R}$, an element $\alpha X \in \mathscr{S}$ is uniquely specified.

III. The operations defined in (I) and (II) satisfy Properties 1 to 8; that is, for all $X, Y, Z \in \mathscr{S}$ and $\alpha, \beta \in \mathbb{R}$, we have:

1. Addition is associative: $(X + Y) + Z = X + (Y + Z)$.
2. Addition is commutative: $X + Y = Y + X$.
3. Existence of zero: $\mathscr{S}$ contains an element O satisfying $X + O = X$.
4. Existence of additive inverses: for each $X \in \mathscr{S}$ there is an element $-X$ satisfying $X + (-X) = O$.
5. $\alpha(\beta X) = (\alpha\beta)X$.
6. $1X = X$.
7. $(\alpha + \beta)X = \alpha X + \beta X$.
8. $\alpha(X + Y) = \alpha X + \alpha Y$.

We already know that $\mathbb{R}^2$ and $\mathbb{R}^n$ are vector spaces. Also, the set $\mathbb{R}$ of real numbers is itself a vector space (one-dimensional, as we shall see). If these were the only examples we might encounter, we would not need the abstract definition of vector space. But we will be dealing with many more vector spaces. A humble but very useful specimen is the space consisting of the single element O—the origin of $\mathbb{R}^2$, for example. You might check this. Moreover, we shall see that such natural objects as (unbounded) straight lines through the origin in $\mathbb{R}^2$, lines and planes through the origin in $\mathbb{R}^3$, and so on, are all vector spaces themselves. Thus, the notion of "vector space" serves to describe a variety of mathematical phenomena.

1.2d Pictures

In the preceding section we viewed $\mathbb{R}^2$ as a plane. Is there a similar representation for $\mathbb{R}^n$ with $n > 2$?

If $n = 3$, the answer is yes. The standard picture for $\mathbb{R}^3$, with $X = (x_1, x_2, x_3)$ a typical vector, is as given here. See Fig. 1.6.

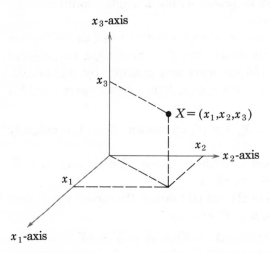

FIGURE 1.6

If $n > 3$, most of us (a few mathematicians excepted) cannot visualize $\mathbb{R}^n$. Nonetheless, it is possible and sometimes necessary to deal with such "higher-dimensional" spaces. In fact, much of modern mathematics occurs in "infinite-dimensional" spaces by scientific necessity, in order to understand aspects of our three-dimensional world (or is it four?). In dealing with these higher-dimensional spaces, we must rely on two things: our spatial intuition in $\mathbb{R}^2$ and $\mathbb{R}^3$, used with care, and our algebraic precision. For you should note that we have defined these spaces $\mathbb{R}^n$ as *algebraic* structures. We will reduce problems therein to algebra and, later, calculus, not geometric argument. Now geometrically it is natural for us to picture three dimensions and impossible for us to picture five. But algebraically we can solve linear equations in x_1, x_2, x_3, x_4, x_5 almost as easily as in x_1, x_2, x_3; there is no essential difference.

REMARKS: 1. In the terminology of the Introduction we are still in stage one, acquiring the basic vector algebra we will need for our purposes. Recall that eventually we wish to study functions $f(X)$ of a *vector* variable $X = (x_1, \ldots, x_n)$ for some n.

2. We have defined the sum, but not the product, of two vectors in $\mathbb{R}^n$; we have no rule for multiplying X by Y to get a third vector. In general, there is no such well-behaved product, although there *is* a natural product in certain very special cases. We shall soon encounter, however, the "inner" product $\langle X, Y \rangle$ of vectors X and Y, but this gives a number, not a new vector.

Exercises

Some basic calisthenics in vector algebra.

1. Let $X = (1, 2, -2)$, $Y = (0, 2, 1)$, $Z = (1, -2, 2)$ in $\mathbb{R}^3$. Compute:

    (a) $X + Y$,   (e) $2X + 4Y$,

    (b) $X - Y$,   (f) $3X - 2Y + Z$,

    (c) $X + Z$,   (g) $7(X - Y)$.

    (d) $X - Y - Z$,

2. Let $Y = (1, 2, -1, 1)$, $Z = (0, 2, -1, 0)$ in $\mathbb{R}^4$. Compute a vector X satisfying:

    (a) $X = Y + Z$,   (d) $Z - 3X - Y = 0$,

    (b) $2X + Y - Z = 0$,   (e) $7X + 7Y + 7Z = 0$.

    (c) $Y - X = 2Z$,

3. Sketch a portion of $\mathbb{R}^3$ as in the text (horizontal x_1x_2-plane, vertical x_3-axis, seen from the first octant $x_1 \geq 0$, $x_2 \geq 0$, $x_3 \geq 0$). Locate the vectors $O = (0, 0, 0)$, $(3, 0, 0)$, $(0, 1, 0)$, $(0, 0, -1)$, $(1, 1, 0)$, $(0, 1, 2)$, $(1, 1, 1)$, $(1, -1, 1)$.

4. (a) Define vectors

$$e_1 = (1, 0, 0), \qquad e_2 = (0, 1, 0), \qquad e_3 = (0, 0, 1).$$

Show that given any $X = (x_1, x_2, x_3)$, there exist scalars $\alpha_1, \alpha_2, \alpha_3$ such that $X = \alpha_1 e_1 + \alpha_2 e_2 + \alpha_3 e_3$. Note that these scalars are uniquely determined by X.

    (b) Let $X = (1, -2, 3)$. Find a, b, c so that $X = ae_1 + be_2 + ce_3$.

5. Let $Y_1 = (1, 2, 0)$, $Y_2 = (1, 1, 0)$, $Y_3 = (-1, 4, 0)$. Is it possible to find for every $X \in \mathbb{R}^3$ scalars $\alpha_1, \alpha_2, \alpha_3$ so that $X = \alpha_1 Y_1 + \alpha_2 Y_2 + \alpha_3 Y_3$? For instance, what if $X = (1, 3, 1)$?

6. (a) Let $\mathscr{S} = \{X = (x_1, x_2, x_3) \mid x_1 - x_2 = 0\}$. Show that the sum of any two vectors in $\mathscr{S}$ is also in $\mathscr{S}$.

(b) Is this true if the equation determining the set is $x_1 + 2x_2 - x_3 = 0$? State your reasons.

(c) If the equation is $x_1 = 4$?

(d) For what kinds of linear equations of the general form $ax_1 + bx_2 + cx_3 = d$ is it true that if the coordinates of two vectors X, X' satisfy this equation, then the coordinates of the sum $X + X'$ satisfy the same equation? Give conditions on a, b, c, d.

7. Sketch in $\mathbb{R}^2$ the set of all points $X = (x_1, x_2)$ such that $x_1 + x_2 = 0$. Likewise for $x_1 + x_2 = 1$.

8. Let e_1, e_2, e_3 be as in Exercise 4. Let $\mathscr{S}$ be the set of all vectors X of the form $X = \alpha_1 e_1 + \alpha_2 e_2$.

(a) Sketch $\mathbb{R}^3$ and locate the set $\mathscr{S}$ in your sketch.

(b) If $X = (x_1, x_2, x_3)$ is a typical vector of $\mathscr{S}$, what simple equation must its coordinates satisfy?

(c) Locate in your sketch of $\mathbb{R}^3$ the set of all vectors satisfying $x_2 = 0$.

(d) Likewise for $x_1 = 0$, then $x_1 = 1$, then $x_3 = 5$.

1.3 SUBSPACES

1.3a Introduction

Now we delve into the internal anatomy of the vector spaces $\mathbb{R}^n$. The notion of "subspace," which we are about to make precise, is very basic and will be used often in what follows.

We begin with a simple example. Let $\mathbb{R}^2$ be the plane as usual, and consider a straight line (extended infinitely in both directions) through the origin $O = (0, 0)$. We denote this line by the script letter $\mathscr{S}$. See Fig. 1.7. Now each point (see X and Y) on $\mathscr{S}$ is a vector, since we are in $\mathbb{R}^2$.

It is crucial to note the following:

1. The sum of any two vectors on $\mathscr{S}$ is also on $\mathscr{S}$. Thus $X, Y \in \mathscr{S}$ implies $X + Y \in \mathscr{S}$. (We say that $\mathscr{S}$ is **closed** under the operation of addition.)

2. If $X \in \mathscr{S}$ and $\alpha \in \mathbb{R}$, then $\alpha X \in \mathscr{S}$. (We say that $\mathscr{S}$ is closed under multiplication by scalars.) It follows that if $X \in \mathscr{S}$, then $-X = -1X \in \mathscr{S}$. Therefore, if $Y \in \mathscr{S}$, then $Y + (-X) = Y - X$ is in $\mathscr{S}$. We describe this situation as follows: We may add, subtract, or

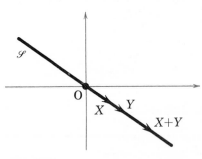

FIGURE 1.7

multiply by scalars any vectors in $\mathscr{S}$ and rest assured that the sum, difference, or product is again a vector in $\mathscr{S}$. In fact, $\mathscr{S}$ itself is a vector space; not only are vector addition and multiplication by scalars possible in $\mathscr{S}$, but also Properties 1 to 8 (the vector-space axioms) are satisfied. (Note that we were careful that $\mathscr{S}$ contained the zero vector O.) We say that $\mathscr{S}$ is a "subspace" of $\mathbb{R}^2$.

DEFINITION: The subset $\mathscr{S}$ of the vector space $\mathbb{R}^n$ is a **subspace (vector subspace, linear subspace)** of $\mathbb{R}^n$ if and only if $\mathscr{S}$ is itself a vector space (see Sec. 1) under the addition and multiplication by scalars defined in $\mathbb{R}^n$.

EXAMPLES: 1. As we have seen, any line through the origin in $\mathbb{R}^2$ is a subspace of $\mathbb{R}^2$.

2. Any line or plane through the origin in $\mathbb{R}^3$ is a subspace of $\mathbb{R}^3$.

For note that if we add two vectors X and Y that are in some plane (an unbounded plane, of course, not just a piece of one), then the sum $X + Y$ is also in that plane. See Fig. 1.8.

3. The subset of $\mathbb{R}^n$ consisting of $O = (0, \ldots, 0)$ alone is a subspace, the **zero subspace**, often denoted {O}.

4. We should note that since $\mathbb{R}^n$ is a subset of itself—not a **proper** subset, to be sure—then $\mathbb{R}^n$ is a subspace of $\mathbb{R}^n$. These subspaces—just

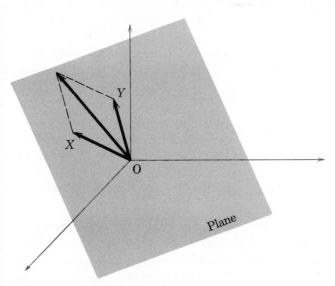

FIGURE 1.8

O and all of $\mathbb{R}^n$—are the trivial or **improper** subspaces of $\mathbb{R}^n$. All others are **proper**.

5. Now an important example motivated by algebra. Let $\mathscr{S}$ be the set of all vectors $X = (x_1, x_2, x_3, x_4)$ in $\mathbb{R}^4$ whose coordinates satisfy some homogeneous linear equation, say $2x_1 - x_2 + x_3 + 5x_4 = 0$. (Here "homogeneous" means that the right-hand side is zero.) Thus $(0, 1, 1, 0)$ and $(5, 1, 1, -2)$ are among the vectors in $\mathscr{S}$. You should check that $\mathscr{S}$ is a vector space, the "space of solutions" to the linear equation. For instance, to see that $\mathscr{S}$ is closed under addition, let $X = (x_1, x_2, x_3, x_4)$ and $Y = (y_1, y_2, y_3, y_4)$ be in $\mathscr{S}$. We ask: is

$$X + Y = (x_1 + y_1, x_2 + y_2, x_3 + y_3, x_4 + y_4)$$

also in $\mathscr{S}$? To see that it is, note that

$$2(x_1 + y_1) - (x_2 + y_2) + (x_3 + y_3) + 5(x_4 + y_4)$$
$$= (2x_1 - x_2 + x_3 + 5x_4) + (2y_1 - y_2 + y_3 + 5y_4) = 0 + 0 = 0,$$

since both X and Y are in $\mathscr{S}$.

REMARK: Example 5 is the forerunner of an important set of ideas: the relations between equations, or systems of equations, of the form $\alpha_1 x_1 + \cdots + \alpha_n x_n = 0$ and subspaces of $\mathbb{R}^n$. In fact, you should become aware that most of the basic computational problems in the vector algebra of $\mathbb{R}^n$ reduce to solving some system of linear equations. Actually, this connection between subspaces and equations should not be completely surprising. If the straight line in Example 1 has slope m in the sense of calculus, then it is the graph of $x_2 = mx_1$, that is, the set of solutions (x_1, x_2) to the equation $mx_1 - x_2 = 0$.

1.3b A criterion for subspace

Suppose that we are given, in some way, a subset $\mathscr{S}$ of $\mathbb{R}^n$. How can we tell if it is actually a subspace? We could, of course, check to see if $\mathscr{S}$ satisfies the definition of vector space. This involves checking the associativity and commutativity of addition, the existence of a zero vector in $\mathscr{S}$, and so on, through the eight properties. But reflection should convince us that much of this is superfluous. Since addition in $\mathscr{S}$ is addition in $\mathbb{R}^n$, it is certainly associative and commutative and well behaved with respect to multiplication by scalars. The real issue is whether the vectors in $\mathscr{S}$ comprise a self-contained algebraic system themselves, that is, whether sums and scalar multiples of vectors in $\mathscr{S}$ are themselves in $\mathscr{S}$.

THEOREM 1: Let $\mathscr{S}$ be a nonempty subset of $\mathbb{R}^n$. Then $\mathscr{S}$ is a subspace $\Leftrightarrow \mathscr{S}$ is closed under addition and closed under multiplication by scalars.

REMARK: We recall that "$\mathscr{S}$ is **closed** under addition" means that whenever X and Y are in $\mathscr{S}$, then so is their sum $X + Y$. Likewise for multiplication by scalars.

Proof: We must verify that each statement implies the other. It is best to begin with the easier.

($\Rightarrow$) This is simply a matter of definitions. Given that $\mathscr{S}$ is a subspace, it is certainly a vector space and thereby closed under both operations. Done.

($\Leftarrow$) We are given that $\mathscr{S}$ is closed under both operations. We check the vector-space axioms 1 to 8 in turn:

1. Associativity is clear, because addition in $\mathscr{S}$ is addition in $\mathbb{R}^n$.
2. Commutativity is clear for the same reason.
3. Why is $O \in \mathscr{S}$? Reason: $\mathscr{S}$ is nonempty, and so there is some X in $\mathscr{S}$. Also $\mathscr{S}$ is closed under multiplication by scalars. Therefore $0 \cdot X$ is in $\mathscr{S}$. But $0 \cdot X = O$, the zero vector.
4. Additive inverses: If $X \in \mathscr{S}$, then $-1X \in \mathscr{S}$. But $-1X = -X$.
5, 6, 7, 8. These hold in $\mathscr{S}$, because they hold in $\mathbb{R}^n$. $\ll$

REMARK: We now mention a very useful necessary criterion that $\mathscr{S}$ be a subspace. First an example: Let $\mathscr{S}$ be the vertical straight line $x_1 = 2$ in $\mathbb{R}^2$. Is $\mathscr{S}$ a subspace of $\mathbb{R}^2$? Perhaps the easiest way to convince oneself that $\mathscr{S}$ is *not* a subspace is to observe that the zero vector is not in $\mathscr{S}$. This line does not contain the origin. Such a line cannot be a subspace. Very often in practice we can decide immediately that some candidate for subspace-hood is not qualified by observing that it does not contain O. This will be useful in the exercises.

1.3c Where we are

In the preceding three sections we defined the concepts of *vector space* and *subspace* and presented several examples. Much of this was abstract and axiomatic: language building. The word "subspace," for example, is a wonderfully useful bag in which we will place the diverse notions of "line," "plane," "set of solutions of a system of homogeneous linear equations," and so on. From now on we will be moving gradually toward dealing with "honest" problems—problems arising in geometry or linear equations, say, not those prompted by our own definitions. Nonetheless, it is to this careful process of definition, observation, and theorem that the diverse problems mentioned above will yield their basic similarity and, ultimately, their solutions. We continue this process in the following section.

Exercises

Some practice in verifying whether a given subset of $\mathbb{R}^n$ is actually a subspace.

1. Sketch the following subsets of $\mathbb{R}^2$. Which are subspaces? Here $X = (x_1, x_2)$, as usual, and the sets are defined by way of coordinates:

(a) $\{X \in \mathbb{R}^2 \,|\, x_1 = 0\}$,

(b) $\{X \in \mathbb{R}^2 \,|\, x_1 \neq 0\}$,

(c) $\{X \in \mathbb{R}^2 \,|\, x_1 - x_2 = 0\}$,

(d) $\{X \in \mathbb{R}^2 \,|\, x_1^2 - x_2 = 0\}$,

(e) $\{X \in \mathbb{R}^2 \,|\, x_1 \geq 0\}$.

(f) $\{X \in \mathbb{R}^2 \,|\, x_1 = 1\}$,

(g) $\{X \in \mathbb{R}^2 \,|\, x_1 - x_2 = 2\}$.

2. Decide which of the following subsets of $\mathbb{R}^3$ are subspaces:

(a) $\{X \in \mathbb{R}^3 \,|\, x_1^2 + x_2^2 + x_3^2 = 1\}$,

(b) $\{X \in \mathbb{R}^3 \,|\, x_1 + x_2 + x_3 = 1\}$,

(c) $\{X \in \mathbb{R}^3 \,|\, x_1 + x_2 + x_3 = 0\}$,

(d) $\{X \in \mathbb{R}^3 \,|\, x_1 \leq x_2 \leq x_3\}$,

(e) $\{X \in \mathbb{R}^3 \,|\, x_1 - x_2 = 0, x_3 = 0\}$,

(f) $\{X \in \mathbb{R}^3 \,|\, x_1 - x_2 = 0\}$,

(g) $\{X \in \mathbb{R}^3 \,|\, x_2 = 0\}$,

(h) $\{X \in \mathbb{R}^3 \,|\, x_1 = x_2 = x_3 = 0\}$.

3. Let X_1, X_2 be given fixed vectors in $\mathbb{R}^3$. Is the set

$$\mathscr{S} = \{X \in \mathbb{R}^3 \,|\, X = \alpha_1 X_1 + \alpha_2 X_2, \alpha_1, \alpha_2 \text{ arbitrary scalars}\}$$

a subspace of $\mathbb{R}^3$? Say why or why not. Does your reasoning depend essentially on $\mathbb{R}^3$?

4. True or false:

(a) Every straight line in $\mathbb{R}^2$ is a subspace of $\mathbb{R}^2$.

(b) Every subspace of $\mathbb{R}^n$ contains the origin of $\mathbb{R}^n$.

(c) Every subspace of $\mathbb{R}^3$ that contains a nonzero vector X also contains the straight line through O and X.

(d) The intersection (part in common) of two subspaces of $\mathbb{R}^n$ is also a subspace.

(e) The union of two subspaces of $\mathbb{R}^n$ is also a subspace.

(f) $\mathbb{R}^1$ has some proper subspace.

(g) All proper subspaces of $\mathbb{R}^3$ are planes through the origin.

(h) Every proper subspace of $\mathbb{R}^2$ is the set of all solutions $X = (x_1, x_2)$ to some homogeneous linear equation $\alpha_1 x_1 + \alpha_2 x_2 = 0$.

(i) The set of all $X = (x_1, x_2, x_3)$ satisfying $x_3 = 0$ is a line through the origin in $\mathbb{R}^3$.

(j) If a subspace $\mathscr{S}$ of $\mathbb{R}^n$ contains two different straight lines through the origin, then $\mathscr{S}$ contains the plane containing both lines.

(k) $\mathbb{R}^2$ has a proper subspace $\mathscr{S}$ that consists of a straight line through the origin and certain points not on this line as well.

(l) A subset $\mathscr{S}$ of $\mathbb{R}^2$ that can be enclosed in the unit circle is not a proper subspace of $\mathbb{R}^2$.

(m) There exist proper subspaces of $\mathbb{R}^n$ with finitely many elements (vectors).

5. Suppose that $X \in \mathbb{R}^n$ may be written $X = \alpha_1 X_1 + \alpha_2 X_2 = \beta_1 X_1 + \beta_2 X_2$ with $\alpha_1 \neq \beta_1$, $\alpha_2 \neq \beta_2$. Find γ_1, γ_2 nonzero such that $0 = \gamma_1 X_1 + \gamma_2 X_2$ (note also $0 = 0X_1 + 0X_2$).

6. One way to specify a particular subspace of $\mathbb{R}^n$ is to give a homogeneous linear equation (or equations) of which the subspace is the set of solutions. Can you describe another way of specifying or building subspaces? See Exercise **3**.

7. a) Describe geometrically all the proper subspaces of $\mathbb{R}^2$.
b) Likewise for $\mathbb{R}^3$. Can you prove your assertions? See the following section.

1.4 AN EXTENDED EXERCISE: THE SUBSPACES OF $\mathbb{R}^3$

1.4a Introduction

In this section we grapple with the following problem: "Classify all subspaces of the vector space $\mathbb{R}^3$." This will afford us some useful exercise in working with the concepts in the preceding sections. In addition, it will enable us to introduce some new ideas and, finally, will yield some information about $\mathbb{R}^3$ we will use later.

1.4b The standard basis

First let us look at $\mathbb{R}^3$ itself. Consider the vectors

$$e_1 = (1, 0, 0), \qquad e_2 = (0, 1, 0), \qquad e_3 = (0, 0, 1).$$

If $X = (x_1, x_2, x_3)$ is any vector in $\mathbb{R}^3$, then note that we may write $X = x_1 e_1 + x_2 e_2 + x_3 e_3$. See Fig. 1.9. For instance, if $X = (2, -1, -7)$, then $X = 2e_1 - e_2 - 7e_3$ also; if $Y = (1, 0, 4)$, then $Y = e_1 + 4e_3$ also. The important thing here is that *every* vector X in $\mathbb{R}^3$ may be written uniquely using e_1, e_2, e_3 and scalars as coefficients. These three vectors comprise the **standard basis** of $\mathbb{R}^3$.

REMARK: Roughly speaking, $\mathbb{R}^3$ is "three-dimensional," because the standard basis consists of three vectors, no more, no less.

Throughout this discussion, we will be using expressions like $x_1 e_1 + x_2 e_2 + x_3 e_3$. Hence, let us make the following general definition. Let $X_1, \ldots, X_k$ be vectors in $\mathbb{R}^n$. Then $Y \in \mathbb{R}^n$ is a **linear combination** of $X_1, \ldots, X_k$ if and only if $Y = \alpha_1 X_1 + \cdots + \alpha_k X_k$ for some scalars $\alpha_1, \ldots, \alpha_k$.

EXAMPLES: 1. If $X_1 = (2, 1)$, $X_2 = (1, 4)$ in $\mathbb{R}^2$, then $Y = (3, 5) = X_1 + X_2$ is a linear combination of X_1, X_2, as is $(0, -7) = X_1 - 2X_2$.

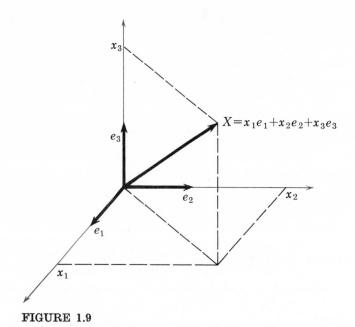

$X = x_1 e_1 + x_2 e_2 + x_3 e_3$

FIGURE 1.9

2. Note that $X_1 = X_1 + 0X_2 + \cdots + 0X_k$ and likewise with X_2, $\ldots, X_k$, so that each X_i is a linear combination of $X_1, \ldots, X_k$.

3. We saw above that *every* vector X in $\mathbb{R}^3$ is a unique linear combination of the standard basis vectors e_1, e_2, e_3.

1.4c   Lines and planes revisited

When asked to list *all* the subspaces of $\mathbb{R}^3$, we are likely to begin, "Lines through the origin, planes through the origin, . . ." We might then recall that the easily overlooked set {O} consisting of the zero vector (origin) alone is a subspace, as is the entire space $\mathbb{R}^3$. Are there any others?

As we ponder this question, we are brought to the realization that its answer requires that we be more precise about the notions we have just used. What is a line? How can we tell if a subset given by such-and-such a condition or property is a plane? In the preceding section, we used lines and planes pictorially, in $\mathbb{R}^2$ and $\mathbb{R}^3$. But what is a plane in $\mathbb{R}^n$ when n is arbitrary?

Lines through the origin.   In the special case of $\mathbb{R}^2$, a line through the origin may be described as the set of all (x_1, x_2) such that $\alpha_1 x_1 + \alpha_2 x_2 = 0$, where α_1, α_2 are scalars not both zero. This definition involving one equation, however, does not work in $\mathbb{R}^n$; for example, you can see that the set of solutions (x_1, x_2, x_3) to the single equation $x_3 = 0$ in $\mathbb{R}^3$ is the *plane* consisting of all points $(x_1, x_2, 0)$ and *not* a line.

Here is a natural definition of the line through the origin whose charm is that it works in an arbitrary $\mathbb{R}^n$. A subset $\mathscr{S}$ of $\mathbb{R}^n$ is a **line through the origin (one-dimensional subspace)** if and only if there is a fixed nonzero vector X in $\mathscr{S}$ such that

$$\mathscr{S} = \{Y \in \mathbb{R}^n \mid Y = \alpha X, \alpha \in \mathbb{R}\}.$$

Note that $0 = 0X$ is in $\mathscr{S}$. See Fig. 1.10. You should convince yourself that any such set $\mathscr{S}$ actually does satisfy the definition of subspace.

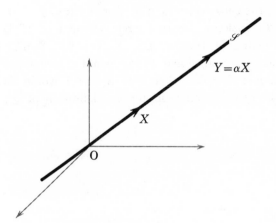

A one-dimensional subspace of $\mathbb{R}^3$

FIGURE 1.10

REMARK: In the language introduced above, we may say that a line $\mathscr{S}$ through the origin is the set of all linear combinations, αX, of one of its nonzero elements X. The line $\mathscr{S}$ is "one-dimensional," precisely because it can be built up by taking *one* vector X and forming the set of all linear combinations thereof.

Planes through the origin. This should not be surprising. A subset $\mathscr{S}$ of $\mathbb{R}^n$ is a **plane through the origin (two-dimensional subspace)** if and only if (1) $\mathscr{S}$ contains two vectors X, Y that do not lie on the same line through the origin, that is, they are *non-collinear* with the origin, and (2) $\mathscr{S}$ is the set of all linear combinations of X and Y, that is,

$$\mathscr{S} = \{Z \in \mathbb{R}^n \mid Z = \alpha X + \beta Y \quad \text{with} \quad \alpha, \beta \in \mathbb{R}\}.$$

We say also that $\mathscr{S}$ is **spanned** by or **generated** by the vectors X, Y.

REMARKS: 1. The set $\mathscr{S}$ so defined *is* a subspace. Reason: By Theorem 1 it suffices to prove that $\mathscr{S}$ is closed under addition and multiplication by scalars. We will check addition and leave multiplication by scalars to you. Thus, let $Z = \alpha X + \beta Y$, $Z' = \alpha' X + \beta' Y$ be in $\mathscr{S}$. To check whether $Z + Z'$ is in $\mathscr{S}$, note that $Z + Z' = (\alpha X + \beta Y) + (\alpha' X + \beta' Y) = (\alpha + \alpha')X + (\beta + \beta')Y$, where $\alpha, \alpha', \beta, \beta'$ are scalars. Now this expression is a linear combination of X and Y and, therefore, an element of $\mathscr{S}$. It follows that $\mathscr{S}$ is closed under addition.

2. You should verify that the zero vector is in $\mathscr{S}$. Just let $\alpha = \beta = 0$.

3. The plane $\mathscr{S}$ is two-dimensional, because every element of $\mathscr{S}$ is a linear combination of *two* vectors (examples: $X + Y$, $X - 2Y$, X, $6X$), and, crucially, *both* vectors are required. For if Y, say, could be written as a linear combination of X alone, $Y = \alpha X$, then both X and Y would lie on the same line through the origin, contradicting the definition of $\mathscr{S}$.

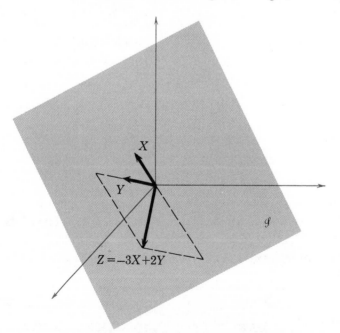

Every $Z \in \mathscr{S}$ is a linear combination of X, Y

FIGURE 1.11

4. Continuing in this vein, you ought to be able to define the notion of k-dimensional subspace of $\mathbb{R}^n$. See Fig. 1.11.

1.4d Classifying the subspaces of $\mathbb{R}^3$

Let $\mathscr{S}$ be a subspace of $\mathbb{R}^3$. What must $\mathscr{S}$ look like? Must $\mathscr{S}$ be one of the "known" subspaces, namely $\{O\}$, a line through the origin, a plane through the origin, or $\mathbb{R}^3$ itself?

We deal with these possibilities in turn:

1. The given subspace $\mathscr{S}$ consists of the zero vector alone, $\mathscr{S} = \{O\}$.

2. If $\mathscr{S} \neq \{O\}$, then it contains a nonzero vector X. Being a subspace, $\mathscr{S}$ therefore contains all vectors of the form αX, $\alpha \in \mathbb{R}$. These form a line through the origin. Thus, $\mathscr{S}$ might actually equal this line through the origin.

3. On the other hand, if $\mathscr{S}$ contains a vector Y not of the form αX, then $\mathscr{S}$ also contains all vectors of the form $\alpha X + \beta Y$, since $\mathscr{S}$ is a subspace and therefore closed under the formation of linear combinations of its elements. Hence $\mathscr{S}$ might be a plane through the origin.

4. But $\mathscr{S}$ might contain a vector Z *not* of the form $\alpha X + \beta Y$. Hence $\mathscr{S}$ is not a plane. See Fig. 1.12. How may we describe $\mathscr{S}$? Can there exist subspaces $\mathscr{S}$ of $\mathbb{R}^3$ which contain planes through the origin as proper subsets but which themselves are proper subsets of $\mathbb{R}^3$? Or is the only subspace $\mathscr{S}$ of $\mathbb{R}^3$ that contains a plane but is "larger" than that plane actually all of $\mathbb{R}^3$, $\mathscr{S} = \mathbb{R}^3$?

Let us consider the following unglamorous but useful lemma.

LEMMA 2: Let X, Y, Z be vectors in $\mathbb{R}^3$ that do not lie on any plane through the origin; that is, they are non-coplanar with the origin. Then it is possible to express each of the standard basis vectors e_1, e_2, e_3 as a linear combination of X, Y, Z.

REMARK: A **lemma** is a fact used in establishing some theorem.

Proof: The method of proof is that of elimination of variables in linear equations. It is absolutely basic to linear algebra. Thus, let

$$X = x_1 e_1 + x_2 e_2 + x_3 e_3$$
$$Y = y_1 e_1 + y_2 e_2 + y_3 e_3$$
$$Z = z_1 e_1 + z_2 e_2 + z_3 e_3$$

be the given vectors, expressed as linear combinations of e_1, e_2, e_3. (We

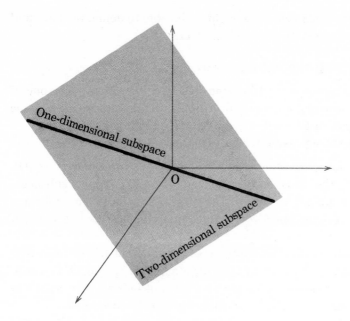

Which subspaces of $\mathbb{R}^3$ contain a plane?

FIGURE 1.12

know that this is possible.) We must show that these equations may be solved for e_1, e_2, e_3 in terms of X, Y, Z. We proceed much as if X, Y, Z and e_1, e_2, e_3 were numbers and not vectors.

Let us show that e_3 is a linear combination of X, Y, Z. An entirely similar argument works with e_1 and e_2.

First, note that at least one of the coefficients x_1, y_1, z_1 must be different from zero; otherwise X, Y, Z would all lie in the plane subspace determined by e_2 and e_3, contrary to the hypothesis. We may as well suppose that $x_1 \neq 0$, since otherwise we could rename our vectors to make this the case.

Now we subtract off the e_1 part from Y and Z; that is, let $Y' = Y - (y_1/x_1)X$ and $Z' = Z - (z_1/x_1)X$. Then, for example, $Y' = y_2'e_2 + y_3'e_3$, where $y_2' = y_2 - (y_1/x_1)x_2$ and similarly for y_3'. The important thing is that Y' and Z' are linear combinations of e_2 and e_3 alone; e_1 has been eliminated:

$$Y' = y_2'e_2 + y_3'e_3,$$
$$Z' = z_2'e_2 + z_3'e_3.$$

This is a smaller system of equations.

Now we argue that at least one of y_2', z_2' is different from zero. For if not, then both Y' and Z' would lie on the line through the origin determined by e_3. It would follow that both Y and Z would lie in the plane determined by X and e_3, since we can reconstruct Y and Z using Y', Z' and X, namely, $Y = Y' + (y_1/x_1)X$, $Z = Z' + (z_1/x_1)X$. But then Y, Z and surely X will lie on the plane determined by X and e_3, contrary to our hypothesis.

Thus, as above, we may suppose that $y_2' \neq 0$. Again we eliminate a vector, this time forming $Z'' = Z' - (z_2'/y_2')Y'$, which has neither e_1 nor e_2 in its makeup. Hence we have

$$Z'' = z_3'' e_3.$$

Just as before, z_3'' is not zero, or else $Z'' = Z' - (z_2'/y_2')Y' = 0$, which would imply that Z' and Y' lie on the same line through the origin, whence X, Y, Z would lie on one plane through the origin. (Check this!)

Since $z_3'' \neq 0$, we may divide by it, obtaining

$$e_3 = \frac{1}{z_3''}Z''$$

$$= \frac{1}{z_3''}\left(Z' - \frac{z_2'}{y_2'}Y'\right)$$

$$= \frac{1}{z_3''}\left(Z - \frac{z_1}{x_1}X - \frac{z_2'}{y_2'}\left(Y - \frac{y_1}{x_1}X\right)\right),$$

where we have gone backwards from Z'' to Z' and Y' to X, Y, Z. This ungainly expression shows that e_3 is a linear combination of X, Y, Z, as claimed. Do *not* simplify it! We don't care about the exact value of coefficients.

You might show in a similar fashion that e_1 and e_2 are also linear combinations of the non-coplanar X, Y, Z. Done. $\ll$

Having proved the lemma, let us apply it to the subspace question. If we know that $\mathscr{S}$ contains vectors X, Y, Z, none of which lies on the plane through the origin determined by the other two, then we conclude that the standard basis vectors e_1, e_2, e_3, being linear combinations of vectors X, Y, Z in $\mathscr{S}$, are themselves in $\mathscr{S}$; this is because $\mathscr{S}$ is a subspace.

But again, since the subspace $\mathscr{S}$ contains e_1, e_2, e_3, then it contains every vector in $\mathbb{R}^3$, since every vector in $\mathbb{R}^3$ is a linear combination of e_1, e_2, e_3. Thus $\mathscr{S} = \mathbb{R}^3$; the only subspace of $\mathbb{R}^3$ larger than a plane is the entire space $\mathbb{R}^3$.

Let us state our findings as a theorem.

THEOREM 3: Let $\mathscr{S}$ be a subspace of $\mathbb{R}^3$. Then $\mathscr{S}$ must be one of the following: the zero subspace $\{O\}$, a (one-dimensional) line through the origin, a (two-dimensional) plane through the origin, or the entire space $\mathbb{R}^3$ itself.

REMARKS: 1. The important aspects of the argument above were the careful definition of line and plane (pictures were not enough) and the lemma, whose proof was a glorious exercise in solving three linear equations in three "unknowns," as taught in high school.

2. You may note that, in proving the lemma, it was important to have x_1, y_2', z_3'' nonzero. We then proceeded to divide by each of them. This often happens in mathematics. If you are told that some number or function is never zero, then you will soon be dividing by it.

3. The theorem indicates that every subspace, at least in $\mathbb{R}^3$, can be built by carefully choosing some vectors from the subspace and then forming all linear combinations of them. The resulting set of all linear combinations is the subspace itself. However, it raises a question for now. It was easy to see (Example 5 in the preceding section) that the set of all vectors $X = (x_1, x_2, x_3)$ such that, for example, $x_1 - x_2 + 5x_3 = 0$ form a subspace of $\mathbb{R}^3$. Now that we claim to know all the subspaces of $\mathbb{R}^3$, we ought to be able to decide which subspace is determined by $x_1 - x_2 + 5x_3 = 0$.

1.4e Subspaces and homogeneous linear equations

Let us consider the problem raised in the remarks above. We are given the homogeneous linear equation $x_1 - x_2 + 5x_3 = 0$. What can we say about the set of solutions (x_1, x_2, x_3), considered as a subset of $\mathbb{R}^3$? Let us denote the set by $\mathscr{S}$.

First, $\mathscr{S}$ is not empty. For we observe that $O = (0, 0, 0)$ and $(5, 0, -1)$, among other vectors, are in $\mathscr{S}$.

Second, $\mathscr{S}$ is a subspace of $\mathbb{R}^3$. We leave it to you to prove that it is.

Third, we ask, "Is $\mathscr{S}$ a line, a plane, or all of $\mathbb{R}^3$?"

Fourth, we observe that $\mathscr{S} \neq \mathbb{R}^3$, since, for example, the vector $(0, 0, 1)$ does not give a solution to $x_1 - x_2 + 5x_3 = 0$. Hence $\mathscr{S}$ must be either a line or a plane.

Fifth, we have already seen that $X_1 = (5, 0, -1)$ is in $\mathscr{S}$. Are all other vectors Y in $\mathscr{S}$ scalar multiples of this vector X_1? Is $\mathscr{S}$ a line?

The answer is no. For we readily find another vector $X_2 = (0, 5, 1)$ in $\mathscr{S}$ such that $X_2 \neq \alpha X_1$. To check this inequality, note that $(0, 5, 1)$

$= X_2 = \alpha X_1 = (5\alpha, 0, -\alpha)$ would imply $5\alpha = 0$, whence $\alpha = 0$. But $X_2 = 0X_1$ would imply $X_2 = O$, which is false.

And this does it. $\mathscr{S}$ must be the plane consisting of all vectors X of the form $\alpha X_1 + \beta X_2$, where $X_1 = (5, 0, -1)$ and $X_2 = (0, 5, 1)$.

REMARK: This gives a good description of the space $\mathscr{S}$ of solutions. We know that $X = (x_1, x_2, x_3)$ is a solution of $x_1 - x_2 + 5x_3 = 0$ if and only if $x_1 = 5\alpha$, $x_2 = 5\beta$, $x_3 = -\alpha + \beta$ for scalars α, β. For each choice of α, β we get a unique solution to the homogeneous equation. The fact that there are two choices (α and β) underscores the two-dimensionality of the space $\mathscr{S}$ of solutions. Other ways of saying this: "The space of solutions has two degrees of freedom" or "There is a two-parameter, namely, α and β, family of solutions."

Although we considered the special case $x_1 - x_2 + 5x_3 = 0$, you should be convinced that the following theorem is true.

THEOREM 4: Let $\alpha_1 x_1 + \alpha_2 x_2 + \alpha_3 x_3 = 0$ be a homogeneous linear equation with at least one of the coefficients $\alpha_1, \alpha_2, \alpha_3$ different from zero. Then the set of solutions $X = (x_1, x_2, x_3)$ is a (two-dimensional) plane subspace in $\mathbb{R}^3$.

Proof: Left to you. Note that the key in the special case above was our ability to find the non-collinear solutions X_1 and X_2. See the examples below also. X_1 and X_2 are called a *basis* for the plane. $\ll$

EXAMPLES: 1. Consider $3x_1 + 4x_2 + 5x_3 = 0$. Then one solution is $X_1 = (5, 0, -3)$. This was easy. To get a second solution X_2 not of the form αX_1, we begin by putting the zero in the first slot this time: $X_2 = (0, ?, ?)$. *Any* vector of the form $(0, ?, ?)$, except, of course, $(0, 0, 0)$, that satisfies the homogeneous equation will *not* be of the form αX_1. Think about this. Hence, $X_2 = (0, 5, -4)$ is a solution and not collinear with X_1.

2. Consider $x_1 + x_3 = 0$. Then $X_1 = (1, 0, -1)$ and $X_2 = (0, 1, 0)$ are non-collinear solutions. Any other solution is of the form $\alpha X_1 + \beta X_2$.

The converse problem. We have just seen that every equation $\alpha_1 x_1 + \alpha_2 x_2 + \alpha_3 x_3 = 0$ determines a plane. The converse question is "Does every plane determine an equation?" That is, suppose that $\mathscr{S}$ is the two-dimensional subspace of $\mathbb{R}^3$ consisting of all linear combinations of given vectors X_1, X_2; that is, $\mathscr{S}$ is spanned by X_1, X_2. Is $\mathscr{S}$ actually the space of solutions to some equation $\alpha_1 x_1 + \alpha_2 x_2 + \alpha_3 x_3 = 0$?

Let us consider a typical example. Let $X_1 = (1, 2, 3)$ and $X_2 = (1, 0, 2)$ determine a two-dimensional subspace in $\mathbb{R}^3$; that is, $\mathscr{S}$ consists of all vectors $\alpha X_1 + \beta X_2$ with $\alpha, \beta \in \mathbb{R}$.

Now let us find an equation for which X_1 and X_2 are both solutions. The equation will be of the form $\alpha_1 x_1 + \alpha_2 x_2 + \alpha_3 x_3 = 0$. We must find suitable coefficients $\alpha_1, \alpha_2, \alpha_3$.

The method of determining $\alpha_1, \alpha_2, \alpha_3$ is easy. Just plug $X_1 = (1, 2, 3)$ and $X_2 = (1, 0, 2)$ into the formal equation $\alpha_1 x_1 + \alpha_2 x_2 + \alpha_3 x_3 = 0$. This gives two equations in unknowns $\alpha_1, \alpha_2, \alpha_3$ as follows:

$$\alpha_1 + 2\alpha_2 + 3\alpha_3 = 0,$$
$$\alpha_1 \qquad\; + 2\alpha_3 = 0.$$

Subtracting the second from the first eliminates α_1 and leads to $\alpha_2 = -\frac{1}{2}\alpha_3$. Thus we obtain the relations

$$\alpha_1 = -2\alpha_3, \qquad \alpha_2 = -\tfrac{1}{2}\alpha_3.$$

For any nonzero value of α_3, these relations yield the coefficients of a homogeneous equation satisfied by X_1 and X_2, namely,

$$(-2\alpha_3)x_1 + (-\tfrac{1}{2}\alpha_3)x_2 + \alpha_3 x_3 = 0,$$

or, showing the dependence on α_3,

$$\alpha_3(-2x_1 - \tfrac{1}{2}x_2 + x_3) = 0.$$

For instance, when $\alpha_3 = -2$, we get

$$4x_1 + x_2 - 2x_3 = 0.$$

Each of these equations has been constructed so that X_1 and X_2 are solutions. You may check this by plugging in X_1 and X_2.

What have we done? Starting with the non-collinear vectors X_1 and X_2, we have constructed an equation, or, better, a family of equations depending on α_3, that has among its solutions the vectors X_1 and X_2. You should convince yourself of these two facts:

1. For any nonzero choice of α_3, the equation

$$\alpha_3(-2x_1 - \tfrac{1}{2}x_2 + x_3) = 0$$

has precisely the same solutions (x_1, x_2, x_3) independent of α_3.

2. Since X_1, X_2 are solutions and since the equation is homogeneous, that is, right-hand side equal to zero, any linear combination of X_1, X_2 is also a solution of the equation.

From this we may conclude that the space of solutions of any equation of the form $\alpha_3(-2x_1 - \tfrac{1}{2}x_2 + x_3) = 0$ is the original plane $\mathscr{S}$ spanned by the given X_1, X_2.

REMARK: The key to this is the fact that we can solve two homogeneous equations in three unknowns $\alpha_1, \alpha_2, \alpha_3$.

What we have done in the special case of $X_1 = (1, 2, 3)$, $X_2 = (1, 0, 2)$ can, of course, be done with arbitrary non-collinear vectors X_1, X_2. Let us state this as a theorem about the relation between planes and homogeneous equations that incorporates Theorem 4.

THEOREM 5: A subset $\mathscr{S}$ of $\mathbb{R}^3$ is a plane through the origin (two-dimensional subspace) $\Leftrightarrow \mathscr{S}$ is the set of all solutions $X = (x_1, x_2, x_3)$ of a homogeneous linear equation $\alpha_1 x_1 + \alpha_2 x_2 + \alpha_3 x_3 = 0$ with at least one coefficient different from zero.

REMARKS: 1. The statement given by ($\Leftarrow$) is Theorem 4.

2. This theorem is worthwhile in that it relates a geometric object, a plane through the origin, to an algebraic object, a homogeneous linear equation. It is thereby in the great tradition of geometry (coordinate geometry, analytic geometry) since Descartes. We will see several generalizations of this theorem in what follows, notably in the Hyperplane Theorem.

3. Again we remark that the proof of the theorem depended on a careful definition of "plane through the origin" and the observation that certain linear equations (or systems) are solvable by the method of eliminating variables. Very many of the problems one meets in linear algebra boil down to finding the system of equations at the heart of the problem and solving it or showing that a solution is possible.

4. The geometry of $\mathbb{R}^3$ will be broadened when we introduce the notions of affine subspace (example: a plane *not* containing the origin) and perpendicularity using the inner product.

Exercises

These give you practice in relating the basic linear algebra (subspaces, bases) of $\mathbb{R}^3$ with linear equations and computations.

1. Which pairs $X, Y \in \mathbb{R}^3$ lie on the same straight line through the origin $O = (0, 0, 0)$? If they do, write Y as αX for a suitable scalar α:

(a) $X = (1, 2, 4)$, $Y = (-1, -4, -2)$;   (d) $X = (1, 0, 2)$, $Y = (2, 1, 4)$;
(b) $X = (1, 2, 4)$, $Y = (3, 6, 12)$;   (e) $X = (1, 0, 0)$, $Y = (0, 0, 0)$;
(c) $X = (1, 0, 1)$, $Y = (2, 0, -2)$;   (f) $X = (\tfrac{1}{3}, \tfrac{2}{7}, \tfrac{9}{5})$, $Y = (\tfrac{1}{4}, \tfrac{3}{14}, \tfrac{27}{20})$.

2. Observe that $X_1 = (1, 1, 0)$, $X_2 = (0, 1, 1)$ span a two-dimensional subspace $\mathscr{S}$ of $\mathbb{R}^3$. Which of the following vectors are in $\mathscr{S}$? You should learn to do these by inspection:

(a) $(2, 2, 0)$,   (f) $(0, 0, 0)$,

(b) $(1, 2, 1)$,   (g) $(0, -1, -1)$,

(c) $(1, 0, 0)$,   (h) $(0, 1, -1)$,

(d) $(0, 0, 1)$,   (i) $(1, -1, -2)$.

(e) $(1, 1, 1)$,

3. Express each of those vectors in Exercise 2 which *is* in $\mathscr{S}$ as a linear combination $\alpha_1 X_1 + \alpha_2 X_2$ for suitable α_1, α_2. Thus $X = (1, -1, -2)$ in Exercise 2*i* is $X = X_1 - 2X_2$. Of course, if a vector is not in $\mathscr{S}$, then such a representation is impossible. This problem may be worked by solving three linear equations for the two unknowns α_1, α_2, but inspection is preferable.

4. Verify that the three given vectors all lie on the same line through the origin in $\mathbb{R}^3$; that is, show that two of them, Y and Z, are of the form $Y = \alpha X$, $Z = \beta X$ for the third vector X:

(a) $(0, 0, 0)$, $(1, -1, 2)$, $(-2, 2, -4)$;

(b) $(4, -4, 4)$, $(-3, 3, -3)$, $(5, -5, 5)$;

(c) $(2, 0, 8)$, $(1, 0, 4)$, $(-4, 0, -16)$.

5. Verify that the three given vectors all lie in some two-dimensional subspace (a plane through the origin) in $\mathbb{R}^3$; that is, write one of them as a linear combination of the other two:

(a) $(1, 1, 0)$, $(1, 0, 1)$, $(3, 1, 2)$;

(b) $(0, 0, 0)$, $(1, 1, 1)$, $(1, 1, -1)$;

(c) $(3, 3, 1)$, $(1, 1, 1)$, $(1, 1, -1)$;

(d) $(3, -1, 5)$, $(6, 4, -17)$, $(12, 2, -7)$.

6. In each case below let $\mathscr{S}$ be the subspace of $\mathbb{R}^3$ spanned by the given X_1, X_2, X_3. Decide whether $\mathscr{S}$ is a line through the origin, a plane, and so on:

(a) $X_1 = (1, 0, 0)$, $X_2 = (1, 1, 0)$, $X_3 = (1, 1, 1)$;

(b) $X_1 = (0, 0, 0)$, $X_2 = (0, 0, 0)$, $X_3 = (0, 0, 0)$;

(c) $X_1 = (1, 2, 0)$, $X_2 = (2, 1, 0)$, $X_3 = (1, -1, 0)$;

(d) $X_1 = (1, 2, 3)$, $X_2 = (-3, -6, -9)$, $X_3 = (2, 4, 6)$;

(e) $X_1 = (1, 2, 3)$, $X_2 = (1, 2, -3)$, $X_3 = (1, 4, 0)$;

(f) $X_1 = (0, 1, 0)$, $X_2 = (0, 2, 0)$, $X_3 = (0, 3, 0)$.

7. From equations to subspaces. Given the homogeneous linear equation below, compute two vectors X_1, X_2 such that every solution $X = (x_1, x_2, x_3)$ may be written uniquely as $X = \alpha_1 X_1 + \alpha_2 X_2$ for suitable α_1, α_2. Note that X_1, X_2 are not uniquely determined by the equation:

(a) $x_3 = 0$,   (c) $x_1 - x_2 + x_3 = 0$,

(b) $x_1 + x_2 = 0$,   (d) $x_1 + x_2 + x_3 = 0$.

8. Decide whether the subspaces of solutions to each of the systems of equations below is a line (one-dimensional) or a plane in $\mathbb{R}^3$. Accordingly, in each case compute a vector X_1 or vectors X_1, X_2 that span the subspace:

(a) $x_1 = 0,$
$\quad x_3 = 0;$

(d) $x_1 - x_3 = 0,$
$\quad x_1 + x_3 = 0;$

(b) $x_1 - x_2 \quad\quad = 0,$
$\quad x_1 + x_2 + x_3 = 0;$

(e) $\quad x_1 - x_2 = 0,$
$\quad -x_1 + x_2 = 0;$

(c) $2x_1 - 4x_2 - 6x_3 = 0,$
$\quad -x_1 + 2x_2 - 3x_3 = 0;$

(f) $2x_1 - 4x_2 - 6x_3 = 0,$
$\quad -x_1 + 2x_2 + 3x_3 = 0.$

9. Solve these systems for the α's in each case (see Exercise 10 also):

(a) $\alpha_1 - \alpha_2 + \alpha_3 = 0,$
$\quad 2\alpha_1 + \alpha_2 + \alpha_3 = 0;$

(b) $3\alpha_1 + \alpha_2 \quad\quad = 0,$
$\quad \alpha_1 \quad\quad + \alpha_3 = 0.$

10. From subspaces to equations. Find a linear equation $\alpha_1 x_1 + \alpha_2 x_2 + \alpha_3 x_3 = 0$ (compute the α's) that determines the plane subspace spanned by the given X_1, X_2 (See Exercise 9):

(a) $X_1 = (1, -1, 1),$
$\quad X_2 = (2, 1, 1);$

(b) $X_1 = (3, 1, 0),$
$\quad X_2 = (1, 0, 1).$

11. For each equation, compute two "independent" solutions $(\alpha_1, \alpha_2, \alpha_3)$ and $(\alpha_1', \alpha_2', \alpha_3')$; that is, $(\alpha_1, \alpha_2, \alpha_3) \neq \beta(\alpha_1', \alpha_2', \alpha_3')$ for any β (see Exercise 12 also):

(a) $\alpha_1 + 2\alpha_2 + \alpha_3 = 0,$

(b) $\alpha_1 - \alpha_2 - 3\alpha_3 = 0.$

12. From lines to equations. Given the vectors

(a) $X_0 = (1, 2, 1),$

(b) $X_0 = (1, -1, -3),$

in each case find a system of *two* homogeneous linear equations in x_1, x_2, x_3 whose space of solutions is the one-dimensional subspace spanned by X_0 (use Exercise 11).

13. Some inhomogeneous equations. Invent systems of linear equations with the following properties:

(a) Two inhomogeneous equations, three unknowns, no solutions;
(b) Two inhomogeneous equations, three unknowns, having solutions;
(c) Three inhomogeneous equations, unknowns x_1, x_2, having solutions;
(d) One inhomogeneous equation in x_1, x_2, x_3 with no solutions;
(e) Three homogeneous equations, unknowns x_1, x_2, having an infinite number of distinct solutions.

Appendix 1. A very useful theorem

It depends only on the simplest algebra. It is this:

THEOREM: Let

$$\alpha_{11}x_1 + \cdots + \alpha_{1n}x_n = 0$$
$$\vdots \quad\quad\quad \vdots \quad\quad\quad \vdots$$
$$\alpha_{q1}x_1 + \cdots + \alpha_{qn}x_n = 0$$

be a system of q homogeneous linear equations in n unknowns (the x's) with $n > q$. Then the system has a nontrivial simultaneous solution $x_1, \ldots, x_n$.

NOTE: More unknowns than equations.

TERMINOLOGY: "Homogeneous" means that the right-hand side of each equation is zero. The solution $x_1, \ldots, x_n$ is "nontrivial" provided that at least one of the numbers x_i is different from zero. The trivial solution is $(0, 0, \ldots, 0)$, that is, $x_1 = x_2 = \ldots = x_n = 0$.

Remarkably enough, this result is essentially equivalent to some readily visualized geometric facts; for example, a special case says that the collection of all vectors in $\mathbb{R}^3$ perpendicular to a given line through the origin forms a plane through the origin (see Chap. 2). And, moreover, the method of proof we shall give tells us how to construct the plane in question (that is, an equation determining the plane). See Fig. 1.13.

The family of vectors perpendicular
to the line forms a plane through
the origin

FIGURE 1.13

How do we prove this basic theorem? Actually, you may have seen the key idea already. First, however, let us look at the very special simple case of a single equation, $q = 1$. We have

$$\alpha_{11}x_1 + \cdots + \alpha_{1n}x_n = 0.$$

If all α's are zero, then clearly any choice of $x_1, \ldots, x_n$, zero or not, would make the equation true. If, say $\alpha_{11} \neq 0$, then we may choose $x_2, \ldots, x_n$

arbitrarily (not all zero). We might choose $x_2 = \cdots = x_n = 100$, but no matter. This leads to an equation in one unknown x_1 (we have not chosen a particular x_1 yet):

$$\alpha_{11}x_1 + c = 0.$$

Here c is obtained from the values of $x_2, \ldots, x_n$ we chose earlier. But now we need only define $x_1 = -c/\alpha_{11}$ to obtain a nontrivial solution $x_1, x_2, \ldots, x_n$.

EXAMPLE: Let us consider $2x_1 - x_2 + 3x_3 + 0x_4 - 2x_5 = 0$. We choose $x_2 = 3$, $x_3 = 1$, $x_4 = 1$, $x_5 = 0$. This leads to

$$2x_1 + 0 = 0,$$

whence $x_1 = 0$ necessarily. Our nontrivial solution is therefore

$$(x_1, x_2, x_3, x_4, x_5) = (0, 3, 1, 1, 0).$$

There are others. You might compute one.

Now we turn to the general case of $n > q \geq 1$.

THE KEY IDEA: ELIMINATION OF VARIABLES: Let us work with an example. Consider the simultaneous equations E_1, E_2, E_3.

E_1: $\qquad\qquad 2x_1 + x_2 - x_3 + x_4 + 0x_5 = 0,$

E_2: $\qquad\qquad 3x_1 + \frac{5}{2}x_2 + 0x_3 + 2x_4 - 2x_5 = 0,$

E_3: $\qquad\qquad -x_1 + x_2 + x_3 - x_4 - x_5 = 0.$

First we eliminate x_1 from E_2. Subtracting $\frac{3}{2}E_1$ from E_2 gives $E_2' = E_2 - \frac{3}{2}E_1$:

E_2': $\qquad\qquad 0x_1 + x_2 + \frac{3}{2}x_3 + \frac{1}{2}x_4 - 2x_5 = 0.$

We do likewise with E_3, obtaining $E_3' = E_3 + \frac{1}{2}E_1$:

E_3': $\qquad\qquad 0x_1 + \frac{3}{2}x_2 + \frac{1}{2}x_3 - \frac{1}{2}x_4 - x_5 = 0.$

This leads to the system

E_1: $\qquad\qquad 2x_1 + x_2 - x_3 + x_4 + 0x_5 = 0,$

E_2': $\qquad\qquad x_2 + \frac{3}{2}x_3 + \frac{1}{2}x_4 - 2x_5 = 0,$

E_3': $\qquad\qquad \frac{3}{2}x_2 + \frac{1}{2}x_3 - \frac{1}{2}x_4 - x_5 = 0.$

It is crucial to realize that the systems E_1, E_2, E_3 and E_1, E_2', E_3' have precisely the same solution. This is because if $x_1, \ldots, x_5$ satisfy equation E_1, then they satisfy E_2 if and only if they satisfy $E_2' = E_2 - \frac{3}{2}E_1$.

Let us continue. We have eliminated x_1 from the last two equations.

Now let us eliminate x_2 from the third equation by forming $E_3'' = E_3'$ $- \frac{3}{2}E_2'$. This yields a new system:

E_1: $2x_1 + x_2 - x_3 + x_4 + 0x_5 = 0,$

E_2': $x_2 + \frac{3}{2}x_3 + \frac{1}{2}x_4 - 2x_5 = 0,$

E_3'': $-\frac{7}{4}x_3 - \frac{5}{4}x_4 + 2x_5 = 0.$

Now each equation has one fewer variable than the one preceding. To get a solution for the single equation E_3'', we may choose $x_3 = 5$, $x_4 = -7$, $x_5 = 0$. We did this kind of thing earlier. Plugging these values into E_2', we obtain

$$x_2 + \tfrac{15}{2} - \tfrac{7}{2} - 0 = 0,$$

so that $x_2 = -4$. Now plugging our values for x_2, x_3, x_4, x_5 into E_1, we obtain

$$2x_1 - 4 - 5 - 7 + 0 = 0,$$

so that $x_1 = 8$. Thus we have constructed a solution

$$(x_1, x_2, x_3, x_4, x_5) = (8, -4, 5, -7, 0)$$

to the original system E_1, E_2, E_3. ≪

REMARKS: 1. We can obtain a different solution to the full system by choosing x_3, x_4, x_5 differently as solutions to E_3''.

2. The procedure used here is sometimes called **gaussian elimination**, after the immortal German mathematician Karl Friedrich Gauss (1777–1855).

3. Did we use $n > q$ (here $5 > 3$)? Where?

4. Gaussian elimination may be used to obtain solutions in more general situations, *provided that the solutions exist*. Here is an example of an inhomogeneous system worth remembering:

$$x_1 + x_2 + x_3 = 1,$$
$$3x_1 + 3x_2 + 3x_3 = 0.$$

Subtracting three times the first equation from the second (to eliminate x_1) yields a new second equation:

$$0x_1 + 0x_2 + 0x_3 = -3.$$

And this equation (hence the system) clearly has no solutions.

Appendix 2.   The hyperplane theorem

Theorems 4 and 5 assure us that a two-dimensional plane subspace of $\mathbb{R}^3$ is the same as the set of solutions to a single linear equation of the form

$\alpha_1 x_1 + \alpha_2 x_2 + \alpha_3 x_3 = 0$. Suppose now that we are in $\mathbb{R}^n$. How should we describe the set of solutions to a single equation $\alpha_1 x_1 + \cdots + \alpha_n x_n = 0$, with not all the α's equal to zero? It is not hard to see that this set of solutions is a subspace of $\mathbb{R}^n$. How big is it? How big, for example, is the set of all solutions to the equation $x_4 = 0$ in $\mathbb{R}^4$?

To deal with these questions in $\mathbb{R}^n$, let us extend the concept of dimension to more general situations. All the needed ingredients occur in the definition of two-dimensional subspace (see the preceding section). The first condition there states that two vectors X and Y are not on the same straight line through the origin. If they were, then they would be "dependent," and we could write Y, say, as αX for some scalar α (possibly zero). The generalized definition is: The system of vectors X_1, X_2, $\ldots$, X_r is **linearly dependent** if and only if there exist scalars α_1, α_2, $\ldots$, α_r not all zero such that $\alpha_1 X_1 + \alpha_2 X_2 + \cdots + \alpha_r X_r = O$. Note that if X_1, X_2, $\ldots$, X_r are dependent with, say, $\alpha_1 \neq 0$, then we may write $X_1 = (-1/\alpha_1)(\alpha_2 X_2 + \cdots + \alpha_r X_r)$. Compare this with $Y = \alpha X$ in the case of two vectors. If X_1, X_2, $\ldots$, X_r are not dependent, then we say they are **linearly independent**. It is not hard to see that the standard basis vectors $e_1 = (1, 0, \ldots, 0), \ldots, e_n = (0, \ldots, 0, 1)$ are independent in $\mathbb{R}^n$.

EXERCISE: Show that $X_1 = (1, 1, 0)$, $X_2 = (2, 0, 3)$, $X_3 = (1, -1, 3)$ are linearly dependent.

The second condition in the definition of two-dimensional plane subspace involved the idea of "span": *Every* vector could be written in the form $\alpha X + \beta Y$. We generalize it now. The vector space $\mathscr{S}$ (possibly a subspace of a larger space) is **spanned** by the system X_1, X_2, $\ldots$, X_r of vectors in $\mathscr{S}$ provided that every $X \in \mathscr{S}$ can be written as a linear combination of the X_j, that is,

$$X = \alpha_1 X_1 + \alpha_2 X_2 + \cdots + \alpha_r X_r$$

for suitable scalars $\alpha_1, \alpha_2, \ldots, \alpha_r$. Note that the standard basis vectors $e_1, e_2, \ldots, e_n$ do span $\mathbb{R}^n$, while in $\mathbb{R}^3$ the vectors e_1, e_2 span the $x_1 x_2$-plane ($x_3 = 0$).

In order to prove anything about the size of a subspace, we must introduce some numerology into our discussion of linear dependence and span.

MAIN LEMMA ON LINEAR DEPENDENCE: Say that $X_1, \ldots, X_r$ span a vector space $\mathscr{S}$ and $Y_1, \ldots, Y_s$ are in $\mathscr{S}$. If $s > r$, then $Y_1, \ldots, Y_s$ are linearly dependent. Thus, if $Y_1, \ldots, Y_s$ are known to be independent, then we must have $s \leq r$.

Proof: Given $s > r$. We want to find scalars $\gamma_1, \ldots, \gamma_s$, *not all zero*, such that $\gamma_1 Y_1 + \cdots + \gamma_s Y_s = 0$. We know each $Y_j = \alpha_{1j} X_1 + \cdots + \alpha_{rj} X_r$, for some constants $\alpha_{1j}, \ldots, \alpha_{rj}$ (why?), so that replacing the Y's by the X's and collecting terms yield the following condition on $\gamma_1, \ldots, \gamma_s$:

$$(\alpha_{11}\gamma_1 + \alpha_{12}\gamma_2 + \cdots + \alpha_{1s}\gamma_s)X_1 + \cdots + (\alpha_{r1}\gamma_1 + \cdots + \alpha_{rs}\gamma_s)X_r = 0.$$

This is the same as the original equation in the Y's. We now ask, naturally, can we solve this long equation (recall that we want a nontrivial solution, that is, at least one $\gamma_j \neq 0$)? We can find a solution if the γ_j's, not all zero, satisfy

$$\alpha_{11}\gamma_1 + \alpha_{12}\gamma_2 + \cdots + \alpha_{1s}\gamma_s = 0$$
$$\cdot \qquad \cdot \qquad \qquad \cdot \qquad \cdot$$
$$\cdot \qquad \cdot \qquad \qquad \cdot \qquad \cdot$$
$$\cdot \qquad \cdot \qquad \qquad \cdot \qquad \cdot$$
$$\alpha_{r1}\gamma_1 + \alpha_{r2}\gamma_2 + \cdots + \alpha_{rs}\gamma_s = 0.$$

The answer is *yes*, because the number of unknowns is greater than the number of equations, $s > r$ (by Appendix 1). Since we can find the γ's (not all zero), the Y's are dependent and the lemma is proved. $\ll$

EXERCISE: Use the lemma to prove that any six vectors in $\mathbb{R}^5$ are linearly dependent.

Now another definition that builds on the preceding. Given a vector space $\mathscr{S}$ and vectors $X_1, \ldots, X_n$ in $\mathscr{S}$ such that (1) $X_1, \ldots, X_n$ are linearly independent and (2) they span $\mathscr{S}$, we call $X_1, \ldots, X_n$ a **basis** for $\mathscr{S}$. Note that $e_1, \ldots, e_n$ does give a basis for $\mathbb{R}^n$ in this new sense.

Now if $X_1, \ldots, X_n$ is a basis for $\mathscr{S}$, then we say that the **dimension** of $\mathscr{S}$ is n and write dim $\mathscr{S} = n$. Note that $\mathbb{R}^5$ *is* five-dimensional.

One trouble. Maybe we can find two bases $X_1, \ldots, X_n$ and $Y_1, \ldots, Y_q$ for $\mathscr{S}$ with $q \neq n$. Then dim $\mathscr{S} = n$ and also dim $\mathscr{S} = q$; nonsense. Fortunately, it is not hard to see that *any two bases for $\mathscr{S}$ must have the same number of elements*, $n = q$. Here's how. Since $X_1, \ldots, X_n$ span $\mathscr{S}$ (why?) and $Y_1, \ldots, Y_q$ are independent (why?), we must have $q \leq n$. Reversing the roles of the X's and Y's, we conclude also that $n \leq q$. Thus $n = q$. Done. $\ll$

EXERCISE: Show that $X_1 = (1, 1)$, $X_2 = (1, -1)$ gives a basis for $\mathbb{R}^2$. Can you find another basis?

At last we have a measure, dimension, of the size of a vector space. Now we can generalize Theorem 5.

THE HYPERPLANE THEOREM: A subspace $\mathscr{S}$ in $\mathbb{R}^n$ is a hyperplane (that is, dim $\mathscr{S} = n - 1$) $\Leftrightarrow \mathscr{S}$ is precisely the set of solutions $X = (x_1, \ldots, x_n)$ in $\mathbb{R}^n$ of a homogeneous linear equation $\alpha_1 x_1 + \cdots + \alpha_n x_n = 0$ with some α's nonzero.

NOTE: Thus a hyperplane in $\mathbb{R}^3$ is an ordinary plane, while in $\mathbb{R}^2$ a hyperplane is a straight line.

Proof:

($\Rightarrow$) Since dim $\mathscr{S} = n - 1$, there is a basis $X_1, \ldots, X_{n-1}$ for $\mathscr{S}$. Say that $X_i = (c_{i1}, c_{i2}, \ldots, c_{in})$, $i = 1, \ldots, n - 1$. To get the α's, that is, to construct the equation $\alpha_1 x_1 + \cdots + \alpha_n x_n = 0$, form the system

$$c_{11}\alpha_1 \quad + c_{12}\alpha_2 \quad + \cdots + c_{1n}\alpha_n \quad = 0$$
$$\vdots \qquad\qquad \vdots \qquad\qquad\qquad \vdots \qquad\qquad \vdots$$
$$c_{n-1,1}\alpha_1 + c_{n-1,1}\alpha_2 + \cdots + c_{n-1,n}\alpha_n = 0.$$

This is just what we did when given two vectors X_1, X_2 in $\mathbb{R}^3$. Again we have more unknowns than equations, and so we know that there exist $\alpha_1, \ldots, \alpha_n$, not all zero, satisfying the system. It follows that each X_i gives a solution to the single equation thus found:

$$\alpha_1 x_1 + \cdots + \alpha_n x_n = 0.$$

Since $X_1, \ldots, X_{n-1}$ is a basis for $\mathscr{S}$, every $X = (x_1, \ldots, x_n)$ in $\mathscr{S}$ satisfies this equation. Since the space of solutions to the single equation is not all of $\mathbb{R}^n$ (why not?) and since it contains the $(n - 1)$-dimensional space $\mathscr{S}$, it must be $\mathscr{S}$. This proves half of the theorem.

($\Leftarrow$) Here is a constructive proof. After renumbering the variables, we may suppose that the single linear equation has the form

$$\alpha_1 x_1 + \cdots + \alpha_q x_q + 0 x_{q+1} + \cdots + 0 x_n = 0,$$

with $\alpha_1, \ldots, \alpha_q$ nonzero. It is easy to see that the vectors

$$X_2 = (\alpha_2, -\alpha_1, 0, 0, \ldots \ldots, 0)$$
$$X_3 = (\alpha_3, 0, -\alpha_1, 0, \ldots \ldots, 0)$$
$$\vdots$$
$$X_q = (\alpha_q, 0, \ldots, 0, -\alpha_1, 0, \ldots, 0),$$

together with the vectors

$$X_{q+1} = (0, \ldots \ldots, 0, 1, 0, \ldots, 0) \qquad \text{here only } x_{q+1} = 1$$
$$\phantom{X_{q+1}} \quad . \qquad\qquad . \qquad\qquad\qquad\qquad .$$
$$\phantom{X_{q+1}} \quad . \qquad\qquad . \qquad\qquad\qquad\qquad .$$
$$\phantom{X_{q+1}} \quad . \qquad\qquad . \qquad\qquad\qquad\qquad .$$
$$X_n \quad = (0, \ldots \ldots \ldots \ldots, 0, 1),$$

form a set of $n - 1$ *independent* solutions to the single equation. Thus the space of solutions (call it $\mathscr{S}$) satisfies dim $\mathscr{S} \geq n - 1$.

Can dim $\mathscr{S} = n$? To see that it cannot, observe first that $\mathscr{S} \neq \mathbb{R}^n$, because we can easily construct a vector in $\mathbb{R}^n$ that does not satisfy $\alpha_1 x_1 + \cdots + \alpha_n x_n = 0$. Second, if dim $\mathscr{S} = n$, then we could take a vector Y in R^n but not in $\mathscr{S}$ that, together with a basis for $\mathscr{S}$, would give $n + 1$ linearly independent vectors in R^n. This is impossible, since dim $\mathbb{R}^n = n$. Thus we have a sandwich

$$n - 1 \leq \dim \mathscr{S} \leq n - 1.$$

We can only conclude dim $\mathscr{S} = n - 1$. Done. $\ll$

REMARKS: 1. We have translated "dimension," a spatial concept, into algebra, using the idea of a basis.

2. The key algebraic idea in the Hyperplane Theorem is the fact that a system of homogeneous linear equations in sufficiently many unknowns always has a nontrivial solution. Again and again in linear algebra, questions about subspaces, planes, dimension, and so forth, are settled by translating the question into one about the solvability of such a system of equations.

FUNCTION SPACES

Throughout this book, the formal discussion of vector spaces is limited to $\mathbb{R}^n$ and its subspaces. The power and significance of the vector space idea, however, is due to the fact that many different sets in mathematics satisfy Properties 1 to 8 and hence are vector spaces.

EXAMPLES: 1. $\mathscr{P}_n$, the space of polynomials

$$p(x) = a_0 + a_1 x + \cdots + a_n x^n$$

of degree at most n. Observe that if p, q are in $\mathscr{P}_n$, then so is $p + q$ and cp, where c is any scalar. Properties 1 to 8 are evidently satisfied, and so

$\mathscr{P}_n$ is a vector space. Can we find a "basis" for $\mathscr{P}_n$ (see Appendix 2)? Sure. Notice that a polynomial $p \in \mathscr{P}_n$ is uniquely determined by its coefficients $a_0, a_1, \ldots, a_n$. Thus we can associate p with the vector $(a_0, a_1, \ldots, a_n)$. If we let $e_0 = 1$, $e_1 = x$, $e_2 = x^2, \ldots, e_n = x^n$, then any $p \in \mathscr{P}_n$ can be written uniquely as

$$p = a_0e_0 + a_1e_1 + \cdots + a_ne_n.$$

Thus $e_0, \ldots, e_n$ is a "basis" for $\mathscr{P}_n$. Since each p in $\mathscr{P}_n$ is determined by $n + 1$ numbers $a_0, a_1, \ldots, a_n$, we see that the dimension of $\mathscr{P}_n$ is $n + 1$. Note: In considering $\mathscr{P}_n$ as a vector space, we *ignore* the fact that we can multiply polynomials as well as add them.

2. $\mathscr{C}[-7, 3]$, the set of functions f continuous on the interval $-7 \leq x \leq 3$. Now, if f and g are in $\mathscr{C}[-7, 3]$, then so is $f + g$ (the sum of continuous functions is also continuous), as is cf, where c is any scalar. Properties 1 to 8 of a vector space are the standard algebraic rules for functions. Thus $\mathscr{C}[-7, 3]$ is a vector space whose elements (the "vectors") are continuous functions. Remarkable!

Since every polynomial is a continuous function, it is clear that $\mathscr{P}_n \subset \mathscr{C}[-7, 3]$. But $\mathscr{P}_n$ is also a vector space. Therefore $\mathscr{P}_n$ is a subspace of $\mathscr{C}[-7, 3]$.

Now the dimension of $\mathscr{P}_n$ is $n + 1$. Since n can be any, possibly huge, integer, we are led to the fact that $\mathscr{C}[-7, 3]$ is *infinite-dimensional*. Although this is a bit terrifying, one must not get upset. It just means that the space has lots of room. The virtue of considering it as a vector space is that many features of $\mathbb{R}^n$ depend only on the fact that it is a vector space and *not* on its dimension. These vector-space results therefore carry over immediately to the more complicated space $\mathscr{C}[-7, 3]$, and so we prove theorems about a complicated object by merely thinking of pictures in the much simpler $\mathbb{R}^2$ or $\mathbb{R}^3$. Both $\mathscr{C}[-7, 3]$ and $\mathscr{P}_n$ are examples of vector spaces of functions, that is, *function spaces*.

Only in this century have mathematicians realized that many seemingly unrelated sets of objects were all vector spaces, and they have grown to know and cherish them. Physicists and many other applied scientists often use infinite-dimensional vector spaces (they are fundamental to quantum mechanics). Anyone going on to use higher mathematics will soon find that he must make his peace with many kinds of vector spaces. $\mathbb{R}^n$ is fine, but it is just a beginning.

1.5 AFFINE SUBSPACES

1.5a Introduction

This brief section will be useful when we study the very important affine functions.

So far the lines, planes, and so on that we have discussed have all contained the origin of $\mathbb{R}^n$; they have all been subspaces in the sense of vector space theory. We know from calculus, however, that we must often consider more general lines that do not necessarily contain the origin. The same is true of planes, and so on, in higher dimensions.

The following example contains most of the points we will raise. Let $\mathscr{S}$ be a line through the origin in $\mathbb{R}^2$, and $\mathscr{A}$ a line parallel to $\mathscr{S}$ but different from $\mathscr{S}$. See Fig. 1.14. We know that $\mathscr{S}$ is a subspace and $\mathscr{A}$ is not (why?). First question: How can we obtain the set $\mathscr{A}$ from the set $\mathscr{S}$? That is, characterize the points of $\mathscr{A}$ in terms of the points of $\mathscr{S}$.

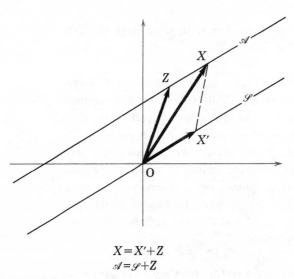

$$X = X' + Z$$
$$\mathscr{A} = \mathscr{S} + Z$$

FIGURE 1.14

An answer: Let Z be a *fixed* vector in $\mathscr{A}$. Then if X is any vector in $\mathscr{A}$, there exists a unique X' in $\mathscr{S}$ such that $X = X' + Z$. In fact, X' is just $X - Z$.

If we use the illuminating notation $\mathscr{S} + Z$ to denote the set

$\{X' + Z \mid X' \in \mathscr{S}\}$, then we have that $\mathscr{A} = \mathscr{S} + Z$. The two sets have precisely the same elements.

Second question: Is Z unique?

The answer: Clearly not. If Z_1 is *any* vector in $\mathscr{A}$, then $\mathscr{S} + Z_1$ is the same set as $\mathscr{S} + Z$, namely, $\mathscr{A}$.

We can say more, however, and we should. Although Z and Z_1 need not be equal, that is, their difference $Z_1 - Z$ need not be O, yet they are still related, for their difference $Z_1 - Z$ must be in $\mathscr{S}$. This is because Z_1 can be written $X' + Z$ with $X' \in \mathscr{S}$, whence $Z_1 - Z \in \mathscr{S}$.

1.5b A definition

True to form, we use these specific observations in $\mathbb{R}^2$ to make a general definition. A subset $\mathscr{A}$ of $\mathbb{R}^n$ is an **affine subspace** if and only if $\mathscr{A}$ is of the form $\mathscr{S} + Z$, where $\mathscr{S}$ is a subspace of $\mathbb{R}^n$ in the usual sense and Z is a vector in $\mathbb{R}^n$. In this case, $\mathscr{A}$ and $\mathscr{S}$ are said to be **parallel**.

Further, it is natural to call $\mathscr{A}$ a **line** if $\mathscr{A} = \mathscr{S} + Z$, where $\mathscr{S}$ is a line through the origin in $\mathbb{R}^n$, or a *plane* if $\mathscr{S}$ is a plane through the origin. Lines and planes are again one- and two-dimensional, respectively.

Examples of affine subspaces in $\mathbb{R}^2$ and $\mathbb{R}^3$ abound. Some examples in $\mathbb{R}^3$ are $x_3 = 1$ and $2x_1 + x_2 - x_3 = 4$. See Fig. 1.15.

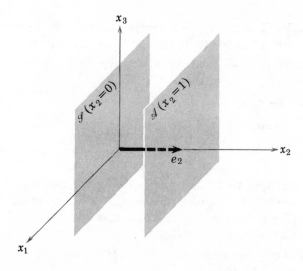

$$\mathscr{A} = \mathscr{S} + e_2$$

FIGURE 1.15

REMARK: Here are some consequences of the definition $\mathscr{A} = \mathscr{S} + Z$:

1. If Z happens to be in $\mathscr{S}$, then $\mathscr{A} = \mathscr{S} + Z = \mathscr{S}$.

2. If Z is not in $\mathscr{S}$, then $\mathscr{A} \neq \mathscr{S}$ as sets, because $Z = O + Z$ *is* in $\mathscr{A}$, and the sets do not have the same elements. In fact, they have no elements in common (they *are* parallel). In this case, moreover, $\mathscr{A}$ *is not a subspace in the usual sense*, since O is not in $\mathscr{A}$. This is one of those accidents of language which plague pedants but not reasonable men: An affine subspace is seldom a subspace in the sense of vector space theory. It is a "subspace" that has been moved (translated) to one side or another, away from the origin.

Here is a question you might think about. Suppose that we are told that a certain subset $\mathscr{A}$ of $\mathbb{R}^n$ is, in fact, an affine subspace. Thus there is a subspace $\mathscr{S}$ such that $\mathscr{A} = \mathscr{S} + Z$ for some $Z \in \mathscr{A}$. Describe $\mathscr{S}$ in terms of $\mathscr{A}$. What vectors are in $\mathscr{S}$? Is $\mathscr{S}$ uniquely determined by $\mathscr{A}$?

1.5c Affine subspaces and linear equations

To see what might be true with regard to equations, let us return to $\mathbb{R}^2$. In the example above, the line $\mathscr{S}$ through the origin is given by an equation $\alpha_1 x_1 + \alpha_2 x_2 = 0$, where α_1, α_2 are known scalars:

$$\mathscr{S} = \{(x_1, x_2) \,|\, \alpha_1 x_1 + \alpha_2 x_2 = 0\}.$$

Now let $\mathscr{A} = \mathscr{S} + Z$ be an affine subspace. Problem: Find an equation that determines $\mathscr{A}$.

To solve this problem, let $X = (x_1, x_2)$ be a typical point on $\mathscr{A}$. Then we know $X = X' + Z$ with $X' \in \mathscr{S}$. We write $X' = (x_1', x_2')$ and $Z = (z_1, z_2)$, so that $(x_1, x_2) = (x_1' + z_1, x_2' + z_2)$. Now we plug this into the equation for $\mathscr{S}$:

$$\begin{aligned}
\alpha_1 x_1 + \alpha_2 x_2 &= \alpha_1(x_1' + z_1) + \alpha_2(x_2' + z_2) \\
&= \alpha_1 x_1' + \alpha_2 x_2' + \alpha_1 z_1 + \alpha_2 z_2 \\
&= \alpha_1 z_1 + \alpha_2 z_2,
\end{aligned}$$

since $\alpha_1 x_1' + \alpha_2 x_2' = 0$.

The constant $\alpha_1 z_1 + \alpha_2 z_2$ appears at first glance to depend on which fixed vector Z we chose from $\mathscr{A}$. This is false, however. If Z' is any other vector in $\mathscr{A}$, say $Z' = (z_1', z_2')$, then $\alpha_1 z_1' + \alpha_2 z_2' = \alpha_1 z_1 + \alpha_2 z_2$, because $Z - Z'$ is in $\mathscr{S}$. Check this!

We summarize our observations in $\mathbb{R}^2$ as follows: *If the subspace $\mathscr{S}$ consists of all vectors $X = (x_1, x_2)$ satisfying*

$$\alpha_1 x_1 + \alpha_2 x_2 = 0,$$

then the affine subspace $\mathscr{A} = \mathscr{S} + Z$ consists of all $X = (x_1, x_2)$ satisfying

$$\alpha_1 x_1 + \alpha_2 x_2 = \beta,$$

where the constant $\beta = \alpha_1 z_1 + \alpha_2 z_2$, $Z = (z_1, z_2)$. If also $\mathscr{A} = \mathscr{S} + Z$ $= \mathscr{S} + Z'$ with $Z' = (z_1', z_2')$, then $\beta = \alpha_1 z_1 + \alpha_2 z_2 = \alpha_1 z_1' + \alpha_2 z_2'$.

A numerical example in the same vein: Let $\mathscr{S}$ be the line through the origin in $\mathbb{R}^2$ given by $3x_1 - 4x_2 = 0$. If $\mathscr{A}$ is the line parallel to $\mathscr{S}$ through the point $Z = (6, 11)$, then $\mathscr{A} = \mathscr{S} + Z$ consists of those (x_1, x_2) satisfying

$$3x_1 - 4x_2 = 3 \cdot 6 - 4 \cdot 11 = -26.$$

Draw a picture! Other points on $\mathscr{A}$ are $(0, 13/2)$, $(-26/3, 0)$, $(-10, -1)$. You can check that the difference of any two of these lies on $\mathscr{S}$.

1.5d Three points determine a plane

Let us examine this familiar phrase in $\mathbb{R}^3$. Recall first that we defined a plane subspace to be the set spanned by vectors X_1, X_2 not collinear with the origin in $\mathbb{R}^3$. In this case the "three points" are X_1, X_2 (heads of the arrows) and the origin O.

Here is a problem: Imagine a plane $\mathscr{A}$ in $\mathbb{R}^3$ balanced on three given points Z_1, Z_2, Z_3. See Fig. 1.16. Determine a subspace $\mathscr{S}$ and vector Z such that $\mathscr{A} = \mathscr{S} + Z$. Also, find equations for the parallel planes $\mathscr{A}$ and $\mathscr{S}$.

Actually, we have accomplished all the preliminaries needed to handle this problem. Let us solve it in the special case where $Z_1 = (1, 2, 4)$, $Z_2 = (3, 1, -2)$, $Z_3 = (0, 2, 2)$.

1. The main problem is the determination of the subspace $\mathscr{S}$. We must find vectors X_1, X_2 that span $\mathscr{S}$. See Fig. 1.17. To do this, recall that $\mathscr{A}$ must equal $\mathscr{S} + Z_3$, since Z_3 is in $\mathscr{A}$. It follows that the vectors $Z_1 - Z_3$ and $Z_2 - Z_3$ are in $\mathscr{S}$. So we form $X_1 = Z_1 - Z_3 = (1, 0, 2)$ and $X_2 = (3, -1, -4)$. These are in $\mathscr{S}$ and do not lie on the same line through the origin. It follows that they span a plane subspace that must be $\mathscr{S}$.

2. Now let us find an equation $\alpha_1 x_1 + \alpha_2 x_2 + \alpha_3 x_3 = 0$ whose space of solutions $X = (x_1, x_2, x_3)$ is precisely the plane $\mathscr{S}$. We saw this in Sec. 1.4e. To determine $\alpha_1, \alpha_2, \alpha_3$, we plug X_1 and X_2 successively into the formal equation, obtaining two simultaneous equations:

$$\alpha_1 + \qquad 2\alpha_3 = 0,$$

$$3\alpha_1 - \alpha_2 - 4\alpha_3 = 0.$$

The general solution here is $\alpha_1 = -2\alpha_3$, $\alpha_2 = -10\alpha_3$.

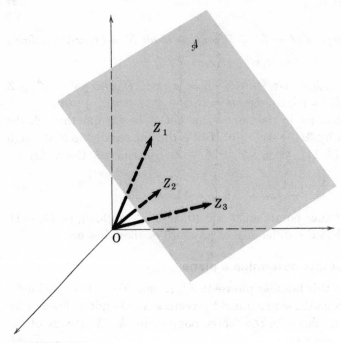

FIGURE 1.16

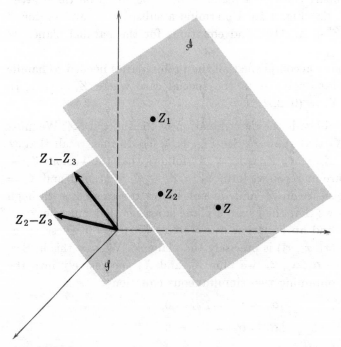

FIGURE 1.17

48

Thus, $\mathscr{S}$ is the space of solutions to any equation of the form

$$\alpha_3(-2x_1 - 10x_2 + x_3) = 0$$

provided that $\alpha_3 \neq 0$, since all these equations clearly have the same space of solutions.

3. Now we show that every vector $X = (x_1, x_2, x_3)$ in the original plane $\mathscr{A}$ satisfies an equation of the form $\alpha_1 x_1 + \alpha_2 x_2 + \alpha_3 x_3 = \beta$, with $\beta \neq 0$ and $\alpha_1, \alpha_2, \alpha_3$, as found above.

We know, for instance, that every vector in the subspace $\mathscr{S}$ satisfies the equation

$$2x_1 + 10x_2 - x_3 = 0,$$

obtained by letting $\alpha_3 = -1$ in the general equation above.

Now since every vector X in $\mathscr{A}$ is of the form $X = X' + Z_3$, with X' in $\mathscr{S}$ (for instance, $Z_1 = X_1 + Z_3$), we see that each such X satisfies $2x_1 + 10x_2 - x_3 = \beta$, where β is obtained by substituting Z_3 into $2x_1 + 10x_2 - x_3$. Since $Z_3 = (0, 2, 2)$, $\beta = 18$. Thus $\mathscr{A}$ is the affine subspace of solutions to the inhomogeneous equation

$$2x_1 + 10x_2 - x_3 = 18.$$

You might check that, in particular, Z_1 and Z_2 satisfy this equation.

Let us recapitulate. We were given the points Z_1, Z_2, Z_3. Note that these three did not lie on any one line. They thereby determined a unique plane $\mathscr{A}$ in 3-space:

1. We formed the subspace $\mathscr{S}$ spanned by $X_1 = Z_1 - Z_3$, $X_2 = Z_2 - Z_3$. Then $\mathscr{A} = \mathscr{S} + Z_3$.

2. Using X_1, X_2, we obtained the equation (homogeneous) $2x_1 + 10x_2 - x_3 = 0$, whose space of solutions was $\mathscr{S}$.

3. We then observed that every element X of $\mathscr{A}$ satisfied $2x_1 + 10x_2 - x_3 = 18$, using the fact that Z_3 satisfied this equation.

Now we state a general theorem whose proof is accessible to us.

THEOREM 6: The subset $\mathscr{A}$ of $\mathbb{R}^3$ is a plane $\Leftrightarrow \mathscr{A}$ is the affine subspace of solutions to a linear equation $\alpha_1 x_1 + \alpha_2 x_2 + \alpha_3 x_3 = \beta$, with at least one of $\alpha_1, \alpha_2, \alpha_3$ different from zero. In this case, moreover, $\mathscr{A}$ is parallel to the two-dimensional subspace $\mathscr{S}$ of solutions to the homogeneous equation $\alpha_1 x_1 + \alpha_2 x_2 + \alpha_3 x_3 = 0$; that is, $\mathscr{A} = \mathscr{S} + Z$, where Z is any solution to $\alpha_1 x_1 + \alpha_2 x_2 + \alpha_3 x_3 = \beta$. Thus, if $Z = (z_1, z_2, z_3)$, then $\beta = \alpha_1 z_1 + \alpha_2 z_2 + \alpha_3 z_3$.

ANOTHER EXAMPLE: Let $\mathscr{A}$ be the affine plane in $\mathbb{R}^3$ that intersects the three coordinate axes at the points e_1, e_2, e_3. See Fig. 1.18. Problem:

(1) Find the subspace $\mathscr{S}$ of $\mathbb{R}^3$ parallel to $\mathscr{A}$. (2) Find a linear equation that has $\mathscr{A}$ as its set of solutions.

We handle this just as we did the previous example:

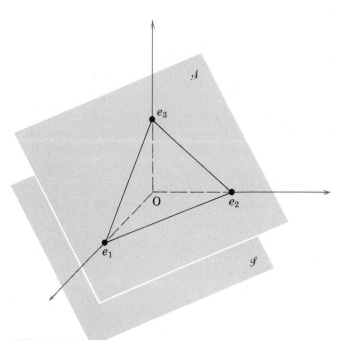

FIGURE 1.18

1. The subspace $\mathscr{S}$ is spanned by the two vectors X_1, X_2 defined by $X_1 = e_1 - e_3$, $X_2 = e_2 - e_3$. (We could have chosen $e_2 - e_1$, $e_3 - e_1$, and so on, but these two vectors span the same $\mathscr{S}$.) This settles (1).

2. Now we obtain a linear equation $\alpha_1 x_1 + \alpha_2 x_2 + \alpha_3 x_3 = 0$ that determines $\mathscr{S}$. Since $X_1 = (1, 0, -1)$ and $X_2 = (0, 1, -1)$ must be solutions to this linear equation (why?), we see that

$$\alpha_1 - \alpha_3 = 0, \qquad \alpha_2 - \alpha_3 = 0,$$

whence

$$\alpha_1 = \alpha_2 = \alpha_3.$$

Let us choose all three coefficients equal to 1. This yields a homogeneous equation for $\mathscr{S}$,

$$x_1 + x_2 + x_3 = 0.$$

3. An equation determining the original plane $\mathscr{A}$ will thereby have the form

$$x_1 + x_2 + x_3 = \beta.$$

Since $e_1 = (1, 0, 0)$ must satisfy this equation (why?), we conclude $\beta = 1$, and so $\mathscr{A}$ is the set of solutions to

$$x_1 + x_2 + x_3 = 1.$$

Note that e_2 and e_3 also satisfy the last equation.

Exercises

Some elementary computations with affine subspaces in the same spirit as the preceding sections.

1. Lines in $\mathbb{R}^2$. Let $\mathscr{S}$ be the subspace of $\mathbb{R}^2$ spanned by the vector $X = (2, -1)$. Let $\mathscr{A}$ be the affine subspace parallel to $\mathscr{S}$ through the point $Z = (3, 0)$:

    (a) Compute a homogeneous linear equation in x_1, x_2 determining $\mathscr{S}$.
    (b) Compute a linear equation determining $\mathscr{A}$.
    (c) Does $\mathscr{A} = \mathscr{S} + Z$?
    (d) Find two other points on $\mathscr{A}$.
    (e) Sketch $\mathbb{R}^2$, $\mathscr{S}$, $\mathscr{A}$, X, Z and the points found in Exercise 1(d).

2. Planes in $\mathbb{R}^3$. Let $\mathscr{S}$ be the subspace (strict sense) of $\mathbb{R}^3$ spanned by $X_1 = (1, 1, 0)$, $X_2 = (0, 1, 1)$. Let $\mathscr{A}$ be the affine subspace parallel to $\mathscr{S}$ through the point $Z = (2, 2, 2)$:

    (a) Compute a homogeneous linear equation in x_1, x_2, x_3 determining $\mathscr{S}$.
    (b) Compute a linear equation determining $\mathscr{A}$.
    (c) Does $\mathscr{A} = \mathscr{S} + Z$?
    (d) Find some other points on $\mathscr{A}$.
    (e) Sketch.

3. Let $\mathscr{S}$ be the two-dimensional vector subspace of $\mathbb{R}^3$ spanned by the standard basis vectors e_1, e_3:

    (a) Compute an equation determining $\mathscr{S}$.
    (b) Compute an equation determining the affine subspace $\mathscr{A}$ parallel to $\mathscr{S}$ through the point $Z = (-3, 7, 4)$; through the point $Z_1 = (4, 7, -6)$; through the point $Z_2 = (4, 8, -5)$.

4. Let $\mathscr{A}$ be the (affine) plane in $\mathbb{R}^3$ determined by the three points $e_1, e_2, 2e_3$ in space. Let $\mathscr{S}$ be the two-dimensional subspace of $\mathbb{R}^3$ parallel to $\mathscr{A}$, and containing the origin, of course:

    (a) Compute a homogeneous equation in x_1, x_2, x_3 determining $\mathscr{S}$.
    (b) Compute an equation determining $\mathscr{A}$.
    (c) Do the same for the affine plane $\mathscr{A}'$ determined by $2e_1, 2e_2, 2e_3$.

5.   (a) Let $\mathscr{A}$ be the affine plane determined by $x_2 = 5$ in $\mathbb{R}^3$. What is an equation determining the subspace $\mathscr{S}$ parallel to $\mathscr{A}$?

(b) Same question where the equation for $\mathscr{A}$ is $x_1 + x_2 + x_3 = 5$.

6.   Prove Theorem 6.

7.   (a) Let $\mathscr{A} \subset \mathbb{R}^3$ be an affine subspace and X_1, X_2, X_3 be in $\mathscr{A}$. If $a + b + c = 1$ and $X = aX_1 + bX_2 + cX_3$, prove that X is in $\mathscr{A}$.

(b) Let $X_1, X_2, X_3 \in \mathbb{R}^3$ be nonzero vectors. Show that the set

$$\mathscr{A} = \{X \mid X = aX_1 + bX_2 + cX_3, \text{ where } a + b + c = 1\}$$

is an affine subspace.

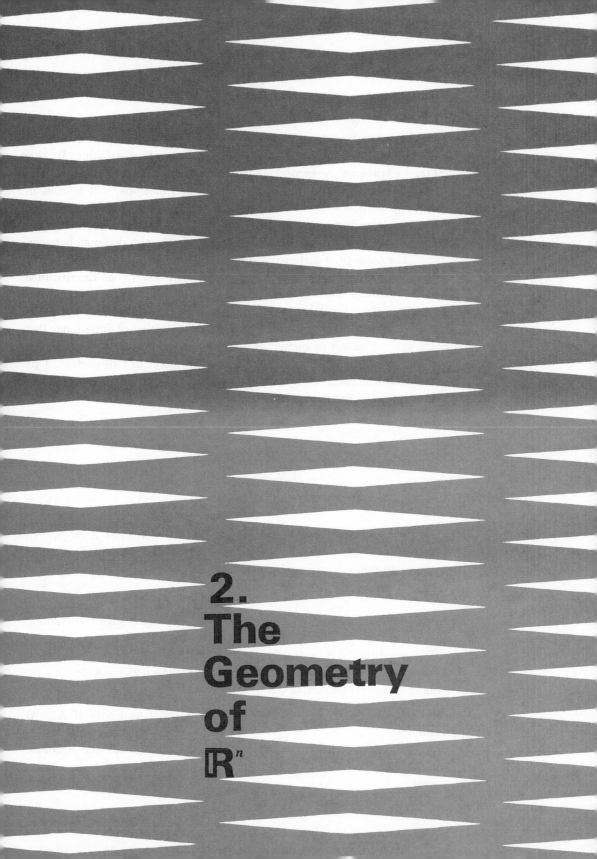

2.
The
Geometry
of
$\mathbb{R}^n$

2.0 INTRODUCTION

With our treatment of affine subspaces of $\mathbb{R}^n$ we completed the first stage of our four-stage approach to differential vector calculus. In this chapter we traverse the second stage, which involves distance and geometry. The last two stages, linear and nonlinear functions, will be met in subsequent chapters.

We now introduce geometry by developing the notion of the length (or "norm") of a vector. Using this, we may define the distance between vectors. The concept of distance will be useful later in obtaining a theory of limits in $\mathbb{R}^n$; the phrase "X approaches X_0" will mean that the distance between these two vectors (here X is a variable vector) approaches zero in the sense of real numbers. Just as in calculus, we need the notion of limit to deal with the continuity and differentiability of functions. But we are getting ahead of ourselves.

In this chapter we also study the "inner product" of two vectors. This will be useful in several ways, as we shall see. Let us begin by considering the notion of the length of a vector.

2.1 THE NORM OF A VECTOR

2.1a Introduction

We work in $\mathbb{R}^2$ first, because things are familiar here. Let $X = (x_1, x_2)$. If we think of X as a directed line segment or arrow in the plane, then it is natural to define the **length** or **norm** of X, denoted $\|X\|$, as follows:

$$\|X\| = \sqrt{x_1^2 + x_2^2}.$$

Figure 2.1 should convince you that this is in agreement with Pythagoras' famous theorem on right triangles: $x_1^2 + x_2^2 = \|X\|^2$.

The quantity $\|X\|$ is easy to compute: thus if $X = (1, -1)$, then $\|X\| = \sqrt{(1^2 + (-1)^2)} = \sqrt{2}$; if $Y = (5, 2)$, then $\|Y\| = \sqrt{29}$.

Three basic properties of $\|X\|$ are:

1. Positivity: $\|X\| \geq 0$ and $\|X\| = 0$ (scalar) if and only if $X = O$ (vector).

2. Homogeneity: $\|\alpha X\| = |\alpha| \|X\|$, where $|\alpha|$ denotes the absolute value of the scalar α.

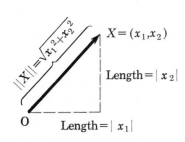

$X = (x_1, x_2)$

Length $= |x_2|$

Length $= |x_1|$

$\|X\| = \sqrt{x_1^2 + x_2^2}$

FIGURE 2.1

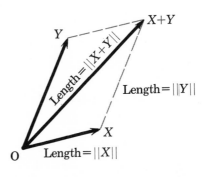

$X+Y$

Y

Length $= \|X+Y\|$

Length $= \|Y\|$

X

Length $= \|X\|$

Observe $\|X+Y\| \le \|X\| + \|Y\|$

FIGURE 2.2

3. Triangle inequality: $\|X + Y\| \le \|X\| + \|Y\|$.

The first property is immediate. The second says that "stretching" the vector X by the factor α changes its length as expected. The third property says that the shortest distance between two points, namely, O and $X + Y$, is along the straight line. See Fig. 2.2.

Although the triangle inequality seems self-evident, a picture is not a proof. We could cook up an elementary proof involving certain magical algebraic identities at crucial steps or by invoking the law of cosines, but we refrain. We will give a proof after introducing the inner product.

2.1b $\|X\|$ in $\mathbb{R}^n$

Now we generalize the notion of length to vectors in n-space. If $X = (x_1, \ldots, x_n)$, we define the **norm** of X as

$$\|X\| = \sqrt{x_1^2 + \cdots + x_n^2}.$$

EXAMPLES: 1. If $X = (1, -1, 1, -1, 1)$ in $\mathbb{R}^5$, then $\|X\| = \sqrt{5}$.

2. If $X = (2, 3, 4)$ in $\mathbb{R}^3$, then $\|X\| = \sqrt{29}$.

This definition of $\|X\|$ clearly generalizes that given for $\mathbb{R}^2$ in the paragraph above. It is sometimes called the **Pythagorean** norm, since $\|X\|^2 = x_1^2 + \cdots + x_n^2$, in analogy with Pythagoras' "sums of squares." See Fig. 2.3.

The space $\mathbb{R}^n$, equipped with the Pythagorean norm, is usually called **Euclidean n-space**, in honor of the famous geometer. Euclidean n-space is sometimes denoted $\mathbb{E}^n$, but we will continue to use $\mathbb{R}^n$.

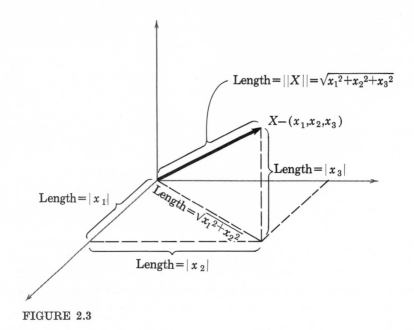

$$\text{Length} = ||X|| = \sqrt{x_1^2 + x_2^2 + x_3^2}$$

$$X - (x_1, x_2, x_3)$$

$$\text{Length} = |x_3|$$

$$\text{Length} = |x_1|$$

$$\text{Length} = \sqrt{x_1^2 + x_2^2}$$

$$\text{Length} = |x_2|$$

FIGURE 2.3

2.1c Distance

What is the **distance** between two points X and Y in Euclidean n-space? We define it to be the norm of their difference, that is, $||X - Y||$. Figure 2.4 shows that this is reasonable. Note $||X||$ *is* the distance from X to 0; that is, $||X|| = ||X - 0||$.

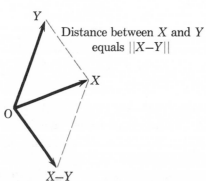

Distance between X and Y equals $||X-Y||$

FIGURE 2.4

An immediate consequence of the definition is the "distance formula." If $X = (x_1, \ldots, x_n)$, $Y = (y_1, \ldots, y_n)$, then the distance from X to Y is given by

$$||X - Y|| = \sqrt{(x_1 - y_1)^2 + \cdots + (x_n - y_n)^2}.$$

EXAMPLE: The distance from $(1, 2, 3, 4)$ to $(2, 3, 1, 4)$ in $\mathbb{R}^4$ is $\sqrt{6}$.

REMARK: It is worth noting that again we have reduced a geometric notion ("length" or "distance") to algebra (the distance formula), just as we did with the geometric notions of "space" and "dimension."

Exercises

These provide practice in computing the norm of a vector and constructing unit vectors with a given direction.

1. Verify (see the examples in Sec. 2.1b) that $\|X\| = \sqrt{5}$ for $X = (1, -1, 1, -1, 1)$ and $\|X\| = \sqrt{29}$ if $X = (2, 3, 4)$.

2. Compute the norms of the following vectors. In $\mathbb{R}^2$:

    (a) $(0, -4)$,   (c) $(1, -1)$,

    (b) $(1, 1)$,   (d) $(3, 4)$;

    In $\mathbb{R}^3$:

    (e) $(0, 0, -4)$,   (g) $(1, 1, 1)$,

    (f) $(1, 2, -2)$,   (h) $\left(\dfrac{1}{\sqrt{3}}, \dfrac{1}{\sqrt{3}}, \dfrac{1}{\sqrt{3}}\right)$;

    In $\mathbb{R}^4$:

    (i) $(0, 0, 0, 4)$,   (k) $(1, 1, 1, 1)$.

    (j) $(\tfrac{1}{2}, \tfrac{1}{2}, \tfrac{1}{2}, \tfrac{1}{2})$,

3. Which property of $\|X\|$ makes the computation of Exercise 2(h) an instant corollary of that of Exercise 2(g) and likewise for Exercises 2(j) and (k)?

4. Compute the distance between X and Y, where:

    (a) $X = (1, 5)$, $Y = (1, 1)$ in $\mathbb{R}^2$;

    (b) $X = (3, 4, 5)$, $Y = (2, 3, 4)$ in $\mathbb{R}^3$;

    (c) $X = (1, 0, 0, 0)$, $Y = (0, 0, 1, 0)$ in $\mathbb{R}^4$.

5. Given nonzero $X \in \mathbb{R}^n$, we wish to find a positive scalar α so that the vector αX, which points in the same direction as X, is a *unit* vector, $\|\alpha X\| = 1$. Solve for α in this equation, thereby obtaining a general formula for α in terms of $\|X\|$.

6. Find unit vectors that have the same direction as the vectors in Exercises 2(e), (f), (g), and (h) (observe immediately that the vector in Exercise 2(h) *is* a unit vector).

7. (a) Show that Property 2 of a norm implies $\|O\| = 0$.

    (b) Show that Properties 2 and 3 of a norm imply $\|X\| \geq 0$. (Thus, Property 1 is really needed only to assert $\|X\| = 0 \Rightarrow X = O$.)

8. Give a direct algebraic proof of the triangle inequality for $\mathbb{R}^2$. Some algebraic fortitude may be needed.

9. (a) If $X \in \mathbb{R}^2$, show that $\|X\|_1 = |x_1| + |x_2|$ defines another norm for $\mathbb{R}^2$ by verifying Properties 1 to 3. (This is sometimes called the "taxicab norm" for $\mathbb{R}^2$. It is what you would use in measuring the distances between two points in a city in terms of city blocks.)

    (b) Show that $\|X\| \leq \|X\|_1 \leq \sqrt{2}\,\|X\|$.

2.2 THE INNER PRODUCT

2.2a Introduction

Let $X = (x_1, \ldots, x_n)$, $Y = (y_1, \ldots, y_n)$ in $\mathbb{R}^n$. We define their **inner product** (denoted $\langle X, Y \rangle$) to be the following number:

$$\langle X, Y \rangle = x_1 y_1 + x_2 y_2 + \cdots + x_n y_n.$$

Thus, if $X = (1, -2, 1)$ and $Y = (-3, 1, 4)$ in $\mathbb{R}^3$, we compute immediately that $\langle X, Y \rangle = (1)(-3) + (-2)(1) + (1)(4) = -1$. Also, $\langle X, X \rangle = 6$ and $\langle Y, Y \rangle = 26$.

This definition of $\langle X, Y \rangle$ may be motivated by the frequency with which the expression $x_1 y_1 + \cdots + x_n y_n$ appears in vector computations. Note that $\langle X, X \rangle = x_1^2 + \cdots + x_n^2 = \|X\|^2$, for example, whence $\|X\| = \sqrt{\langle X, X \rangle}$. This is commonly used.

REMARK: You are warned that our inner product is also widely known as the "dot product" (denoted $X \cdot Y$) or "scalar product." The notation $X \cdot Y$ *is* reasonable; however, we use "inner product" and $\langle X, Y \rangle$ throughout except in Chap. 10, where we use $X \cdot Y$.

FACT OF LIFE: The inner product in $\mathbb{R}^n$ has been defined here in a way that makes it easy to compute, namely,

$$\langle X, Y \rangle = x_1 y_1 + \cdots + x_n y_n.$$

This definition, however, has the disadvantage that it is not clear what the inner product $\langle X, Y \rangle$ *means* (geometrically, say) or what algebraic properties or axioms it satisfies. Now we must work to obtain these geometric and algebraic insights. With work we shall show, for instance, that $\langle X, Y \rangle = \|X\| \|Y\| \cos \theta$, where θ is the angle between X and Y; this is a useful and readily interpretable geometric result.

The above is by no means an apology for an ill-conceived definition. Rather, there is a fact of mathematical life to be acquired here. It is this: First, a new mathematical notion is generally presented in one of three ways:

1. It is given as a computation, such as $\langle X, Y \rangle$ above.
2. It is required to satisfy certain axioms, such as "vector space."
3. It is given so as to be readily visualized or interpreted meaningfully, perhaps in terms of previous concepts, perhaps in terms of geometry, such as "line through the origin."

Admittedly, the three styles of definition may overlap or be open to argument. The second remark, however, is that, for any interesting mathematical object *all three* kinds of information, the computational, the axiomatic, the interpretive, will be necessary. Given one, we must labor to obtain the other two.

Let us give an example. Suppose we had *defined*

$$\langle X, Y \rangle = \| X \| \| Y \| \cos \theta$$

(this is to be our Theorem 1). This inner product cries out with geometric meaning, lengths and angles, yet given two explicit vectors X and Y in some $\mathbb{R}^n$, it is not clear how to compute the number $\langle X, Y \rangle$ in terms of $\| X \| \| Y \| \cos \theta$. Even after computing the lengths $\| X \|$ and $\| Y \|$, we must somehow obtain the cosine of the angle between them. This is by no means immediate. The inner product of $X = (1, 2, -2, -2)$ and $Y = (4, 6, 8, 10)$ in $\mathbb{R}^4$ becomes a major computation if we know only $\langle X, Y \rangle = \| X \| \| Y \| \cos \theta$. We then would be delighted to have the *theorem* that

$$\langle X, Y \rangle = x_1 y_1 + \cdots + x_n y_n$$

(or some such formula). And, as always, discovering and establishing the correct formula may involve insight, logic, faith, luck, and plain hard work in varying amounts.

We close this discussion by suggesting that, when studying a theorem, you ask yourself, "What do I already know about this situation, and what further question is this theorem answering? Will it enable me to compute a number or interpret something geometrically, or will it provide access to another, more distant theorem?"

2.2b Algebraic properties of $\langle X, Y \rangle$

These follow readily from the definition $\langle X, Y \rangle = x_1 y_1 + \cdots + x_n y_n$ and basic properties of the real numbers. We have, for all $X, Y, Z \in \mathbb{R}^n$ and $\alpha \in \mathbb{R}$:

1. Positivity: $\langle X, X \rangle \geq 0$ and $\langle X, X \rangle = 0$ if and only if $X = O$.
2. Symmetry: $\langle X, Y \rangle = \langle Y, X \rangle$.
3. Homogeneity: $\langle \alpha X, Y \rangle = \alpha \langle X, Y \rangle$.
4. Distributivity: $\langle X, Y + Z \rangle = \langle X, Y \rangle + \langle X, Z \rangle$.

The first statement follows from the fact that $x_1^2 + \cdots + x_n^2$ is zero if and only if each of the x_i is zero. The rest are just as immediate.

Is it true, as we would hope, that

$$\langle X + Y, Z \rangle = \langle X, Z \rangle + \langle Y, Z \rangle? \quad \text{(Compare Property 4 as above.)}$$

It is, because $\langle X + Y, Z \rangle = \langle Z, X + Y \rangle$ (by symmetry), which equals $\langle Z, X \rangle + \langle Z, Y \rangle$ (by Property 4), which equals $\langle X, Z \rangle + \langle Y, Z \rangle$, as claimed. You can prove similarly that $\langle X, \alpha Y \rangle = \alpha \langle X, Y \rangle$.

We mention that 1 to 4 are the essential properties of $\langle X, Y \rangle$ and that many theorems are proved by appealing to these properties, rather than to the definition itself.

For example:

1. Prove that if $\langle Z, X \rangle = 0$ for *all* $X \in \mathbb{R}^n$, then $Z = 0$.

Proof: This is easier than it looks. Let $X = Z$. Then $\langle Z, Z \rangle = 0$, whence $Z = 0$ by Property 1. Done. $\ll$

2. Prove that if $\langle Z_1, X \rangle = \langle Z_2, X \rangle$ for *all* $X \in \mathbb{R}^n$, then $Z_1 = Z_2$.

Proof: We have $\langle Z_1, X \rangle - \langle Z_2, X \rangle = 0$. Using distributivity and homogeneity, we have $\langle Z_1 - Z_2, X \rangle = 0$ for all X. By Example 1, $Z_1 - Z_2 = 0$; thus $Z_1 = Z_2$. Done. $\ll$

NOTE: As in Example 2, often the neatest way to prove that two objects are equal is to prove that their *difference* is zero.

2.2c An interpretation in $\mathbb{R}^2$

Now we relate $\langle X, Y \rangle$ to the geometry (and trigonometry) of the plane $\mathbb{R}^2$. Our result will enable us to prove the triangle inequality $\| X + Y \| \leq \| X \| + \| Y \|$, which, although visually obvious, was beyond our algebraic resources a few paragraphs above. The following theorem is the key. It gives an expression for $\langle X, Y \rangle$ involving geometric concepts, *not* coordinates (a "coordinate-free" expression.)

THEOREM 1: Let X, Y be vectors in $\mathbb{R}^2$. Then

$$\langle X, Y \rangle = \| X \| \| Y \| \cos \theta,$$

where θ is the angle between X and Y. (Since $\cos \theta = \cos(-\theta)$, the sense in which we take the angle does not matter.)

Proof: We know $\langle X, Y \rangle = x_1 y_1 + x_2 y_2$, where $X = (x_1, x_2)$, $Y = (y_1, y_2)$ as usual. We will translate x_1, y_1, x_2, y_2 into trigonometry. Let ω and φ be the angles from the horizontal axis to X and Y, respectively. Now we note crucially that

$$\cos \omega = \frac{x_1}{\| X \|}, \qquad \sin \omega = \frac{x_2}{\| X \|},$$

whence $x_1 = \| X \| \cos \omega$, $x_2 = \| X \| \sin \omega$. Likewise $y_1 = \| Y \| \cos \varphi$, $y_2 = \| Y \| \sin \varphi$. See Fig. 2.5.

Thus $\langle X,\ Y \rangle = x_1y_1 + x_2y_2 = \|X\|\ \|Y\|\ (\cos \omega \cos \varphi + \sin \omega \sin \varphi)$. We recall $\cos \theta = \cos (\varphi - \omega) = \cos \omega \cos \varphi + \sin \omega \sin \varphi$. Thus $\langle X,\ Y \rangle = \|X\|\|Y\| \cos \theta$, as claimed. $\ll$

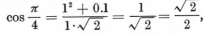

FIGURE 2.5

APPLICATIONS: Theorem 1 has several geometric consequences, as follows:

1. Computation of cosines. We have $\cos \theta = \langle X,\ Y \rangle / \|X\|\ \|Y\|$. To obtain $\cos \pi/4$ (recall that $\pi/4$ radians $= 45°$), let $X = e_1 = (1,\ 0)$ and $Y = (1,\ 1)$. Then the angle θ between X and Y is $\pi/4$, and hence

$$\cos \frac{\pi}{4} = \frac{1^2 + 0.1}{1 \cdot \sqrt{2}} = \frac{1}{\sqrt{2}} = \frac{\sqrt{2}}{2},$$

as is well known.

Also, given *any* vectors X and Y, we may compute $\cos \theta$ as above and, using a table of cosines, determine the angle θ between them.

2. Orthogonality in $\mathbb{R}^2$. The vectors (arrows) X and Y are at right angles (perpendicular or, as we shall say, **orthogonal**) if and only if $\cos \theta = 0$, where θ is as above or either of them is O. Hence X is orthogonal to Y, written $X \perp Y$, if and only if $\langle X,\ Y \rangle = 0$. (You should note the connection here with the "slopes are negative reciprocals" requirement for the perpendicularity of lines in plane analytic geometry.)

3. The *Cauchy-Schwarz inequality* in $\mathbb{R}^2$. This is very useful and far-reaching. It will give us the triangle inequality. It says that, for all $X,\ Y$ in $\mathbb{R}^2$,

$$|\langle X,\ Y \rangle| \le \|X\|\ \|Y\|.$$

Given Theorem 1, it follows immediately from the fact that $|\cos \theta| \le 1$; just take the absolute value of both sides in the expression $\langle X,\ Y \rangle = \|X\|\ \|Y\| \cos \theta$.

We will see that "Cauchy-Schwarz" is true in $\mathbb{R}^n$ also.

4. The triangle inequality in $\mathbb{R}^2$. We wish to prove that, for all $X,\ Y$ in $\mathbb{R}^2$,

$$\|X + Y\| \le \|X\| + \|Y\|.$$

Now this is true if and only if it is true with both sides squared; that is,

$$\|X + Y\|^2 \le (\|X\| + \|Y\|)^2.$$

Now we have $\|X + Y\|^2 = \langle X + Y,\ X + Y \rangle = \langle X,\ X \rangle + \langle X,\ Y \rangle$

$+ \langle Y, X \rangle + \langle Y, Y \rangle = \langle X, X \rangle + 2\langle X, Y \rangle + \langle Y, Y \rangle$. Here we have used the symmetry and distributivity of the inner product. Note: Squaring gets rid of the troublesome square roots.

On the other hand, $(\|X\| + \|Y\|)^2 = \|X\|^2 + 2\|X\|\|Y\| + \|Y\|^2 = \langle X, X \rangle + 2\|X\|\|Y\| + \langle Y, Y \rangle$. Hence $\|X + Y\| \leq \|X\| + \|Y\|$ is true provided that $\langle X, Y \rangle \leq \|X\|\|Y\|$. But this is immediate from the Cauchy-Schwarz inequality. The triangle inequality in $\mathbb{R}^2$ is proved at last.

REMARK: You may object that we expend so much effort to prove a visible fact. We reply that (1) the triangle inequality is not quite so obvious in the higher-dimensional spaces $\mathbb{R}^n$ (or in infinite-dimensional space). (2) Such spaces are regularly encountered by physicists, engineers, economists, and certainly mathematicians and statisticians. (3) Our proof generalizes to these spaces verbatim, although some care is necessary in proving $\langle X, Y \rangle = \|X\|\|Y\|\cos\theta$ in $\mathbb{R}^n$ when $n \geq 2$. In the following paragraphs we discuss the inner product in dimensions higher than two.

2.2d The inner product in $\mathbb{R}^n$

In the paragraph above, we obtained the following results for all X, Y in $\mathbb{R}^2$:

Theorem 1: $\langle X, Y \rangle = \|X\|\|Y\|\cos\theta$,
Cauchy-Schwarz inequality: $|\langle X, Y \rangle| \leq \|X\|\|Y\|$,
Triangle inequality: $\|X + Y\| \leq \|X\| + \|Y\|$.

We note that "Cauchy-Schwarz" follows from Theorem 1 by taking absolute values on both sides and that the triangle inequality follows readily from Cauchy-Schwarz. Moreover, we observe that if Theorem 1 were true for X, Y in $\mathbb{R}^n$, $n \geq 2$, then the two inequalities would follow in n-space word for word as in 2-space.

We now claim that Theorem 1 is true in $\mathbb{R}^n$ for all $n \geq 1$; that is,

THEOREM 2: Let X, Y be vectors in $\mathbb{R}^n$. Then

$$\langle X, Y \rangle = \|X\|\|Y\|\cos\theta,$$

where θ is the angle between X and Y.

REMARK: This is true because it is essentially "two-dimensional" in the sense that our concern is with only two vectors X and Y. These vectors, although each has n coordinates, together span a plane subspace $\mathscr{S}$ inside $\mathbb{R}^n$. See Fig. 2.6. We may regard this subspace $\mathscr{S}$ as basically the same as the

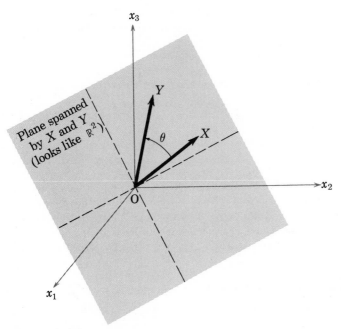

FIGURE 2.6

standard plane $\mathbb{R}^2$; $\mathscr{S}$ is just a "copy" of $\mathbb{R}^2$ tilted inside some larger $\mathbb{R}^n$. Since X, Y are in a two-dimensional space, we have $\langle X,\ Y \rangle = \|X\|\|Y\| \cos \theta$, as desired. We omit details.

APPLICATION: Theorem 2 allows us to conclude that the Cauchy-Schwarz and triangle inequalities are true in all spaces $\mathbb{R}^n$. As mentioned above, these inequalities are proved exactly as in $\mathbb{R}^2$.

In addition, the exercises contain a proof of the Cauchy-Schwarz inequality that uses no trigonometry whatever but rather the famous quadratic formula of high school algebra.

Exercises

1.   Compute $\langle X,\ Y \rangle$ for the given vectors:

(a) $X = (1, 3)$, $Y = (6, -2)$ in $\mathbb{R}^2$;

(b) $X = (1, 3, 1)$, $Y = (2, 0, 5)$ in $\mathbb{R}^3$;

(c) $X = (1, 4, 0, 1)$, $Y = (0, 1, 1, -1)$ in $\mathbb{R}^4$;

(d) $X = (1, 1, -1, 2, 0, 1)$, $Y = (1, -1, 1, 0, 4, 1)$ in $\mathbb{R}^6$.

2.   (a) Compute $\langle X, X \rangle$, where $X = (1, -3, 1)$ in $\mathbb{R}^3$.

(b) Now compute $\|X\|$, using Exercise 2(a).

3.   Compute $\cos \theta$ where θ is the angle between $X = (1, 1)$ and $Y = (0, 2)$ in $\mathbb{R}^2$. Note that $\cos \theta$ is unchanged, as we would expect, if we replace X by $2X$.

4.   (a) Compute a nonzero vector $Y = (y_1, y_2)$ that is orthogonal to $X = (1, 5)$. Is Y unique?

(b) How many *unit* vectors are orthogonal to X? Draw them.

5.   Think of $f(x_1, x_2) = 3x_1 + 4x_2$ as a function of two variables. Let $X = (x_1, x_2)$, and consider only X satisfying $\|X\| \leq 1$. Using the fact that $f(x_1, x_2) = \langle X, Y \rangle$ for $Y = (3, 4)$ and the Cauchy-Schwarz inequality, prove that $|f(x_1, x_2)| \leq 5$.

6.   Let $Y = (2, 4, -1)$ in $\mathbb{R}^3$. Suppose that $X = (x_1, x_2, x_3)$ is orthogonal (perpendicular, at right angles) to Y. What can you say about $\cos \theta$, where θ is the angle between X and Y? What does this and Theorem 2 imply about $\langle X, Y \rangle$? True or false: X and Y are orthogonal if and only if $2x_1 + 4x_2 - x_3 = 0$. Describe the set of all vectors X in $\mathbb{R}^3$ such that X is orthogonal to the fixed vector Y.

7.   High school algebra and Cauchy-Schwarz:

(a) Recall that the quadratic polynomial $f(t) = at^2 + bt + c$, with $a, b, c \in \mathbb{R}$, is ≥ 0 (or ≤ 0) for all t if and only if it has no real roots or a double real root t_0 (for example, $f(t) = t^2$, $t_0 = 0$). From the quadratic formula, these happen if and only if $b^2 - 4ac < 0$ or $b^2 - 4ac = 0$, respectively.

(b) Define $f(t) = \langle X + tY, X + tY \rangle$, where t is a real variable. Observe that $f(t) \geq 0$ (why?).

(c) Using the algebraic properties of the inner product, write $f(t)$ as a polynomial in t with coefficients $a = \langle Y, Y \rangle$, $b = 2\langle X, Y \rangle$, $c = \langle X, X \rangle$.

(d) Using the last sentence of Exercise 7(a), prove $|\langle X, Y \rangle| \leq \|X\| \|Y\|$.

8.   *The law of cosines.* Prove this generalization of Pythagoras' Theorem (see Fig. 2.7):

$$c^2 = \|X\|^2 + \|Y\|^2 - 2\|X\| \|Y\| \cos \theta.$$

Note that $c^2 = \|X - Y\|^2$.

9.   Let X, Y be two vectors in $\mathbb{R}^2$. Prove that X is perpendicular to $Y \Leftrightarrow \|X - Y\|^2 = \|X\|^2 + \|Y\|^2$. (We agree to say that the zero vector is perpendicular to every vector.)

10.   Let $X, Y \in \mathbb{R}^n$. Using only the four properties of $\langle X, Y \rangle$ in Sec. 2.2b, prove:

(a) $4\langle X, Y \rangle = \|X + Y\|^2 - \|X - Y\|^2$

(b) $\|X + Y\|^2 + \|X - Y\|^2 = 2\|X\|^2 + 2\|Y\|^2$

(this states that the sum of the squares of the diagonals of a parallelogram equals the sum of the squares of the sides).

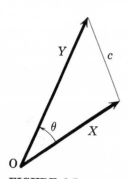

FIGURE 2.7

2.3 ORTHOGONALITY IN $\mathbb{R}^n$

In light of Theorem 2, that is, $\langle X, Y \rangle = \|X\| \|Y\| \cos \theta$, we see that two vectors X, Y in $\mathbb{R}^n$ are perpendicular, or **orthogonal**, denoted $X \perp Y$, if and only if $\langle X, Y \rangle = 0$. Reason: $\langle X, Y \rangle = 0$ means that either $\cos \theta = 0$, and so θ is a right angle, or it means that $X = O$ or $Y = O$. We agree to say that the zero vector is orthogonal to any vector.

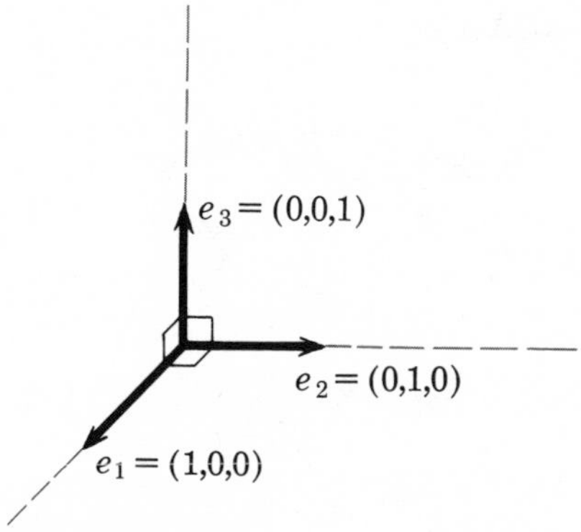

Note that the standard basis vectors e_1, e_2, e_3 in $\mathbb{R}^3$ are pairwise orthogonal, $e_1 \perp e_2$, $e_1 \perp e_3$, $e_2 \perp e_3$. See Fig. 2.8. This agrees with our geometric intuition. Also, in $\mathbb{R}^4$, we readily compute that $X = (1, -1, 2, 0)$ and $Y = (0, 6, 3, -2)$ are orthogonal.

A problem. Given a fixed vector Z in $\mathbb{R}^n$, describe the set of all vectors X such that $X \perp Z$. For instance, is this set a subspace of $\mathbb{R}^n$?

THEOREM 3: The set of all vectors in $\mathbb{R}^n$ orthogonal to a fixed vector Z is a subspace of $\mathbb{R}^n$.

$e_3 = (0,0,1)$

$e_2 = (0,1,0)$

$e_1 = (1,0,0)$

FIGURE 2.8

Proof: Let $\mathscr{S} = \{X \in \mathbb{R}^n \mid X \perp Z\}$. To show that $\mathscr{S}$ is a subspace, it suffices to show that $\mathscr{S}$ is closed under addition and multiplication by scalars. The proof for addition is very clean: Let $X_1, X_2 \in \mathscr{S}$. Then $X_1 + X_2 \in \mathscr{S} \Leftrightarrow (X_1 + X_2) \perp Z \Leftrightarrow \langle X_1 + X_2, Z \rangle = 0$. But $\langle X_1 + X_2, Z \rangle = \langle X_1, Z \rangle + \langle X_2, Z \rangle = 0 + 0 = 0$, and so $\mathscr{S}$ is closed under addition. We leave multiplication by scalars to you. $\ll$

NOTE: If $Z \neq O$, then the subspace $\mathscr{S}$ is not all of $\mathbb{R}^n$, since $Z \notin \mathscr{S}$.

Can we describe $\mathscr{S}$ further? Let us consider this question in $\mathbb{R}^3$, where, happily, our previous work has answered it.

Let $Z = (z_1, z_2, z_3)$ be a nonzero vector and $X = (x_1, x_2, x_3)$. Then $X \perp Z$ if and only if $z_1 x_1 + z_2 x_2 + z_3 x_3 = 0$. Now the set of all solutions $X = (x_1, x_2, x_3)$ to this equation is a plane subspace in $\mathbb{R}^3$, by Theorem 5 in Chap. 1. We state:

THEOREM 4: Let $\mathscr{S}$ be a subset of $\mathbb{R}^3$. Then the following statements are equivalent:

1. $\mathscr{S}$ is a plane subspace of $\mathbb{R}^3$,

2. $\mathscr{S}$ is the set of all solutions $X = (x_1, x_2, x_3)$ to a homogeneous linear equation $z_1 x_1 + z_2 x_2 + z_3 x_3 = 0$ with at least one of the coefficients z_1, z_2, z_3 different from zero,

3. $\mathscr{S}$ is the set of all vectors X orthogonal to the nonzero vector $Z = (z_1, z_2, z_3)$.

EXAMPLE: Let $Z = (1, 1, 1)$. Then the set $\mathscr{S}$ of vectors X such that $X \perp Z$ is the set of solutions to $\langle X, Z \rangle = x_1 + x_2 + x_3 = 0$, a plane through the origin in $\mathbb{R}^3$. See Fig. 2.9. A basis for $\mathscr{S}$ is given by $X_1 = (1, 0, -1)$ and $X_2 = (0, 1, -1)$. Thus $X \perp Z$ if and only if there exist scalars α_1, α_2 such that $X = \alpha_1 X_1 + \alpha_2 X_2$.

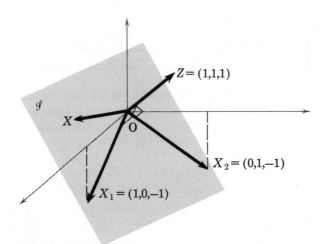

FIGURE 2.9

Exercises

These afford you practice in determining the plane orthogonal to a given vector (or line) and vice versa.

1. (a) Let $Z = (2, -1, 3)$ in $\mathbb{R}^3$. Compute the equation that determines the plane $\mathscr{S}$ of vectors $X = (x_1, x_2, x_3)$ such that $X \perp Z$.
    (b) Find vectors X_1, X_2 in $\mathscr{S}$ that span $\mathscr{S}$. Each of them will, of course, be $\perp Z$.

2. (a) Let X_1, X_2 be as in Exercise 1(b). Describe geometrically the set $\mathscr{S}'$ of all Y such that $Y \perp X_1$ and $Y \perp X_2$. Is this set a plane through the origin; a subspace? Is Z (compare Exercise 1(a)) in this set? What relation has $\mathscr{S}'$ to $\mathscr{S}$?
    (b) Find a system of *two* linear equations whose set of solutions is precisely

the set $\mathscr{S}'$ of those $Y = (y_1, y_2, y_3)$ which satisfy $Y \perp X_1$ and $Y \perp X_2$.

3. Write down immediately vectors perpendicular to the planes through the origin in $\mathbb{R}^3$ determined by each of the following equations:

(a) $2x_1 - 12x_2 + 11x_3 = 0$,   (c) $x_1 - x_2 + x_3 = 0$.

(b) $x_2 = 0$,

4. Given a vector Z in $\mathbb{R}^3$, outline a procedure for obtaining a system of two homogeneous equations in y_1, y_2, y_3 whose set of solutions is the straight line (one-dimensional subspace) determined by Z. Use the methods in Exercises 1(b) and 2(b). Note that this procedure amasses information about the plane $\mathscr{S}$ orthogonal to Z; each linear equation is the statement that some inner product is zero.

5. (a) Given $X_1 = (1, 0, 2)$, $X_2 = (2, 2, 3)$, compute a vector Z (nonzero, of course) orthogonal to both. How unique is Z?

(b) Use Z to find a homogeneous equation determining the plane subspace spanned by X_1, X_2. How unique is this equation?

6. If true, give a proof; if false, give a counterexample:

(a) $Z = (1, 4, -1)$ is orthogonal to the subspace determined by $x_1 + 4x_2 + x_3 = 0$.

(b) Z is not orthogonal to any vectors in the subspace in Exercise 6(a).

(c) If X_1, X_2 are nonzero vectors and if the set of all vectors in $\mathbb{R}^3$ orthogonal to both of them is a plane, then $X_1 = \alpha X_2$ for some scalar α.

(d) $X_1 = (1, 0, 1)$ and $X_2 = (0, -1, 0)$ are orthogonal.

(e) If $Z \perp X$ and $Z \perp Y$, then Z is $\perp$ to any vector $\alpha X + \beta Y$.

(f) If $X \perp Z$ and both are nonzero, then $X = \alpha Z$.

(g) If $X \neq \alpha Z$, then $X \perp Z$.

(h) If $\langle Z, X \rangle = 0$ for all $X \in \mathbb{R}^n$, then $Z = O$.

(i) If $\langle Z_1, X \rangle = \langle Z_2, X \rangle$ for all $X \in \mathbb{R}^n$, then $Z_1 = Z_2$.

7. Complete the proof of Theorem 3.

8. If $X \perp Y$ and $aX + bY = O$, prove that $a = b = 0$ (assume $X \neq O \neq Y$).

9. Let $X, Y \in \mathbb{R}^n$:

(a) If $X \perp Y$, prove that $\|X - aY\| \geq \|X\|$ for all $a \in \mathbb{R}$.

(b) If $\|X - aY\| \geq \|X\|$ for all $a \in \mathbb{R}$, prove that $X \perp Y$.

10. Let $X_0 \in \mathbb{R}^n$ be a given vector, and let $\mathscr{S}$ be a subspace of $\mathbb{R}^n$. Define the numbers α, β by

$$\alpha = \min_{X \in \mathscr{S}} \|X - X_0\|, \qquad \beta = \max_{\substack{Y \in \mathscr{S}^\perp \\ \|Y\| = 1}} |\langle Y, X_0 \rangle|.$$

Show that $\alpha = \beta$. (Suggestion: Picture in $\mathbb{R}^2$? Here $\mathscr{S}^\perp$ is the set of all vectors perpendicular to $\mathscr{S}$.)

EUCLID—USING VECTORS

After a new topic, such as inner products, is learned, it is fun to see what it has to say about an old subject. We shall prove some classical results from Euclidean geometry using vector methods.

PROPOSITION: An angle inscribed in a semicircle is a right angle.

Proof: Say the circle has radius a and is centered at the origin. If one end of a diameter is X, then the other end is $-X$. We must show (see Fig. 2.10) that $X - Z \perp X + Z$. Use the inner product and observe that $\|X\| = a = \|Z\|$:

$$\langle X - Z, X + Z \rangle = \|X\|^2 + \langle X, Z \rangle - \langle Z, X \rangle - \|Z\|^2$$
$$= a^2 - a^2 = 0. \qquad\qquad \text{Q. E. D.}$$

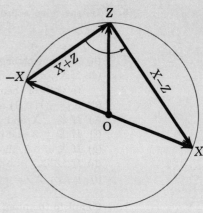

FIGURE 2.10

PROPOSITION: If the medians to two sides of a triangle have equal length, then the triangle is isosceles.

Proof: Place the origin at the vertex that does not have one of the median lines. Then (see Fig. 2.11) we are told that

$$\|X - \tfrac{1}{2}Y\| = \|Y - \tfrac{1}{2}X\|.$$

Now

$$\|X - \tfrac{1}{2}Y\|^2 = \langle X - \tfrac{1}{2}Y, X - \tfrac{1}{2}Y \rangle = \|X\|^2 - \langle X, Y \rangle + \tfrac{1}{4}\|Y\|^2$$

and

$$\| Y - \tfrac{1}{2}X \|^2 = \langle Y - \tfrac{1}{2}X, \, Y - \tfrac{1}{2}X \rangle = \| Y \|^2 - \langle X, Y \rangle + \tfrac{1}{4} \| X \|^2.$$

Equating these two expressions and doing a mental computation, we find that $\| X \| = \| Y \|$.   Q. E. D.

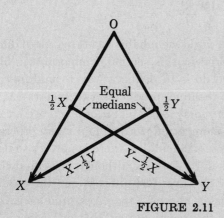

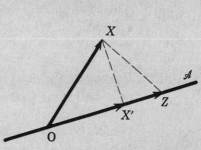

FIGURE 2.11

FIGURE 2.12

PROPOSITION: Given a point X and a line $\mathscr{A}$ in $\mathbb{R}^2$, the shortest distance from X to $\mathscr{A}$ is along the perpendicular dropped from X to $\mathscr{A}$.

Proof: Place the origin on $\mathscr{A}$. Let $X' \in \mathscr{A}$ be the foot of the perpendicular from X to $\mathscr{A}$, so that $X - X' \perp \mathscr{A}$, and let Z be any other vector in $\mathscr{A}$. See Fig. 2.12. Then $X - X' \perp Z$. Therefore, by the Pythagorean Theorem applied to the right triangle $XX'Z$, we find

$$\| X - Z \|^2 = \| X - X' \|^2 + \| X' - Z \|^2,$$

which implies

$$\| X - Z \| \geq \| X - X' \|;$$

that is, X' is closer to X than any other vector Z in $\mathscr{A}$. (Q. E. D.)

Since mathematics is not a spectator sport, we offer some assertions for you to prove:

1. A parallelogram is a rectangle $\Leftrightarrow$ the diagonals have equal length.

2. If a triangle is isosceles, then the medians to the two sides of equal length also have equal length.

3. The sum of the squares of the sides of a parallelogram equals the sum of the squares of the diagonals.

4. Given any quadrilateral Q, construct a new one P by connecting

the midpoints of successive sides of Q. Prove that P is a parallelogram.

5. If the three sides of a triangle with lengths a, b, c satisfy $a^2 + b^2 = c^2$, then the triangle is a right triangle.

2.4 THE CROSS PRODUCT IN $\mathbb{R}^3$

As you grasped years ago, the world we inhabit has three space dimensions. For this reason the material in this section is important in many applications. What we intend to do is to define a way to multiply two vectors X and Y in $\mathbb{R}^3$. Although the inner product of two vectors $\langle X, Y \rangle$, is a *scalar*, this product $X \times Y$, the *cross product* (or *vector product*) is a *vector*. In physics, the ideas of angular momentum, torque, and many others are expressed using the cross product. Incidentally, there is no way to define a "cross product" of two vectors in $\mathbb{R}^n$ if $n > 3$ (although one can define a "product" of $n - 1$ vectors in $\mathbb{R}^n$). Thus, fortuitously, our three-dimensional world is the exact dimension in which the cross product exists.

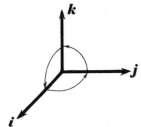

Right-handed basis

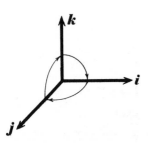

Left-handed basis

FIGURE 2.13

For the remainder of this one section, $\mathbf{i}$, $\mathbf{j}$ and $\mathbf{k}$ (and not e_1, e_2, e_3) will always denote the standard basis, so that $(x_1, x_2, x_3) = x_1\mathbf{i} + x_2\mathbf{j} + x_3\mathbf{k}$, with yet one additional property: the basis is *right-handed*. To explain what right-handed basis means, we use a picture, or rather two pictures. See Fig. 2.13. In a right-handed basis, if we move $\mathbf{i} \rightarrow \mathbf{j}$, $\mathbf{j} \rightarrow \mathbf{k}$, and $\mathbf{k} \rightarrow \mathbf{i}$, the resulting motion viewed from the first octant is counterclockwise, and in a left-handed system it is clockwise. Some people prefer to use a peculiar finger business with the right hand to explain right-handed basis. Others like to think of "right-handed screws." If you like, ask someone to explain them to you. Then adopt your favorite.

Here is the definition of the cross product. It should be memorized (unfortunately).

DEFINITION: The **cross product** of vectors $X = x_1\mathbf{i} + x_2\mathbf{j} + x_3\mathbf{k}$ and $Y = y_1\mathbf{i} + y_2\mathbf{j} + y_3\mathbf{k}$ is defined as the vector

$$X \times Y \equiv (x_2 y_3 - x_3 y_2)\mathbf{i} + (x_3 y_1 - x_1 y_3)\mathbf{j} + (x_1 y_2 - x_2 y_1)\mathbf{k}.$$

NOTE: Be careful with the sign of the $\mathbf{j}$ term. It is the negative of what you might expect.

For those who know 3-by-3 determinants, this can be written in the suggestive notation

$$X \times Y = \begin{vmatrix} \mathbf{i} & \mathbf{j} & \mathbf{k} \\ x_1 & x_2 & x_3 \\ y_1 & y_2 & y_3 \end{vmatrix} = \begin{vmatrix} x_2 & x_3 \\ y_2 & y_3 \end{vmatrix} \mathbf{i} - \begin{vmatrix} x_1 & x_3 \\ y_1 & y_3 \end{vmatrix} \mathbf{j} + \begin{vmatrix} x_1 & x_2 \\ y_1 & y_2 \end{vmatrix} \mathbf{k},$$

where the 3-by-3 determinant is always expanded by the first row. This is a mnemonic device only, since as a determinant it is meaningless because the elements in the first row are vectors and the other elements are numbers.

EXAMPLES: 1. $(1, 0, 0) \times (0, 1, 0) = (0, 0, 1)$, $(0, 0, 1) \times (0, 1, 0) =$ $-(1, 0, 0)$, $(0, 1, 0) \times (0, 1, 0) = O$; that is, $\mathbf{i} \times \mathbf{j} = \mathbf{k}$, $\mathbf{k} \times \mathbf{j} = -\mathbf{i}$, and $\mathbf{j} \times \mathbf{j} = O$.

2. If $X = (1, 2, -1) = \mathbf{i} + 2\mathbf{j} - \mathbf{k}$ and $Y = (-3, 1, 1) = -3\mathbf{i} + \mathbf{j} + \mathbf{k}$, then $X \times Y = (2 + 1)\mathbf{i} + (3 - 1)\mathbf{j} + (1 + 6)\mathbf{k} = 3\mathbf{i} + 2\mathbf{j} + 7\mathbf{k}$, but $Y \times X = (-1 - 2)\mathbf{i} + (1 - 3)\mathbf{j} + (-6 - 1)\mathbf{k} = -3\mathbf{j} - 2\mathbf{j} - 7\mathbf{k}$, and so we immediately see that the cross product is not commutative. We shall see that, in fact, it is always true that $X \times Y = -Y \times X$.

In order to understand the cross product, we shall investigate its algebraic and geometric properties.

THEOREM 5: (Algebraic properties of the cross product).
1. $X \times Y = -(Y \times X)$ (anti-commutative).
2. $X \times (Y + Z) = (X \times Y) + (X \times Z)$ (distributive).
3. $a(X \times Y) = (aX) \times Y$

Proof: These all follow by writing $X = x_1 \mathbf{i} + x_2 \mathbf{j} + x_3 \mathbf{k}$, $Y = y_1 \mathbf{i}$ + etc. and using the definition of the cross product. Thus, Property 1 follows from

$$Y \times X = (y_2 x_3 - y_3 x_2)\mathbf{i} + (y_3 x_1 - y_1 x_3)\mathbf{j} + (y_1 x_2 - y_2 x_1)\mathbf{k}$$
$$= -X \times Y.$$

The proofs of Properties 2 and 3 are just as straightforward and just as tedious. ≪

WARNING: There is (at least) one algebraic rule which you might have anticipated but which was not given above. We are thinking of the associative rule $(X \times Y) \times Z \overset{?}{=} X \times (Y \times Z)$. It is *false* for the cross product. For example,

$$(\mathbf{i} \times \mathbf{j}) \times \mathbf{j} = \mathbf{k} \times \mathbf{j} = -\mathbf{i}, \text{ but } \mathbf{i} \times (\mathbf{j} \times \mathbf{j}) = \mathbf{i} \times O = O.$$

In order to interpret the cross product geometrically, we have to visualize both its direction and its magnitude.

THEOREM 6: (Geometric properties of the cross product).

1. $X \times Y$ is perpendicular to both X and Y and hence to the plane determined by X and Y.

2. $\| X \times Y \| = \| X \| \, \| Y \| \, |\sin \theta|$, where θ is the angle between X and Y (see Fig. 2.14).

3. If $X \neq 0$, $Y \neq 0$, then $X \times Y = 0 \Leftrightarrow X = cY$ for some scalar c; that is, O, X, and Y lie on the same straight line.

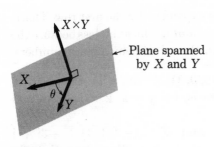

Plane spanned by X and Y

FIGURE 2.14

Proof: 1. To prove that $X \perp (X \times Y)$, we show that $\langle X, X \times Y \rangle = 0$. Thus

$$\langle X, X \times Y \rangle = x_1(x_2y_3 - x_3y_2) + x_2(x_3y_1 - x_1y_3) + x_3(x_1y_2 - x_2y_1) = 0.$$

Similarly, one shows that $\langle Y, X \times Y \rangle = 0$, which implies that $Y \perp (X \times Y)$.

2. We compute $\| X \times Y \|$ by the Pythagorean Theorem; that is, if $Z = z_1\mathbf{i} + z_2\mathbf{j} + z_3\mathbf{k}$, then $\| Z \|^2 = z_1^2 + z_2^2 + z_3^2$. Therefore

$$\begin{aligned} \| X \times Y \|^2 &= (x_2y_3 - x_3y_2)^2 + (x_3y_1 - x_1y_3)^2 + (x_1y_2 - x_2y_1)^2 \\ &= (x_1^2 + x_2^2 + x_3^2)(y_1^2 + y_2^2 + y_3^2) - (x_1y_1 + x_2y_2 + x_3y_3)^2 \\ &= \| X \|^2 \, \| Y \|^2 - \langle X, Y \rangle^2 \end{aligned}$$

(a sharp version of the Schwarz inequality). But

$$\langle X, Y \rangle^2 = \| X \|^2 \, \| Y \|^2 \cos^2 \theta = \| X \|^2 \, \| Y \|^2 (1 - \sin^2 \theta),$$

and so

$$\| X \times Y \|^2 = \| X \|^2 \, \| Y \|^2 \sin^2 \theta.$$

3. This follows from the formula for $\| X \times Y \|$ just proved, for if $X \times Y = 0$, then $0 = \| X \times Y \| = \| X \| \, \| Y \| \, |\sin \theta|$. Since $\| X \| \neq 0$ and $\| Y \| \neq 0$, this means that $\sin \theta = 0$. Thus $\theta = 0$ or $\theta = \pi$. Consequently, X and Y lie along the same straight line, which implies that each of them is a scalar multiple of the other. Done. ≪

COROLLARY: The area of the parallelogram whose adjacent sides are X and Y is $A = \|X \times Y\|$.

Proof: By high school geometry, $A = \|X\| h$ (see Fig. 2.15). But $h = \|Y\| |\sin\theta|$, and so $A = \|X\| \|Y\| |\sin\theta| = \|X \times Y\|$, where the last equality is part 2 of the theorem above. $\ll$

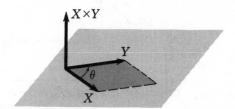

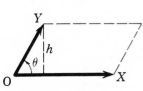

FIGURE 2.15   FIGURE 2.16

To recapitulate the geometric interpretation, $X \times Y$ is a vector perpendicular to the plane determined by X and Y and its magnitude equals the area of the parallelogram whose sides are X and Y. See Fig. 2.16. By convention, the orientation of $X \times Y$ is such that the coordinate system formed by the three vectors X, Y, and $X \times Y$ (in *this order*) is right-handed. Since our basis $\mathbf{i}, \mathbf{j}, \mathbf{k}$ is assumed to be right-handed, we can verify this. It is clear for the special case $X = \mathbf{i}$, $Y = \mathbf{j}$, so that $X \times Y = \mathbf{k}$, but not at all clear otherwise. The difficulty is in attempting to deal with the fuzzy "definition" we gave of right-handed basis. Finger and screw "definitions" suffer the same defect. A more precise definition is needed and is found in Exercise 9.

Undoubtedly the most useful property of the cross product is that, given two vectors X and Y, we can find a vector perpendicular to both of them without having to solve equations. That vector is $X \times Y$.

EXAMPLE: If $X = (1, 1, -1)$ and $Y = (2, 1, 3)$, find all vectors perpendicular to both X and Y in 3-space $\mathbb{R}^3$. Since X and Y span a two-dimensional space in $\mathbb{R}^3$, the set of vectors perpendicular to X and Y is a one-dimensional space in $\mathbb{R}^3$. Since $X \times Y = 4\mathbf{i} - 5\mathbf{j} - \mathbf{k} = (4, -5, -1)$ is perpendicular to both X and Y, it generates the line perpendicular to X and Y, and so every vector Z perpendicular to both X and Y is of the form $Z = c(4, -5, -1)$, where c is a scalar.

Exercises

1. Let $X = i + 2j - 3k$, $Y = i - j$, $Z = -i - 2j + k$. Compute the following:

    (a) $X \times Y$,
    (b) $Y \times X$,
    (c) $X \times Z$,
    (d) $X \times X$,
    (e) $Y \times Y$,
    (f) $Z \times Y$,
    (g) $X \times (Y \times Z)$,

    (h) $(X \times Y) \times Z$,
    (i) $X \times (Z \times Y)$,
    (j) $X \times (X \times Y)$,
    (k) $X \times (2Y + Z)$,
    (l) $(X - 2Z) \times Y$,
    (m) $(X + Y) \times (X - Y)$,
    (n) $(X + Y) \times (Y + Z)$.

2. Complete the following multiplication table:

    $$X \times Y$$

$X \downarrow \quad Y \rightarrow$	i	j	k
i			
j			
k		$-i$	

3. Let X, Y, Z be any vectors in $\mathbb{R}^3$. Prove the following identities:

    (a) $(X + Y) \times Z = X \times Z + Y \times Z$,
    (b) $X \times X = O$,
    (c) $X \times (X + Y) = X \times Y$,
    (d) $X \times (Y \times Z) = \langle X, Z \rangle Y - \langle X, Y \rangle Z$,
    (e) $(X \times Y) \times Z = \langle Z, X \rangle Y - \langle Z, Y \rangle X$,
    (f) $\langle X \times Y, Z \rangle = \langle X, Y \times Z \rangle$,
    (g) $(X \times Y) \times (X \times Z) = \langle X, Y \times Z \rangle X$.

4. By writing $X \times Y = (x_1 i + x_2 j + x_3 k) \times (y_1 i + y_2 j + y_3 k)$ and using the multiplication table in Exercise 2 along with Theorem 5, deduce the formula for $X \times Y$.

5. Find all unit vectors perpendicular to the plane spanned by $X = (0, 3, 2)$ and $Y = (1, 1, -1)$.

6. Let $\mathscr{S}$ be spanned by $(1, 2, 3)$ and $(-1, 0, 4)$. Find a vector that spans $\mathscr{S}^\perp$, the subspace of vectors $\perp \mathscr{S}$.

7. Find all unit vectors in the **ij**-plane that are perpendicular to the vector $X = -3i - 4j + k$.

8. (a) Is there a vector X such that $X \times i = i$? Justify your statement.
    (b) Given the vectors Y and Z, show that there is a vector X such that $X \times Y = Z$ if and only if Z is perpendicular to Y. Is the solution X unique? If not, what is the most general solution?

9. In $\mathbb{R}^3$, we define the vectors X, Y, Z (the order is important) to be a *right-handed coordinate system* if $\langle X \times Y, Z \rangle > 0$, and *left-handed* if $\langle X \times Y, Z \rangle < 0$. Prove the following assertions:

   (a) $\mathbf{i}, \mathbf{j}, \mathbf{k}$ is a right-handed coordinate system.
   (b) $\mathbf{j}, \mathbf{i}, \mathbf{k}$ is a left-handed coordinate system.
   (c) If X, Y, Z is a right-handed coordinate system, then Y, X, Z is a left-handed coordinate system.
   (d) If $X \neq O$, $Y \neq O$, then X, Y, and $Z \equiv X \times Y$ is a right-handed coordinate system, and X, Y, and $Z \equiv Y \times X$ is a left-handed coordinate system.

10. Prove that the vectors X, Y, $Z \in \mathbb{R}^3$ span all of $\mathbb{R}^3 \Leftrightarrow \langle X \times Y, Z \rangle \neq 0$.

11. Prove that $\| X \times Y \| = \| X \| \| Y \| \Leftrightarrow X \perp Y$.

12. Find the area of a parallelogram whose adjacent sides are $X = \mathbf{i} + \mathbf{j} - 2\mathbf{k}$, $Y = 2\mathbf{i} - 3\mathbf{j}$.

13. (a) If $X \times Z = O$ for all vectors $Z \in \mathbb{R}^3$, what can you conclude?
    (b) If $X \times Z = Y \times Z$ for all vectors $Z \in \mathbb{R}^3$, what can you conclude?

14. Find all vectors in the plane spanned by $X = \mathbf{i} + \mathbf{j} - 2\mathbf{k}$ and $Y = -\mathbf{i} + \mathbf{j} + \mathbf{k}$ that are perpendicular to the vector $Z = 2\mathbf{i} + \mathbf{j} + 2\mathbf{k}$.

15. If the vertices of a parallelogram in the xy-plane all have integer coordinates, prove that the area is an integer.

3. Linear Functions

3.0 INTRODUCTION

Having made our acquaintance with Euclidean spaces, we are now
ready to discuss functions defined in such spaces. In this chapter we reexam-
ine the familiar functions $y = mx + b$ of calculus and generalize them to
higher-dimensional spaces. These functions, we recall, were essential to
differential calculus because their graphs (straight lines) were the tangent
lines to the graphs of other, more complicated functions. We were able
to learn a good deal about these more complicated "nonlinear" functions
by a study of the tangent lines. Recall, for example, that maxima and
minima occur where the tangent line is horizontal. Our generalizations of
$y = mx + b$ will be useful in an analogous study of nonlinear functions
in higher dimensions.

Let us write $T(x) = mx + b$. See Fig. 3.1. Here m and b are constants
and x is a real variable. Recall that m is the "slope" of the line $y = mx + b$
and b is the "y-intercept," the height at which the line crosses the y-axis.
The function $T(x)$ is our one-dimensional model for an "affine" function.
A function of the form $L(x) = mx$ is our model of a "linear" function.
Examples of linear functions are $L(x) = 6x$, $L(x) = -x$, $L(x) = (\frac{1}{2})x$,
and examples of affine functions are $T(x) = 6x + 1$, $T(x) = 6x - 5$,
$T(x) = (\frac{1}{2})x + 1$. We note that affine functions are readily built from
linear functions, simply by adding a constant b.

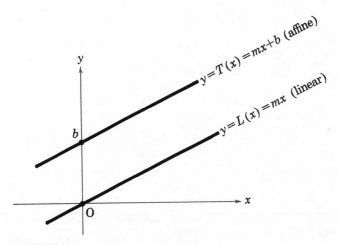

FIGURE 3.1 The case of a single variable x

Linear functions have two essential properties, as follows. If $L(x)$ $= mx$, then for all x, x' and scalars α we have

Additivity: $\qquad\qquad\qquad L(x + x') = L(x) + L(x'),$

Homogeneity: $\qquad\qquad\qquad L(\alpha x) = \alpha L(x).$

These properties, as a moment's reflection shows, are merely the distributive and commutative laws in the real numbers. Furthermore, it can be shown that any function $y = f(x)$ that has the two properties listed must be of the form $f(x) = mx$ for some constant m.

We need functions in higher dimensions that act like the functions $y = L(x) = mx$ in ordinary calculus. Our functions will now be of the form $Y = L(X)$, where X and Y are vectors, not just numbers. How should we define them? We attend to this now.

3.1 DEFINITION AND BASIC PROPERTIES

3.1a Introduction

Let $\mathbb{R}^n$ and $\mathbb{R}^q$ be Euclidean spaces. Perhaps $n = q$, perhaps not. A function F from $\mathbb{R}^n$ to $\mathbb{R}^q$ is a rule that assigns to each vector X in $\mathbb{R}^n$ some vector $F(X)$ in $\mathbb{R}^q$. We write this

$$F\colon \mathbb{R}^n \longrightarrow \mathbb{R}^q.$$

The space $\mathbb{R}^n$ is the **domain** of F, and the space $\mathbb{R}^q$ might be thought of as the **target** of F. The function F "maps" a vector X from $\mathbb{R}^n$ into the target space $\mathbb{R}^q$; X is "mapped" to the vector $F(X)$. Now we single out a special kind of function.

DEFINITION: A function L from $\mathbb{R}^n$ into $\mathbb{R}^q$, that is, $L\colon \mathbb{R}^n \to \mathbb{R}^q$, is **linear** if and only if, for all X, $X' \in \mathbb{R}^n$ and $\alpha \in \mathbb{R}$, it satisfies

Additivity: $\qquad\qquad L(X + X') = L(X) + L(X'),$

Homogeneity: $\qquad\qquad L(\alpha X) = \alpha L(X).$

Thus we require linear functions to satisfy certain behavioral criteria or requirements. Note that homogeneity requires $L(0) = 0$ (just take $\alpha = 0$).

Terminology: Before proceeding to some examples of linear functions, we remark that they are also called linear **mappings**, linear **maps**, linear **transformations**, or linear **operators**. The different names are standard among researchers in different areas of science. Thus a physicist might speak of linear "operators," but a geometer is more likely to say "mapping" or "transformation." We are partial to "linear map." Remember "linear map" = "linear function."

NOTATION: It is customary to drop the parentheses in the expression $L(X)$ whenever possible. Thus we write LX, $L\alpha X = \alpha LX$, $L(X + Y) = LX + LY$, and so on.

EXAMPLES OF LINEAR MAPS: Here are some important linear maps $L: \mathbb{R}^n \longrightarrow \mathbb{R}^q$. Note that in each case we first specify the domain $\mathbb{R}^n$ and target $\mathbb{R}^q$ and then give the rule for LX. Finally we verify that the rule determines a function that is indeed linear.

Whenever feasible we illustrate the map L with a picture showing the domain and target and the geometric transformation accomplished by L.

1. Let $L: \mathbb{R}^n \longrightarrow \mathbb{R}^n$ (note $q = n$) be given by $LX = 2X$ for all $X \in \mathbb{R}^n$. Thus L doubles the length of each X in $\mathbb{R}^n$. Note that $LO = O$. Figure 3.2 is a picture in $\mathbb{R}^2$.

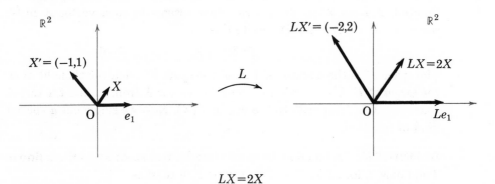

$$LX = 2X$$

FIGURE 3.2

To see that L is linear, we must verify the criteria of the definition. To verify additivity, note that $L(X + X') = 2(X + X') = 2X + 2X' = LX + LX'$, as required.

To verify the homogeneity of L, let α be any scalar. Then $L(\alpha X) = 2(\alpha X) = (\alpha 2)X = \alpha(2X) = \alpha LX$, as required. Thus $LX = 2X$ defines a linear map.

In general, if γ is any fixed scalar, then $LX = \gamma X$ defines a linear map

$L: \mathbb{R}^n \to \mathbb{R}^n$. You should check that this function *is* linear; the proof is just as in the case of $\gamma = 2$.

2. Let $L: \mathbb{R}^n \to \mathbb{R}^q$ (any n, q) be given by $LX = O$, the zero vector in $\mathbb{R}^q$, for *every* X in $\mathbb{R}^n$. Figure 3.3 is a picture in the case of $n = 2$, $q = 2$. This is called the **zero map**.

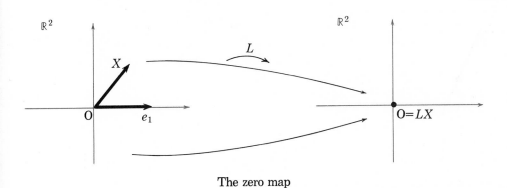

The zero map

FIGURE 3.3

3. Consider the map of $\mathbb{R}^n$ into $\mathbb{R}^n$ that assigns to each vector X itself. This is the **identity map** $I: \mathbb{R}^n \to \mathbb{R}^n$, $IX = X$. It is clearly linear. Figure 3.4 is a picture in $\mathbb{R}^2$. As you can see, the identity map leaves things exactly as they are. Contrast this with the zero map, which sends every vector X in $\mathbb{R}^n$ to the single point O.

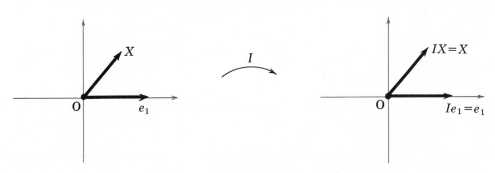

The identity map

FIGURE 3.4

4. Now let $n = 2$, $q = 1$ and define $L: \mathbb{R}^2 \to \mathbb{R}$ as follows. To each $X = (x_1, x_2)$ in $\mathbb{R}^2$, let $LX = x_1$. Thus L maps each vector X to its first coordinate. To see that L is additive, write $X' = (x_1', x_2')$ and note that $L(X + X') = L((x_1 + x_1', x_2 + x_2')) = x_1 + x_1' = LX + LX'$. Homogeneity is verified just as easily, and so L is linear. See Fig. 3.5.

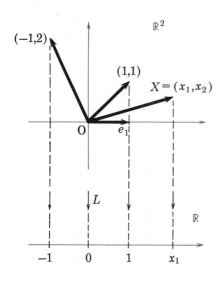

Projection onto the first coordinate

FIGURE 3.5

Note that, in contrast to those above, the next example seems to require the use of coordinates in its definition.

5. This time let $L: \mathbb{R}^2 \to \mathbb{R}$ be given by $LX = x_1 + 2x_2$, where $X = (x_1, x_2)$. If $X = (1, -3)$, then $LX = -5$. Thus LX is given as the "left-hand side" of the linear equation $x_1 + 2x_2 = $ (something).

There is an interesting picture associated with this example. See Fig. 3.6. Note that all points on the line $x_1 + 2x_2 = 0$ are mapped to 0 in the target $\mathbb{R}$ (drawn tilted) and that each line $x_1 + 2x_2 = \beta$ parallel to $x_1 + 2x_2 = 0$ is compressed and projected by L onto the single point β in $\mathbb{R}$.

To prove L linear, note that $L(X + X') = L(x_1 + x_1', x_2 + x_2') = (x_1 + x_1') + 2(x_2 + x_2') = (x_1 + 2x_2) + (x_1' + 2x_2') = LX + LX'$, whence L is additive. You may check homogeneity yourself.

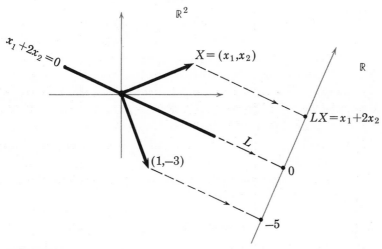

FIGURE 3.6

6. Continuing in the same vein as Example 5, let $L: \mathbb{R}^2 \to \mathbb{R}^2$ be defined by $LX = (x_1 + 2x_2, 3x_1 - x_2)$, where as usual $X = (x_1, x_2)$. See Fig. 3.7. Note that for $e_1 = (1, 0)$, $e_2 = (0, 1)$, we have $Le_1 = (1, 3)$, $Le_2 = (2, -1)$. You should be able to verify that L has the additive and homogeneous properties and so is linear.

Just as the map in Example 5 was related to a single linear equation, the present map L, with target space $\mathbb{R}^2$, is given as the left-hand side of the system of *two* equations

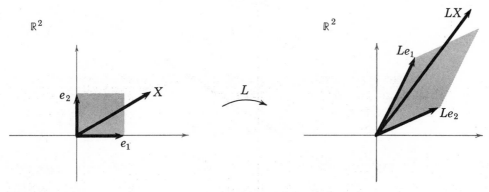

FIGURE 3.7

$$x_1 + 2x_2 = y_1,$$
$$3x_1 - x_2 = y_2,$$

where y_1, y_2 are given numbers and x_1, x_2 are to be found. Roughly speaking, the study of this system of equations is equivalent to the study of the map L. Thus, asking for the solutions x_1, x_2 to this system, with $y_1 = 4$ and $y_2 = -7$, say, is the same as asking for those vectors $X = (x_1, x_2)$ which are mapped by L to the fixed vector $Y = (4, -7)$, $LX = Y$. We encourage you to ponder this; it is basic. For one thing, systems of linear equations are very important in applications and the connection with linear maps allows us to study them using the simpler vector notation and our geometric insight into Euclidean space. We are thereby less likely to become lost in clouds of coordinates.

7. Now we turn to some explicitly geometric examples. Let $L: \mathbb{R} \to \mathbb{R}^2$ as follows. To every real number t, that is, a vector in $\mathbb{R}$, we associate the point $L(t) = (3t, 2t)$ in the plane $\mathbb{R}^2$. See Fig. 3.8. It is easily verified that L is linear. What does the set of points $\{(3t, 2t) \mid t \in \mathbb{R}\}$ look like? Note that $L(1) = (3, 2) \in \mathbb{R}^2$. If we write $Y_1 = (3, 2)$, then $L(t) = tY_1$. As t varies from $-\infty$ to ∞ in $\mathbb{R}$, $L(t)$ varies along the line (one-dimensional subspace) in $\mathbb{R}^2$ determined by the vector Y_1. This example indicates how a line through the origin in $\mathbb{R}^2$ may be represented as the set of "images" $L(t)$ of all numbers $t \in \mathbb{R}$. We will see more of this idea.

8. Let $L: \mathbb{R}^2 \to \mathbb{R}^2$ be given by $LX = (x_1, -x_2)$, where $X = (x_1, x_2)$. See Fig. 3.9.

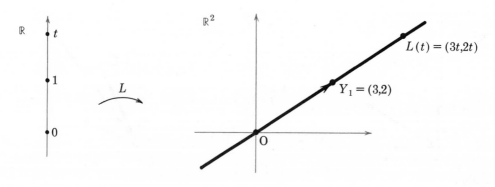

Mapping the t–axis into $\mathbb{R}^2$

FIGURE 3.8

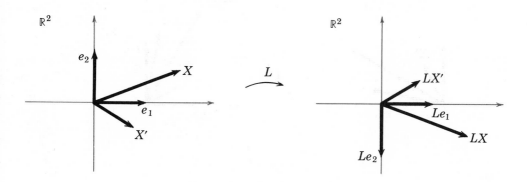

Reflection across x_1–axis

FIGURE 3.9

We see that L reflects every vector X, that is, flips X across the horizontal axis to its image LX.

We leave it to you to verify that L is linear.

9. Here is another map of $\mathbb{R}^2$ into itself that has a geometric definition. (Please do not be disturbed by this proliferation of examples. We will proceed to take a unified point of view that will reveal the order within this seeming chaos.)

Let $L\colon \mathbb{R}^2 \to \mathbb{R}^2$ rotate each vector X through $90°$ counterclockwise. See Fig. 3.10. Note that lengths are not changed, $\|LX\| = \|X\|$, and the angle between LX and LX' equals that between X and X'.

Now let us verify that the function L is indeed linear. To prove additivity, recall that addition of vectors in $\mathbb{R}^2$ obeys the parallelogram rule; that is, $X + X'$ is a diagonal of the parallelogram determined by X and X'. Using Fig. 3.10, you can convince yourself that $L(X + X')$ *is* the diagonal of the parallelogram determined by LX and LX' and therefore $L(X + X') = LX + LX'$.

The homogeneity of L is readily apparent also, and thus the rotation L is a linear map.

EXAMPLES OF NONLINEAR MAPS: Before you conclude that *all* functions are linear, we point out some that are not.

1. Let $f\colon \mathbb{R}^2 \to \mathbb{R}$ be given by $f(X) = x_1^2 + x_2^2$, where $X = (x_1, x_2)$. Then f is neither additive nor homogeneous. For let $X = (1, 1)$, $X' = (2, 1)$. Then $f(X + X') = 3^2 + 2^2 = 13$, while $f(X) + f(X') = 2 + 5 = 7$.

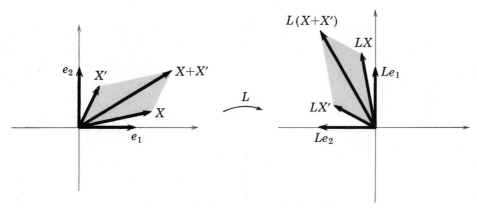

Rotation 90 degrees counterclockwise
Note: $L(X+X')=LX+LX'$

FIGURE 3.10

Hence f is not additive. Also, it is easy to see that $f(\alpha X) = \alpha^2 f(X)$, rather than $\alpha f(X)$, so that f is not homogeneous either. (The property $f(\alpha X) = \alpha^2 f(X)$ is sometimes referred to as **homogeneity of degree two**.)

The important lesson of this example is that f is not linear because the "variables" x_1, x_2 are *squared*. Functions involving squares, cubes, $x_1 x_2$, and the like, are not linear.

2. Another example of a nonlinear map is given by a **translation** function $T: \mathbb{R}^n \rightarrow \mathbb{R}^n$. Let X_0 be a fixed nonzero vector in $\mathbb{R}^n$. Then define $T(X) = X + X_0$. Intuitively, T translates or shifts all of $\mathbb{R}^n$ by adding X_0 to each vector in $\mathbb{R}^n$.

To see that T is not additive, observe that $T(X + X')$ is, by definition, $(X + X') + X_0$, while $T(X) + T(X') = (X + X_0) + (X' + X_0) = X + X' + 2X_0$. Hence T is not linear. You can check that $T(\alpha X) \neq \alpha T(X)$ either. In particular, $T(O) \neq O$.

REMARK: Some remarks on the examples of linear functions are in order. First, they represent a wide variety of mathematical phenomena: linear equations and systems, lines in the plane, projections and rotations. Some of these examples are algebraic, some seem to be devoid of algebra (the concept of rotation, for example.) But our second remark is that by isolating in all these diverse examples the essential property of linearity and studying linear functions in general, we may gain some basic insight into all these phenomena at once and may hope, further, to settle many questions with the one approach. This is the virtue of abstraction.

3.1b   Some consequences of linearity

Going back to the definition of linear function, we derive some immediate consequences. The proofs here will serve to test your grasp of the
definition. The properties will be used constantly.

THEOREM 1:   Let $L: \mathbb{R}^n \to \mathbb{R}^q$ be a linear function. Then:

1. $L(0) = 0$,
2. $L(-X) = -L(X)$, for all $X \in \mathbb{R}^n$,
3. $L(\alpha X + \beta Y) = \alpha L(X) + \beta L(Y)$, for all $X, Y \in \mathbb{R}^n, \alpha, \beta \in \mathbb{R}$.

Proof: 1. First note that the O in the expression $L(O)$ is the zero
vector in $\mathbb{R}^n$, and the zero on the right-hand side of $L(O) = O$ is the zero
vector of $\mathbb{R}^q$ (and q may or may not equal n). Now for the proof, just let
$\alpha = 0$ in $L(\alpha X) = \alpha LX$.

2. We have $L(-X) = L(-1X) = -1L(X) = -L(X)$.

3. We have $L(\alpha X + \beta Y) = L(\alpha X) + L(\beta Y)$ by additivity, and this
equals $\alpha L(X) + \beta L(Y)$ by homogeneity, giving the result. $\ll$

APPLICATION: The easiest way to see that the translation function T:
$\mathbb{R}^n \to \mathbb{R}^n$, defined above by $T(X) = X + X_0$ for a fixed nonzero vector
X_0, is not linear is this: Just observe that $T(O) = X_0 \neq O$ and apply
Theorem 1.

AN IMPORTANT MAPPING PROPERTY: Let $L: \mathbb{R}^n \dashrightarrow \mathbb{R}^q$ be a linear map.
What does L do to a line through the origin in $\mathbb{R}^n$?

From Chap. 1, we know that such a line $\mathscr{S}$ is the set of all points αX,
with $\alpha \in \mathbb{R}$, where X is a nonzero vector on the line. The question asks
us to describe the set of all points $L(\alpha X)$ in $\mathbb{R}^q$.

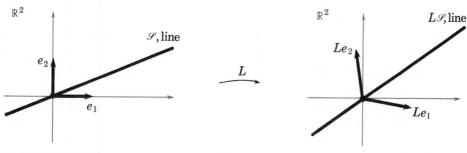

FIGURE 3.11

Since L is linear, $L(\alpha X) = \alpha LX$. Now either $LX = O$ or LX is a non-zero vector Y. In the second case, $L(\alpha X) = \alpha Y$, and the set of all "images" or values $L(\alpha X)$ is another line of points in $\mathbb{R}^q$. We sum up: *The image in $\mathbb{R}^q$ of a line through the origin in $\mathbb{R}^n$ under a linear map is either the single point O or another line through the origin.* In a sense, linear maps preserve lines. They do not transform lines into curves. See Fig. 3.11.

You may use statement 3 in Theorem 1 to verify that *the image in $\mathbb{R}^q$ of a plane through the origin in $\mathbb{R}^n$ under a linear map is either O or a line through the origin or another plane through the origin.*

3.1c Linearity and basis

Now we discuss a remarkable property of linear maps $L: \mathbb{R}^n \to \mathbb{R}^q$. As with many discoveries in mathematics, it is a consequence of relating two ideas.

We have already used the standard basis e_1, e_2, e_3 in $\mathbb{R}^3$. You recall that $e_1 = (1, 0, 0)$, $e_2 = (0, 1, 0)$, $e_3 = (0, 0, 1)$. This system of vectors had the property that every $X = (x_1, x_2, x_3)$ may be written as a linear combination of e_1, e_2, e_3, namely, $X = x_1 e_1 + x_2 e_2 + x_3 e_3$.

The **standard basis** of $\mathbb{R}^n$ consists of the vectors

$$e_1 = (1, 0, \ldots, 0), e_2 = (0, 1, 0, \ldots, 0), \ldots, e_n = (0, \ldots, 0, 1),$$

where e_i has a 1 in the ith slot and 0 in all the other slots. Then, just as in $\mathbb{R}^3$, $X = (x_1, \ldots, x_n)$ may be written $X = x_1 e_1 + \cdots + x_n e_n$.

We remark that, with one or two exceptions, standard basis vectors are the only vectors in $\mathbb{R}^n$, $n > 1$, that we do not denote by capital letters.

Now we turn to linear maps. Suppose that we know that $L: \mathbb{R}^n \to \mathbb{R}^q$ is linear and, moreover, that $Le_1 = Y_1, \ldots, Le_n = Y_n$, where $Y_1, \ldots, Y_n$ are vectors in $\mathbb{R}^q$; that is, we know the "value" of LX at the n points $X = e_1, e_2, \ldots, e_n$. Then we assert the striking fact that we now know LX for *all* vectors X, without exception.

Why is this? It is a simple consequence of the linearity of L. For let $X = (x_1, \ldots, x_n)$ be any vector in $\mathbb{R}^n$. Then, using the standard basis, we have $X = x_1 e_1 + \cdots + x_n e_n$. By the linearity of L, we have

$$LX = L(x_1 e_1 + \cdots + x_n e_n)$$
$$= x_1 Le_1 + \cdots + x_n Le_n$$
$$= x_1 Y_1 + \cdots + x_n Y_n.$$

Thus LX is a linear combination of $Y_1, \ldots, Y_n$ in $\mathbb{R}^q$.

Therefore, if we know the "image" vectors $Le_1, \ldots, Le_n$ in $\mathbb{R}^q$ and are given an explicit vector X in $\mathbb{R}^n$, then we can readily compute LX.

EXAMPLE: Let $L: \mathbb{R}^3 \rightarrow \mathbb{R}^2$ be linear, and denote by e_1, e_2, e_3 the standard basis in $\mathbb{R}^3$. Suppose that

$$Le_1 = (3, 1), \qquad Le_2 = (2, 4), \qquad Le_3 = (-1, 2).$$

These vectors in $\mathbb{R}^2$ are the "values" LX when $X = e_1, e_2, e_3$ respectively. Now we ask, "What is LX in general, when $X = (x_1, x_2, x_3)$?"

To answer this, note that

$$\begin{aligned} LX &= L(x_1e_1 + x_2e_2 + x_3e_3) \\ &= x_1Le_1 + x_2Le_2 + x_3Le_3 \\ &= x_1(3, 1) + x_2(2, 4) + x_3(-1, 2) \\ &= (3x_1 + 2x_2 - x_3, x_1 + 4x_2 + 2x_3). \end{aligned}$$

This is the *coordinate form* of L. If $X = (1, 3, -2)$, then $LX = (11, 9)$, and if $X' = (1, -1, 0)$, then $LX' = (1, -3)$.

CONSEQUENCES: 1. If $L: \mathbb{R}^n \rightarrow \mathbb{R}^q$ and $M: \mathbb{R}^n \rightarrow \mathbb{R}^q$ are linear maps, and they have equal values at $e_1, \ldots, e_n$, that is, $Le_1 = Me_1, \ldots, Le_n = Me_n$, then they are equal as functions, $LX = MX$ for all $X \in \mathbb{R}^n$, and we may write $L = M$.

2. To underscore the last comment, consider the following situation. Suppose that we have a function $L: \mathbb{R}^2 \rightarrow \mathbb{R}^2$ and we know that (*a*) L is linear and (*b*) L operates on the standard basis vectors $e_1 = (1, 0)$ and $e_2 = (0, 1)$ in $\mathbb{R}^2$ by rotating each through one right angle. See Fig. 3.12. Then we may conclude that L rotates *every* vector X in $\mathbb{R}^2$ through a right angle; that is, L is the linear map of Example 9 in the introduction to this section.

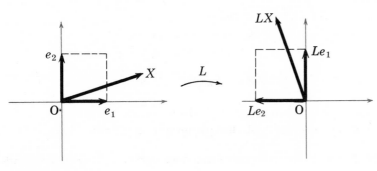

FIGURE 3.12

3. It is now easy to invent linear maps $L\colon \mathbb{R}^n \to \mathbb{R}^q$. Merely choose n vectors $Y_1, \ldots, Y_n$ in $\mathbb{R}^q$, the target space. Then define $Le_1 = Y_1, \ldots,$ $Le_n = Y_n$. This determines a linear map satisfying $LX = x_1Y_1 + \cdots + x_nY_n$, where $X = (x_1, \ldots, x_n)$. And, as mentioned above, the map L so determined is the only linear map such that $Le_1 = Y_1, \ldots, Le_n = Y_n$.

Here is a concrete example.

PROBLEM: Write down the rule for a linear map $L\colon \mathbb{R}^2 \to \mathbb{R}^2$ that rotates each vector $30°$ ($= \pi/6$ radians) counterclockwise.

Solution: It will suffice to calculate Le_1, Le_2, where $e_1 = (1, 0)$, $e_2 = (0, 1)$ in $\mathbb{R}^2$. See Fig. 3.13. We note that $Le_1 = (\cos \pi/6, \sin \pi/6)$ $= (\sqrt{3}/2, 1/2)$, and $Le_2 = (-\sin \pi/6, \cos \pi/6) = (-1/2, \sqrt{3}/2)$, using elementary trigonometry.

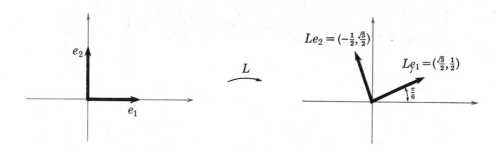

Rotation through $\frac{\pi}{6}$ (thirty degrees) counterclockwise

FIGURE 3.13

It follows that, for $X = (x_1, x_2)$, we have the coordinate form of L

$$LX = x_1Le_1 + x_2Le_2$$

$$= x_1\left(\frac{\sqrt{3}}{2}, \frac{1}{2}\right) + x_2\left(-\frac{1}{2}, \frac{\sqrt{3}}{2}\right)$$

$$= \frac{1}{2}(\sqrt{3}\,x_1 - x_2, x_1 + \sqrt{3}\,x_2).$$

You may check that LX is rotated $30°$ from X.

Let us sum up the result of this paragraph as follows.

THEOREM 2: Let $L\colon \mathbb{R}^n \to \mathbb{R}^q$ be a linear map, and $e_1, \ldots, e_n$ be the standard basis for $\mathbb{R}^n$.

1. If $X = (x_1, \ldots, x_n)$, then $LX = x_1 Le_1 + \cdots + x_n Le_n$.
2. L is uniquely determined by the image vectors $Le_1, \ldots, Le_n$ in $\mathbb{R}^q$.

REMARK: To appreciate this property of linear functions, consider the following. Think of the universe as a huge Euclidean space $\mathbb{R}^3$, designating your own position as the origin O. For each point $X = (x_1, x_2, x_3)$ in the universe, let $T(X)$ be the temperature there, measured in some suitable scale, at a fixed instant in time. Thus

$$T: \mathbb{R}^3 \rightarrow \mathbb{R}, \qquad T(X) = \text{temperature at } X.$$

We ask, "Is T linear?" Well, if it were, then by Theorem 2, it would be possible to pick the three points e_1, e_2, e_3, each quite close to yourself at the origin, and, by taking the temperature $T(e_1)$, $T(e_2)$, $T(e_3)$ at these three points, immediately write down the temperature at any point X of the universe. For we have $X = (x_1, x_2, x_3)$, whence $T(X)$ would equal $x_1 T(e_1) + x_2 T(e_2) + x_3 T(e_3)$.

But this conclusion seems to contradict experience. The temperature at some distant point X, on the moon, say, or in San Francisco Bay, seems quite independent of the temperatures at e_1, e_2, e_3; $T(X)$ may have one value or another without $T(e_1)$, $T(e_2)$, $T(e_3)$ changing. Hence the temperature T is nonlinear.

3.1d   The matrix notation

A linear map L is defined by formal properties, namely, $L(X + X') = LX + LX'$ and $L(\alpha X) = \alpha LX$. Examples involve linear equations, rotations of the plane, projections from one space to another, and so on—a wide variety of phenomena. Faced with this formal definition on the one hand and proliferation of diverse examples on the other, we feel the need for a uniform, concrete, almost mechanical way of representing and computing with all linear maps at once, without regard to special geometric or algebraic interpretations.

Actually, Theorem 2 is a great step in this direction. It assures us that the linear map $L: \mathbb{R}^n \rightarrow \mathbb{R}^q$ is determined by the n vectors $Le_1, \ldots, Le_n$ in $\mathbb{R}^q$. By pushing this insight a bit farther, we come to the idea of a matrix. This, at last, gives us a uniform way of computing with linear maps.

Let us work with an example. Let $L: \mathbb{R}^2 \rightarrow \mathbb{R}^3$ be a linear map. Denote the standard bases in $\mathbb{R}^2$ and $\mathbb{R}^3$ by e_1, e_2 and $\bar{e}_1, \bar{e}_2, \bar{e}_3$, respectively. Remember that $e_1 = (1, 0)$ and $\bar{e}_1 = (1, 0, 0)$, and so on.

Theorem 2 says that L is determined by the vectors Le_1, Le_2. Since these are elements of $\mathbb{R}^3$, they are linear combinations of $\bar{e}_1, \bar{e}_2, \bar{e}_3$, namely,

$$Le_1 = \alpha_1 \bar{e}_1 + \alpha_2 \bar{e}_2 + \alpha_3 \bar{e}_3,$$

$$Le_2 = \beta_1 \bar{e}_1 + \beta_2 \bar{e}_2 + \beta_3 \bar{e}_3,$$

where the α's and β's are scalars. These coefficients determine L. Different coefficients, different map. It is customary to arrange these coefficients into a rectangular array with three horizontal rows and two vertical columns thus:

$$\begin{bmatrix} \alpha_1 & \beta_1 \\ \alpha_2 & \beta_2 \\ \alpha_3 & \beta_3 \end{bmatrix}.$$

This array is termed the **matrix** of L and denoted $[L]$. Since the α's and β's determine the map, the matrix provides an uncluttered and concise summary of L. Observe that its first **column** lists the coefficients in the expression of Le_1:

$$Le_1 = \alpha_1 \bar{e}_1 + \alpha_2 \bar{e}_2 + \alpha_3 \bar{e}_3,$$

and the second column does likewise for Le_2. The reason for writing these coefficients vertically, rather than horizontally, will become clear later, when we define matrix multiplication.

EXAMPLES: 1. Let $L: \mathbb{R}^2 \to \mathbb{R}^3$ be determined by

$$Le_1 = \quad \bar{e}_1 \qquad\quad + 8\bar{e}_3,$$

$$Le_2 = 3\bar{e}_1 - 8\bar{e}_2 + \bar{e}_3.$$

Then the matrix for L is

$$[L] = \begin{bmatrix} 1 & 3 \\ 0 & -8 \\ 8 & 1 \end{bmatrix}.$$

2. Let $L: \mathbb{R}^2 \to \mathbb{R}^2$ be given by $LX = 2X$. We have seen this map already. Thus $Le_1 = 2e_1$, $Le_2 = 2e_2$. It follows that

$$[L] = \begin{bmatrix} 2 & 0 \\ 0 & 2 \end{bmatrix}.$$

3. Let $L: \mathbb{R}^2 \to \mathbb{R}^2$ be the map that rotates the plane counterclockwise through one right angle. Then $Le_1 = e_2$, $Le_2 = -e_1$ (check this!), and so

$$[L] = \begin{bmatrix} 0 & -1 \\ 1 & 0 \end{bmatrix}.$$

4. Suppose that we are told that a linear map L has the following matrix:

$$[L] = \begin{bmatrix} 1 & 3 & 0 & -1 \\ 2 & 1 & 4 & 4 \end{bmatrix}.$$

This means that $L: \mathbb{R}^4 \to \mathbb{R}^2$ is a linear map given by

$$Le_1 = \bar{e}_1 + 2\bar{e}_2,$$
$$Le_2 = 3\bar{e}_1 + \bar{e}_2,$$
$$Le_3 = 4\bar{e}_2,$$
$$Le_4 = -\bar{e}_1 + 4\bar{e}_2.$$

Hence the matrix is a shorthand way of giving a linear map. *Note that its proper interpretation depends on our agreement that the standard bases $e_1, \ldots, e_n$ for $\mathbb{R}^n$ and $\bar{e}_1, \ldots, \bar{e}_q$ for $\mathbb{R}^q$ are to be used.*

Again, we shall soon see that matrices provide not only a concise way of *describing* a linear map but also a useful means of *computing* with these maps. It is for the second reason that they are widely used.

3.1e The image of a linear map

Let $L: \mathbb{R}^n \to \mathbb{R}^q$ be linear. We have called the space $\mathbb{R}^n$ the *domain* of L and the space $\mathbb{R}^q$ the *target*. Let Y be a point in $\mathbb{R}^q$. We ask, "Does there exist X in $\mathbb{R}^n$ such that $LX = Y$?" The answer is "It depends on L." This is best seen by examples.

Thus, if $L: \mathbb{R}^n \to \mathbb{R}^q$ is the zero map, $LX = O$ for all X, then clearly $LX = Y$ has no solution X unless $Y = O$.

Again, the map in Example 7 at the start of this section, given as $L: \mathbb{R} \to \mathbb{R}^2$, $L(t) = (3t, 2t)$, shows that $L(t) = Y$ is solvable for t only if the vector Y is on the line through the origin in $\mathbb{R}^2$ determined by the vector $Y_1 = (3, 2)$. If Y is *not* on this line, no value of t satisfies $L(t) = Y$.

Having observed this phenomenon, we proceed to analyze it.

DEFINITION: The set of all vectors Y in $\mathbb{R}^q$ such that $Y = LX$ for at least one X in $\mathbb{R}^n$ is called the image of L and denoted $\mathscr{I}(L)$. Thus

$$\mathscr{I}(L) = \{Y \in \mathbb{R}^q \mid Y = LX \text{ for some } X \in \mathbb{R}^n\}.$$

Thus, if L is the zero map, then $\mathscr{I}(L) = \{O\}$, the set consisting of the zero vector alone.

If $L: \mathbb{R} \to \mathbb{R}^2$ is the map $L(t) = (3t, 2t)$ mentioned above, then $\mathscr{I}(L)$ is the line determined by the vector $Y_1 = (3, 2)$.

Another example: If $L: \mathbb{R}^2 \to \mathbb{R}^2$ is rotation counterclockwise through a right angle, then $\mathscr{I}(L) = \mathbb{R}^2$. To see this, suppose that Y is given. We show $Y \in \mathscr{I}(L)$ as follows: Let X be the vector of the same length

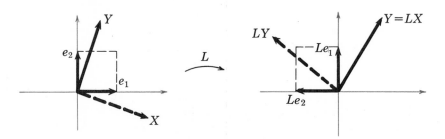

Showing $Y \in \mathscr{I}(L)$

FIGURE 3.14

as Y but pointing $90°$ clockwise from Y (see Fig. 3.14). Then clearly $Y = LX$.

Since Y was arbitrary, every vector is in $\mathscr{I}(L)$.

DEFINITION: If $L: \mathbb{R}^n \to \mathbb{R}^q$ and $\mathscr{I}(L) = \mathbb{R}^q$, then we say that L is **onto** $\mathbb{R}^q$, or **surjective.**

REMARK: We offer two reasons for studying $\mathscr{I}(L)$:

1. The map $L(t) = (3t, 2t)$ shows that interesting *geometric* objects, in this case lines, may be represented as images under linear maps. In this case we say that we have *parametrized* the line $2x_1 - 3x_2 = 0$ by means of the function L. The variable t is called the **parameter**. Parametric representation of lines, curves, surfaces, and so on, is of basic importance in advanced geometric investigations.

2. Many problems in mathematics reduce to this: If $L: \mathbb{R}^n \to \mathbb{R}^q$ and $Y \in \mathbb{R}^q$ is given, can we find a solution X to $LX = Y$? In other words, "Is the given vector Y in $\mathscr{I}(L)$?"

As an example of such a problem, consider the system of equations

$$2x_1 - 3x_2 = 1,$$

$$-6x_1 + 9x_2 = -4.$$

We translate this into the language of maps. Let $L: \mathbb{R}^2 \to \mathbb{R}^2$ be given by $LX = (2x_1 - 3x_2, -6x_1 + 9x_2)$, where $X = (x_1, x_2)$. You should agree that L is linear. Now let $Y = (1, -4)$. Then inquiring about the solvability of the system of equations is the same as asking, "Does there exist $X = (x_1, x_2)$ such that $LX = Y$? In other words, is $Y = (1, -4)$ in $\mathscr{I}(L)$?"

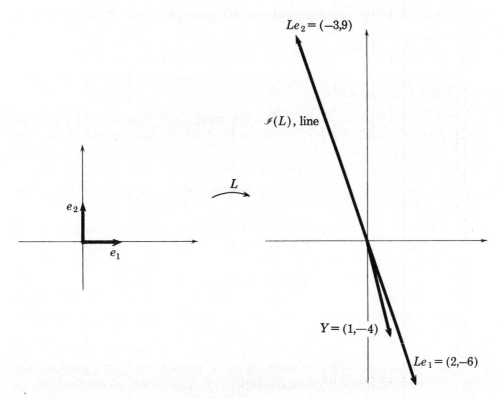

FIGURE 3.15

Let us consider this map. We compute $Le_1 = (2, -6)$, and $Le_2 = (-3, 9)$. Now $LX = x_1Le_1 + x_2Le_2$ and therefore is a linear combination of $(2, -6)$ and $(-3, 9)$. Both these vectors, however, lie on the same line through the origin, since $(2, -6) = -2/3(-3, 9)$. See Fig. 3.15. Thus $\mathscr{I}(L)$ is one-dimensional, and, in particular, the vector Y is not on it.

On the other hand, the vector $Y_1 = (1, -3)$ does lie on the line $\mathscr{I}(L)$. Hence we should be able to solve for x_1, x_2 in

$$2x_1 - 3x_2 = 1,$$

$$-6x_1 + 9x_2 = -3.$$

It is easy to see that $x_1 = 2$, $x_2 = 1$ is a solution. In other words, if $X_1 = (2, 1)$, $LX_1 = Y_1$. Can you find other solutions? In summary, linear equations lead to the study of $\mathscr{I}(L)$.

We will discuss $\mathscr{I}(L)$ again from the points-of-view of linear equations and of geometry. For now, a natural and elementary question to ask is this, "Is the set $\mathscr{I}(L)$ always a vector subspace of $\mathbb{R}^q$?" It is, we note, in the particular examples above. More generally, it is always a subspace.

THEOREM 3: Let $L: \mathbb{R}^n \to \mathbb{R}^q$ be linear. Then the set $\mathscr{I}(L)$ of images is a subspace of $\mathbb{R}^q$. In fact, $\mathscr{I}(L)$ consists of all linear combinations of the vectors $Le_1, \ldots, Le_n$ in $\mathbb{R}^q$.

Proof: Let $Y_1, Y_2 \in \mathscr{I}(L)$. We must show that $Y_1 + Y_2 \in \mathscr{I}(L)$. But $Y_i = LX_i$, $i = 1, 2$ for some $X_i \in \mathbb{R}^n$. It follows, therefore, that $Y_1 + Y_2 = L(X_1 + X_2)$, so that the set $\mathscr{I}(L)$ *is* closed under addition of vectors. We leave it to you to show that if $Y \in \mathscr{I}(L)$ and α is a scalar, then $\alpha Y \in \mathscr{I}(L)$ also. This, of course, completes the proof that $\mathscr{I}(L)$ is a subspace. $\ll$

REMARK: In your studies you may have encountered the **principle of superposition**. In the language we are developing, the principle states the following: *If the equations $LX = Y_1$ and $LX = Y_2$ have the solutions $X = X_1$ and $X = X_2$ respectively, then the equation $LX = \alpha_1 Y_1 + \alpha_2 Y_2$ is solvable and, moreover, has the solution $X = \alpha_1 X_1 + \alpha_2 X_2$.* This statement is roughly equivalent to the fact that $\mathscr{I}(L)$ is a subspace and therefore closed under addition and multiplication by scalars.

"Superposition" is useful in that it reduces the solution of $LX = \alpha_1 Y_1 + \cdots + \alpha_k Y_k$ to the solution of the k possibly simpler equations $LX = Y_1, \ldots, LX = Y_k$. If the solutions of these equations are found to be $X_1, \ldots, X_k$, respectively, then the original equation has a solution given by $X = \alpha_1 X_1 + \cdots + \alpha_k X_k$.

3.1f Further digression on $LX = Y$; the null space

We continue to examine the vector equation $LX = Y$, where Y is given and X is to be found. Suppose that we know that $X = X_1$ is one solution, $LX_1 = Y$. We ask, "Are there any others? What do they look like? In fact, what is the most general solution to $LX = Y$?"

Here is an observation. Suppose that $LZ = 0$. Then $L(X_1 + Z) = LX_1 + LZ = Y + 0 = Y$. Thus the vector $X = X_1 + Z$ is also a solution to $LX = Y$. This prompts a definition.

DEFINITION: Let $L: \mathbb{R}^n \to \mathbb{R}^q$ be linear. Then the **null space** of L, denoted $\mathscr{N}(L)$, is the set in $\mathbb{R}^n$ whose image is zero,

$$\mathscr{N}(L) = \{X \in \mathbb{R}^n \,|\, LX = 0\}.$$

We remark that $\mathscr{N}(L)$ is sometimes referred to as the **kernel** of L.

Thus $\mathscr{N}(L)$ is the set of solutions to the very special equation $LX = 0$.

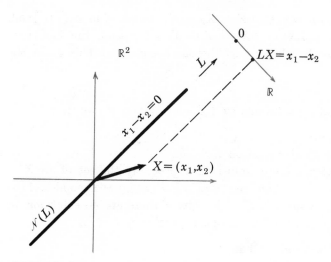

FIGURE 3.16

For example, if $L: \mathbb{R}^2 \rightarrow \mathbb{R}$ is given by $LX = x_1 - x_2$, then $\mathcal{N}(L)$ is the familiar straight line through the origin of $\mathbb{R}^2$ given by $x_1 - x_2 = 0$. See Fig. 3.16.

Note that $\mathcal{N}(L)$ is contained in the domain of L, in contrast to the image $\mathcal{I}(L)$, which is a subspace of the target $\mathbb{R}^q$.

Is the null space a vector subspace of the domain $\mathbb{R}^n$, as the name seems to imply? In the example $LX = x_1 - x_2$, $\mathcal{N}(L)$ is a line through the origin and hence a subspace. We leave it to you to prove

THEOREM 4:   Let $L: \mathbb{R}^n \rightarrow \mathbb{R}^q$ be linear. Then the null space $\mathcal{N}(L)$ is a subspace of $\mathbb{R}^n$. Also, $\mathcal{N}(L) = \mathbb{R}^n \Leftrightarrow L$ is the zero map.

Having this useful notion, we return to the question "What is the most general solution to $LX = Y$?" We state

THEOREM 5:   Let $L: \mathbb{R}^n \rightarrow \mathbb{R}^q$ be linear. Let $Y \in \mathbb{R}^q$ be given, and suppose that $LX_1 = Y$ for a particular $X_1 \in \mathbb{R}^n$. Then every solution to $LX = Y$ is of the form $X = X_1 + Z$, where X_1 is as above and $Z \in \mathcal{N}(L)$.

COROLLARY 6:   If $\mathcal{N}(L) = \{0\}$, then there are either no solutions or exactly one solution X to $LX = Y$ for a given Y.

Proof of Theorem 5:   We saw above that $X_1 + Z$ was a solution to $LX = Y$ if X_1 was. Now suppose that X_1, X_2 are two solutions. We must show that X_2 is of the form $X_1 + Z$ for some $Z \in \mathcal{N}(L)$. The obvious thing to do is subtract: Call $Z = X_2 - X_1$. Is this Z in $\mathcal{N}(L)$? Well, $LZ = L(X_2 - X_1) = LX_2 - LX_1 = Y - Y = 0$, using the fact that X_1 and X_2 were both solutions to the same equation. This gives the theorem. $\ll$

SUMMARY: To find all solutions to $LX = Y$, given Y, we may:

1. Compute $\mathcal{N}(L)$.
2. Compute a particular solution $X = X_1$.
3. Conclude that the set of solutions is precisely the set of all $X = X_1 + Z$, with $Z \in \mathcal{N}(L)$. Thus the "size" of $\mathcal{N}(L)$ gives some idea of the "size" of the solution set of $LX = Y$: Either there are no solutions, or the set of solutions is an affine subspace parallel to $\mathcal{N}(L)$.

Namely, in the notation of Chap. 1,

$$\text{Set of solutions} = \mathcal{N}(L) + X_1.$$

Note that $\mathcal{N}(L)$ depends only on L, not on the choice of Y.

Computations 1 and 2 may require effort (solving linear equations), and we shall not give a full treatment of this. Our chief purpose in this chapter is not the theory of linear equations but rather linear maps as applied to the theory of vector functions. Let us content ourselves with an example in the spirit of the discussion above.

EXAMPLE: Let $LX = x_1 + x_2 + x_3$, $L: \mathbb{R}^3 \to \mathbb{R}$. Find the general solution to $LX = 2$.

1. We realize that $\mathcal{N}(L)$ is the plane of all (x_1, x_2, x_3) such that $x_1 + x_2 + x_3 = 0$. This subspace is spanned by the two vectors $Z_1 = (1, 0, -1)$ and $Z_2 = (0, 1, -1)$.
2. A particular solution to $LX = 2$ is clearly given by $X_1 = (2, 0, 0)$.
3. Thus every solution to $LX = 2$ is of the form

$$X = X_1 + \alpha_1 Z_1 + \alpha_2 Z_2;$$

that is,

$$x_1 = 2 + \alpha_1, \qquad x_2 = \alpha_2, \qquad x_3 = -\alpha_1 - \alpha_2,$$

for any $\alpha_1, \alpha_2 \in \mathbb{R}$.

Some solutions are therefore $(2, 0, 0)$, $(3, 0, -1)$, $(3, 1, -2)$, $(12, 3, -13)$.

Of course, this set of solutions is the affine subspace $x_1 + x_2 + x_3 = 2$, a plane in $\mathbb{R}^3$ that avoids the origin.

3.1g What has been done so far?

This may be answered very simply. First, we defined a vector space as a set in which addition of elements and multiplication by real scalars are possible. A mathematician might say that a vector space is a set endowed with a "linear structure."

Second, we have just discussed functions which *preserve* this structure and which are well behaved with respect to vector addition and multiplication by scalars. For the additivity property

$$L(X + X') = LX + LX'$$

says that a linear map "preserves" vector addition, and a similar connection holds between homogeneity and multiplication by scalars. And it is because of this amicable relation between vector spaces and linear maps that the very important Theorems 2 and 3 are true.

We can see that these theorems are false for more general nonlinear functions. Happily, however, when we launch our study of nonlinear functions, the linear maps will be on hand to guide us and preserve us from chaos. They are the true heros of these chapters.

Exercises

These give you practice in (1) deciding whether a given map is actually linear, (2) writing down a linear map with prescribed properties, (3) using the matix notation. The basic idea of linear maps is simple; it is important, however, to realize that they may be encountered in various forms, involving vectors (say $LX = 7X$), or coordinates, or matrices, or a geometric prescription (rotate $30°$ counterclockwise). Each form has its uses; we must learn to translate from one to another.

1. Which of the following maps L are linear?
$$L: \mathbb{R}^2 \longrightarrow \mathbb{R}, \qquad X = (x_1, x_2)$$

(a) $LX = 2x_1 - x_2$,
(b) $LX = x_1 + x_2 + 2$,

(c) $LX = x_1 x_2$;

$$L: \mathbb{R} \longrightarrow \mathbb{R}^2 \qquad \text{so that the variable is a } scalar \ t$$

(d) $L(t) = (\cos t, \sin t)$,
(e) $L(t) = (0, t)$,

(f) $L(t) = (2t, -t)$;

$$L: \mathbb{R}^2 \longrightarrow \mathbb{R}^2$$

(g) $LX = (x_1^2 - x_2^2, 2x_1 x_2)$,
(h) $LX = (x_1, -x_1)$,
(i) $LX = \|X\|^{-1}X, X \neq 0$,

(j) $LX = (x_1 + x_2, x_1 - x_2)$,
(k) $LX = (x_1, e^{x_2})$;

$$L: \mathbb{R}^3 \longrightarrow \mathbb{R}^3, \qquad X = (x_1, x_2, x_3)$$

(l) $LX = (x_3, x_1, x_2)$,
(m) $LX = (x_1 + x_2, x_2 + x_3, 0)$,

(n) $LX = \|X\| (1, 2, 0)$.

2.　Let $L: \mathbb{R}^2 \longrightarrow \mathbb{R}^2$ be the linear map given by $LX = (x_1, x_1 + x_2)$. Draw a picture of the target space and locate Le_1, Le_2, LX, where $X = (1, 1)$, and locate the images of the x_1-axis and the x_2-axis (lines) under the mapping L. What is the image of the line $x_1 + x_2 - 2 = 0$?

3.　(a) Given that $L: \mathbb{R}^2 \longrightarrow \mathbb{R}^2$ is linear and that $Le_1 = Y_1 = (3, 4)$, $Le_2 = Y_2 = (1, -1)$. Write LX, where $X = (x_1, x_2)$, in terms of Y_1, Y_2. In the case of $X = (13, -2)$, what is LX (coordinates please)?

　　　(b) Where did we use the hypothesis that L was linear in computing the image vector LX in Exercise 3(a)?

4.　Given that $L: \mathbb{R}^2 \longrightarrow \mathbb{R}^2$ is linear and also that $Le_1 = Y_1 = (2, 3)$, $Le_2 = Y_2 = (8, 12)$:

　　　(a) If $Y = (1, 1)$, does there exist X such that $LX = Y$?
　　　(b) Same question, with $Y = (4, 6)$?
　　　(c) Draw a picture of the target space, showing that the set of all LX is a straight line (equation in coordinates y_1, y_2?)
　　　(d) Are there nonzero X such that $LX = O$?

5.　Let $L: \mathbb{R}^2 \longrightarrow \mathbb{R}^2$ be a linear map with matrix

$$[L] = \begin{bmatrix} 1 & -1 \\ 2 & 0 \end{bmatrix}.$$

What is:

　　　(a) Le_1?
　　　(b) Le_2?
　　　(c) LX, where $X = (10, 7) = 10e_1 + 7e_2$?

6.　Given that $L: \mathbb{R}^2 \longrightarrow \mathbb{R}^2$ is linear, $Le_1 = e_1 + 2e_2 = (1, 2)$, $Le_2 = -e_1 = (-1, 0)$. Write down the 2 by 2 matrix $[L]$.

7.　Let $L: \mathbb{R}^2 \longrightarrow \mathbb{R}$ have the matrix $[L] = [2 \quad -3]$. What is:

　　　(a) Le_1?
　　　(b) Le_2?
　　　(c) $L(3e_1 + 2e_2)$?
　　　(d) LX, where $X = (x_1, x_2)$? That is, write LX in coordinate form.

8.　(a) Let $L: \mathbb{R}^2 \longrightarrow \mathbb{R}$ be linear and $Le_1 = 4$, $Le_2 = 3$. Write down the matrix $[L]$.

　　　(b) Let $M: \mathbb{R}^2 \longrightarrow \mathbb{R}$ be given by $MX = 2x_1 - 3x_2$. Write down the matrix $[M]$. (First compute Me_1, Me_2.)

9.　Let $L: \mathbb{R}^3 \longrightarrow \mathbb{R}^2$ be determined by

$$[L] = \begin{bmatrix} 0 & 1 & 1 \\ 1 & 0 & 1 \end{bmatrix}.$$

What is:

　　　(a) Le_1?　　　　　　　　　　(c) Le_3?
　　　(b) Le_2?　　　　　　　　　　(d) $L(e_1 + e_2 - e_3)$?

Write these image vectors Le_1, and so on, in terms of the standard basis $\bar{e}_1, \bar{e}_2$ of $\mathbb{R}^2$.

10. Given a linear map $L: \mathbb{R}^3 \longrightarrow \mathbb{R}^2$ such that $Le_1 = \bar{e}_2$, $Le_2 = \bar{e}_1$, $Le_3 = \bar{e}_1 + \bar{e}_2$. Let $X = (x_1, x_2, x_3)$ in $\mathbb{R}^3$, $Y = (y_1, y_2)$ in $\mathbb{R}^2$:

    (a) If $LX = Y$, what is y_1 in terms of x_1, x_2, x_3?

    (b) Same for y_2. (Hint: $X = x_1e_1 + x_2e_2 + x_3e_3$, $Y = y_1\bar{e}_1 + y_2\bar{e}_2$.)

    (c) Write down $[L]$.

    (d) Do you actually use the fact that L is linear?

11. Draw pictures of the image space $\mathscr{I}(L)$ in $\mathbb{R}^2$ for the following L:

    (a) $L: \mathbb{R} \longrightarrow \mathbb{R}^2$, $L(t) = (2t, t)$.

    (b) $L: \mathbb{R}^2 \longrightarrow \mathbb{R}^2$, $LX = (2x_1 - 6x_2, x_1 - 3x_2)$. (Hint: Think of LX as $Y = (y_1, y_2)$, where y_1, y_2 are coordinates in the target $\mathbb{R}^2$.)

    (c) $L: \mathbb{R}^3 \longrightarrow \mathbb{R}^2$, $X = (x_1, x_2, x_3)$, $LX = (2x_1 - 2x_2 + 6x_3, x_1 - x_2 + 3x_3)$.

12. Using Exercise 11(b), decide for which y_1, y_2 the system

$$2x_1 - 6x_2 = y_1,$$
$$x_1 - 3x_2 = y_2$$

has a solution. Note that you are not asked to find a solution if one exists. But can you?

13. Let $L: \mathbb{R}^2 \longrightarrow \mathbb{R}^2$. Since the null space $\mathscr{N}(L)$ is a subspace (see Sec. 3.1f), it is either $\{O\}$ or some line through O or all of $\mathbb{R}^2$. Describe $\mathscr{N}(L)$ explicitly for the following linear maps L:

    (a) $LX = (x_1, -x_2)$,

    (b) $LX = (x_1, 0)$,

    (c) $LX = (x_1, -x_1)$,

    (d) $LX = (x_1, x_1 - x_2)$,

    (e) $LX = (2x_1 - 2x_2, x_1 - x_2)$.

Note: In each case say for which $X = (x_1, x_2)$ we have $LX = (0, 0)$. A picture might clarify things, especially in Exercise 13(b).

14. (a) If L is linear, can $\mathscr{N}(L)$ be the empty set?

    (b) Given $LX_1 = Y$ and $LX_2 = Y$ also for distinct X_1, X_2, find a nonzero vector in $\mathscr{N}(L)$.

    (c) Suppose that $LX_1 = LX_2$ implies $X_1 = X_2$. What can you say of X if $LX = O$? (This is easy.)

    (d) Given that $\mathscr{N}(L) = \{O\}$. What can you say of X_1, X_2 if $LX_1 = LX_2$? (Hint: Look at $X_1 - X_2$.)

15. True or false. In all cases $L: \mathbb{R}^n \longrightarrow \mathbb{R}^q$ may or may not be linear:

    (a) If $LO = O$, then L is linear.

    (b) L is linear $\Leftrightarrow L(\alpha_1X_1 + \alpha_2X_2) = \alpha_1LX_1 + \alpha_2LX_2$ for all $\alpha_1, \alpha_2 \in \mathbb{R}$, $X_1, X_2 \in \mathbb{R}^n$.

    (c) If L is a rotation of $\mathbb{R}^2$ about the origin O, then L is linear.

    (d) If $L: \mathbb{R}^2 \longrightarrow \mathbb{R}$ is constant, say $LX = 1$ for all X, then L is linear.

    (e) If L is a rotation of $\mathbb{R}^2$ about the point $X_0 = (1, 1)$, then L is linear.

    (f) If L is linear and we know the image vectors $Le_1, \ldots, Le_n$, then we can compute LX for any given $X \in \mathbb{R}^n$.

    (g) If $L: \mathbb{R}^2 \longrightarrow \mathbb{R}^2$ is linear and $Le_1 = e_1$, $Le_2 = e_2$, then $LX = X$ for *all* $X \in \mathbb{R}^2$.

    (h) If $L: \mathbb{R}^2 \longrightarrow \mathbb{R}^2$ is linear and $Le_1 = e_2$, $Le_2 = -e_1$, then L rotates the plane $90°$ counterclockwise about the origin.

(i)   The image $\mathscr{I}(L)$ of a linear map is "flat": the origin, a straight line, a plane, and so on.

(j)   If $L: \mathbb{R}^2 \longrightarrow \mathbb{R}^2$ is linear and the first column of the 2 by 2 matrix $[L]$ is

$$\begin{bmatrix} 1 \\ 0 \end{bmatrix},$$

then $Le_1 = e_1$.

(k)   If L is linear but not the zero map, then the image $\mathscr{I}(L)$ contains a straight line through the origin in $\mathbb{R}^q$.

(l)   The map $L: \mathbb{R}^n \longrightarrow \mathbb{R}$ given by $LX = \|X\|$ is linear.

(m)   To write down the matrix of a linear map L, it is necessary and sufficient to know the image vectors $Le_1, \ldots, Le_n$ in terms of the standard basis $\bar{e}_1, \ldots, \bar{e}_q$ of $\mathbb{R}^q$.

(n)   If $L: \mathbb{R}^n \longrightarrow \mathbb{R}^q$ is linear, then its matrix $[L]$ is square $\Leftrightarrow q = n$.

(o)   Let $e_1 = (1, 1)$ and $e_2 = (0, 1)$, and let $L: \mathbb{R}^2 \longrightarrow \mathbb{R}^3$ satisfy $Le_1 = (1, 2, 3)$, $Le_2 = (1, -2, -1)$. Then $L(2, 3) = (1, 1, 1)$.

(p)   Let $L: \mathbb{R}^n \longrightarrow \mathbb{R}^q$ be linear. If $LX_1 = 0$ and $LX_2 = 0$, then $Z = 3X_1 - 7X_2$ also satisfies $LZ = 0$.

16.   Let $L: \mathbb{R}^3 \longrightarrow \mathbb{R}^3$ be defined by $LX = (x_1, 0, x_3)$. Geometrically describe $\mathscr{N}(L)$ and $\mathscr{I}(L)$. What are the dimensions of $\mathscr{N}(L)$ and $\mathscr{I}(L)$?

17.   Complete the proof of Theorem 3.

18.   Prove Theorem 4.

19.   Prove Corollary 6.

20.   If $L: \mathbb{R}^2 \longrightarrow \mathbb{R}^2$ is linear, prove that the image of a parallelogram is another parallelogram.

LINEAR MAPS ON FUNCTION SPACES

In Chap. 1 we observed that sets of functions may be vector spaces. It turns out that some maps on these spaces are well known to every student of calculus. These maps are the familiar derivative and integral.

To be more precise, let $\mathscr{C}^1(\mathbb{R})$ denote the set of all functions having one continuous derivative for x in $\mathbb{R}$ and $\mathscr{C}(\mathbb{R})$ the set of all continuous functions on $\mathbb{R}$. Of course, $\mathscr{C}^1(\mathbb{R})$ is a subspace of $\mathscr{C}(\mathbb{R})$. Some elements of $\mathscr{C}^1(\mathbb{R})$ are $\sin x$, polynomials such as $x^2 + 1$, e^x, and so on. Then, if $f \in \mathscr{C}^1(\mathbb{R})$, the operator $D: \mathscr{C}^1(\mathbb{R}) \longrightarrow \mathscr{C}(\mathbb{R})$ defined by

$$Df = f',$$

where f' is the derivative of f, is a linear map. For example $Dx^2 = 2x$, $D \sin x = \cos x$. The linearity is just a restatement of the rules for differentiation:

$$D(f + g) = Df + Dg, \qquad D(cf) = c(Df).$$

Similarly, the operator $L: \mathscr{C}(\mathbb{R}) \longrightarrow \mathscr{C}(\mathbb{R})$ defined for $f \in \mathscr{C}(\mathbb{R})$ by

$$(Lf)(x) = \int_0^x f(t)\, dt$$

is linear. For example, $Lx^2 = x^3/3$, $Le^x = e^x - 1$. Note that $L0 = 0$, as required for linear maps.

Because both the derivative and integral are linear maps, we see that *all calculus can be viewed as a study of these two special linear maps on a particular linear space.* For instance, one version of the Fundamental Theorem of Calculus states that

$$\frac{d}{dx} \int_0^x f(t)\, dt = f(x).$$

This can be written in the simple symbolic form

$$DL = I,$$

where I is the identity operator; that is, *$If = f$ for all functions f.*

It is quite simple to get other linear maps. For instance, define T: $\mathscr{C}^1(\mathbb{R}) \to \mathscr{C}(\mathbb{R})$ on $u \in \mathscr{C}^1(\mathbb{R})$ by the rule

$$Tu = u' - u,$$

and so $T(x^2) = 2x - x^2$, $T(7e^x) = 0$, and so on. Since T is the sum of two linear maps,

$$T = D - I,$$

it too is a linear map. The problem of solving the differential equation

$$u' - u = f,$$

where f is a given continuous function, is exactly that of solving $Tu \overset{!}{=} f$, that is, of finding if f is in the image of T. Just as in the case of maps in $\mathbb{R}^n$, and with identical proofs, one sees that the image $\mathscr{I}(T)$ and null space $\mathscr{N}(T)$ are subspaces of $\mathscr{C}(\mathbb{R})$. Moreover, one can see that $\mathscr{N}(T)$ *is one-dimensional* (note that $\mathscr{N}(T)$ is the set of functions $u(x)$ satisfying $u' = u$).

To prove one-dimensionality, first observe that $u(x) = ce^x$ is in $\mathscr{N}(T)$ for any scalar c. It follows that the set of all ce^x for all c is a one-dimensional subspace of $\mathscr{N}(T)$. Call it $\mathscr{S}$; $\mathscr{S} = \{ce^x \mid c \in \mathbb{R}\}$. Clearly $\dim \mathscr{S} = 1$. Now suppose that $v(x) \in \mathscr{N}(T)$, $v'(x) - v(x) = 0$. We will show $v(x) = ce^x$, which will imply $\mathscr{N}(T) = \mathscr{S}$ and so $\dim \mathscr{N}(T) = \dim \mathscr{S} = 1$.

To see that $v(x) = ce^x$, we define a new function $\varphi(x) = v(x)e^{-x}$. Since $v' - v = 0$, we easily obtain

$$\varphi' = v'e^{-x} - ve^{-x} = (v' - v)e^{-x} = 0.$$

Consequently, $\varphi(x)$ is a constant, $\varphi(x) = c$. Thus $v(x) = \varphi(x)e^x = ce^x$, as claimed. We conclude that $\mathcal{N}(T)$, the space of functions satisfying the differential equation $v' - v = 0$, is one-dimensional.

$$dim \; \mathscr{D}(L) = dim \; \mathscr{I}(L) + dim \; \mathscr{N}(L)$$

Let $L: \mathbb{R}^n \to \mathbb{R}^q$ be a linear map. What can we say about the dimension of the image of L, which we write $dim \; \mathscr{I}(L)$? No matter how many dimensions the target space $\mathbb{R}^q$ has, it should be plausible that $dim \; \mathscr{I}(L) \leq n$. This is by "conservation of dimension"; the image cannot be any bigger than the domain $\mathscr{D}(L) = \mathbb{R}^n$.

For example, the map $L: \mathbb{R}^2 \to \mathbb{R}^3$ defined by $L(x_1, x_2) = (x_1, x_2, x_1 + x_2)$ has as its image a two-dimensional subspace of $\mathbb{R}^3$ (it is the plane $y_3 = y_1 + y_2$). Its image could not have taken up more than two dimensions.

Assuming that we have proved this, that is, $dim \; \mathscr{I}(L) \leq dim \; \mathscr{D}(L)$, we ask what happens if $dim \; \mathscr{I}(L) < dim \; \mathscr{D}(L)$. Again an example. Let $L: \mathbb{R}^2 \to \mathbb{R}^3$ be defined by $L(x_1, x_2) = (x_1 - x_2, -x_1 + x_2, 2x_1 - 2x_2)$. The image is the one-dimensional straight line $y_2 = -y_1$, $y_3 = 2y_1$. Can we account geometrically for the difference $dim \; \mathscr{D}(L) - dim \; \mathscr{I}(L) = ?$ Yes. The missing dimensions are in the null space $\mathscr{N}(L)$. In fact, we have the following theorem connecting the numerology. It is both attractive and useful.

DIMENSION THEOREM: Let $L: V \to W$ be a linear map between the vector spaces V and W. Then

$$dim \; \mathscr{D}(L) = dim \; \mathscr{I}(L) + dim \; \mathscr{N}(L).$$

REMARK: Notice that $dim \; W$ does not appear in this, except implicitly in the fact that $\mathscr{I}(L) \subset W$, and so $dim \; \mathscr{I}(L) \leq dim \; W$.

Proof: Say that $dim \; \mathscr{I}(L) = k$ and let $\{Y_1, \ldots, Y_k\} \subset W$ be a basis for $\mathscr{I}(L)$. Then there are vectors $\{X_1, \ldots, X_k\} \subset V$ such that $LX_1 = Y_1, \ldots, LX_k = Y_k$. If $dim \; \mathscr{N}(L) = p$, we let $\{Z_1, \ldots, Z_p\} \subset V$ be a basis for $\mathscr{N}(L)$. We claim that $\{Z_1, \ldots, Z_p, X_1, \ldots, X_k\}$ is a basis for V. Assuming this, we are done, since $p + k = dim \; V$, and so $dim \; \mathscr{N}(L) + dim \; \mathscr{I}(L) = p + k = dim \; V = dim \; \mathscr{D}(L)$.

First we shall prove that $\{Z_1, \ldots, Z_p, X_1, \ldots, X_k\}$ are linearly independent. Assume that $a_1 Z_1 + \cdots + a_p Z_p + b_1 X_1 + \cdots + b_k X_k = 0$. We must show that $a_1 = \cdots = a_p = b_1 = \cdots = b_k = 0$. Applying L to each side and recalling that $LZ_j = 0$, $j = 1, \ldots, p$, we find that

$$O = L(b_1 X_1 + \cdots + b_k X_k) = b_1 LX_1 + \cdots + b_k LX_k$$
$$= b_1 Y_1 + \cdots + b_k Y_k.$$

But the Y_j's are independent, and so $b_1 = b_2 = \cdots = b_k = 0$. Consequently we have $a_1 Z_1 + \cdots + a_p Z_p = O$. Now we use the independence of the Z_j's to conclude that $a_1 = a_2 = \cdots = a_p = 0$.

Next we shall prove that $\{Z_1, \ldots, Z_p, X_1, \ldots, X_k\}$ span V. Say that $X \in V$. Then $LX \in \mathcal{I}(L)$, so that $LX = \alpha_1 Y_1 + \cdots + \alpha_k Y_k$ for some scalars $\alpha_1, \ldots, \alpha_k$. However, $Y_j = LX_j$. Therefore

$$L(X - \alpha_1 X_1 - \alpha_2 X_2 - \cdots - \alpha_k X_k) = O;$$

that is, the vector $X - \alpha_1 X_1 - \cdots - \alpha_k X_k \in \mathcal{N}(L)$. Hence it must be a linear combination of $Z_1, \ldots, Z_p$,

$$X - \alpha_1 X_1 - \cdots - \alpha_k X_k = \beta_1 Z_1 + \cdots + \beta_p Z_p,$$

or, equivalently,

$$X = \alpha_1 X_1 + \cdots + \alpha_k X_k + \beta_1 Z_1 + \cdots + \beta_p Z_p.$$

Thus the X_j's and Z_j's together span V. Done. $\ll$

Several applications of the Dimension Theorem will appear later on. Here we show how it can be used to give a very short proof of one half of the Hyperplane Theorem.

THEOREM: Given numbers $a_1, \ldots, a_n$, not all zero. Let the subspace $\mathcal{S}$ of $\mathbb{R}^n$ consist of all points $X \in \mathbb{R}^n$ that satisfy $a_1 x_1 + \cdots + a_n x_n = 0$. Then $\mathcal{S}$ is a hyperplane; that is, dim $\mathcal{S} = n - 1$.

Proof: Define the linear map $L: \mathbb{R}^n \to \mathbb{R}$ by $LX = a_1 x_1 + \cdots + a_n x_n$. Since some $a_j \neq 0$, we have dim $\mathcal{I}(L) = 1$. Therefore dim $\mathcal{N}(L) = $ dim $\mathcal{D}(L) - $ dim $\mathcal{I}(L) = n - 1$. But $\mathcal{N}(L)$ is precisely $\mathcal{S}$. Done. $\ll$

3.2 A SPECIAL CASE: FUNCTIONALS

A linear map $L: \mathbb{R}^n \to \mathbb{R}$ is commonly called a linear **functional** or linear **form**. We note that the values LX of a functional are *real numbers*, that is, vectors in the one-dimensional space $\mathbb{R}$. Because of this one-dimensionality, functionals have certain nice properties not enjoyed by the generality of linear maps.

AN IMPORTANT EXAMPLE: Let Z be a fixed vector in $\mathbb{R}^n$. For $X \in \mathbb{R}^n$, define $LX = \langle Z, X \rangle$, the inner product of Z and X. Then $L: \mathbb{R}^n \to \mathbb{R}$, since $\langle Z, X \rangle$ is a number and, as you should check, L is linear. Hence, we conclude that *every fixed vector Z in $\mathbb{R}^n$ determines a linear functional*. An illustrative notation for this functional is $L = \langle Z, \quad \rangle$, where we leave a blank space for insertion of the variable X.

For instance, let $L: \mathbb{R}^3 \rightarrow \mathbb{R}$ be given as $L = \langle Z, \quad \rangle$, where $Z = (2, 3, -5)$. Then, if $X = (1, -1, 2)$, say, we have $LX = 2 \cdot 1 + 3 \cdot (-1) - 5 \cdot 2 = -11$.

A QUESTION: We ask about the converse to the above: Given an arbitrary linear functional $L: \mathbb{R}^n \rightarrow \mathbb{R}$, does there exist a vector $Z \in \mathbb{R}^n$ such that $L = \langle Z, \quad \rangle$, that is, $LX = \langle Z, X \rangle$ for all X? If this were the case, then we would have reduced the theory of functionals to that of the inner product.

To answer this question, let us examine a typical linear functional $L: \mathbb{R}^n \rightarrow \mathbb{R}$. If $e_1, \ldots, e_n$ is the standard basis for $\mathbb{R}^n$, then $z_1 = Le_1, \ldots, z_n = Le_n$ are numbers. Since L is linear, if $X = (x_1, \ldots, x_n) = x_1 e_1 + \cdots + x_n e_n$, then $LX = x_1 Le_1 + \cdots + x_n Le_n = x_1 z_1 + \cdots + x_n z_n$. But the last expression is familiar. Suppose that we use the numbers z_j to define the vector Z by $Z = (z_1, \ldots, z_n) = z_1 e_1 + \cdots + z_n e_n$. Then, for any X, we have $\langle Z, X \rangle = x_1 z_1 + \cdots + x_n z_n = LX$. This answers in the affirmative the question raised above: Every functional L may be represented as $\langle Z, \quad \rangle$ for some fixed vector Z.

THEOREM 7: (Representation theorem) Every linear functional $L: \mathbb{R}^n \rightarrow \mathbb{R}$ is of the form $L = \langle Z, \quad \rangle$ for some vector Z in $\mathbb{R}^n$. The vector $Z = (z_1, \ldots, z_n)$ is determined by $z_1 = Le_1, \ldots, z_n = Le_n$. Hence, every functional L may also be expressed in the form

$$LX = z_1 x_1 + \cdots + z_n x_n$$

for $X = (x_1, \ldots, x_n)$, where the coefficients $z_1, \ldots, z_n$ are fixed.

REMARK: Thus every functional looks like the left-hand side of a linear equation.

PROBLEM: Given a functional $L: \mathbb{R}^n \rightarrow \mathbb{R}$, find Z such that $L = \langle Z, \quad \rangle$; that is, find the Z that represents L.

Solution: This is straightforward. Let $n = 3$ for definiteness. Say that L is given by $Le_1 = 4$, $Le_2 = -2$, $Le_3 = 5$. Then we define $Z = (4, -2, 5)$. It is easy to check that for $X = (x_1, x_2, x_3)$ we have

$$LX = \langle Z, X \rangle = 4x_1 - 2x_2 + 5x_3.$$

Exercises

Linear functionals and the closely related scalar-valued affine maps are of basic importance in Chap. 5, because, roughly speaking, their graphs are the tangent planes to "curved" surfaces, the graphs of nonlinear functions (compare straight lines in ordinary calculus). We begin to consider these graphs below.

1. Which of the scalar-valued maps $L: \mathbb{R}^2 \longrightarrow \mathbb{R}$ are linear functionals?

    (a) $LX = x_1$,   (e) $LX = \langle X, X \rangle$,

    (b) $LX = \| X \|$,   (f) $LX = |x_1|$,

    (c) $LX = x_1 - x_2$,   (g) $LX = 1 - x_1 - x_2$,

    (d) $LX = x_1 x_2$,   (h) $LX = (\sin x_1 + \cos x_1)x_2$.

2. The image $\mathscr{I}(L)$ of the zero linear functional is the set consisting of the number 0 only. What is $\mathscr{I}(L)$ if the linear functional L is not identically zero? Describe geometrically.

3. (a) Let $L: \mathbb{R}^2 \longrightarrow \mathbb{R}$ be given by $LX = 2x_1 - 5x_2$. Find a vector Z such that $LX = \langle Z, X \rangle$ for all X.

    (b) Given that $L: \mathbb{R}^2 \longrightarrow \mathbb{R}$ is a linear functional and, moreover, $Le_1 = 2$, $Le_2 = -5$. Compute a vector Z such that $LX = \langle Z, X \rangle$ for all X.

    (c) What is the matrix of the linear map L in Exercise 3(b)? It should look familiar.

4. Let $LX = 2x_1 - 5x_2$, as in Exercise 3(a). Sketch the null space $\mathscr{N}(L)$ as a subspace of $\mathbb{R}^2$.

    (b) What is the relation between $\mathscr{N}(L)$ and the vector Z found in Exercise 3(a)?

    (c) Does there exist a linear functional $L: \mathbb{R}^2 \longrightarrow \mathbb{R}$ whose null space consists of O alone? Why?

    (d) True or false. The null space $\mathscr{N}(L)$ of a nonzero linear functional $L: \mathbb{R}^3 \longrightarrow \mathbb{R}$ is a straight line through the origin in $\mathbb{R}^3$.

5. The **graph** of a linear functional $L: \mathbb{R}^2 \longrightarrow \mathbb{R}$ is, not surprisingly, the set of all points $(X, z) = (x_1, x_2, z)$ in $\mathbb{R}^3$ such that $LX = L(x_1, x_2) = z$. Prove the following:

    (a) The graph of L contains the origin $(0, 0, 0)$ of $\mathbb{R}^3$.

    (b) The graph of L is a (two-dimensional) plane in $\mathbb{R}^3$.

    (c) The graph of L is the unique plane through the three points (O, LO), (e_1, Le_1), (e_2, Le_2), that is, $(0, 0, 0)$, $(1, 0, Le_1)$, $(0, 1, Le_2)$.

    (d) The graph of $z = LX = 2x_1 - 5x_2$ is the plane in $\mathbb{R}^3$ determined by the equation $2x_1 - 5x_2 - z = 0$.

    (e) Any plane in x_1x_2z-space $\mathbb{R}^3$ that does not contain the z-axis is the graph of a linear functional in

$$X = (x_1, x_2).$$

6. Sketch in $\mathbb{R}^3$ the graph of $L: \mathbb{R}^2 \longrightarrow \mathbb{R}$ given by $z = LX = -x_1 - x_2$. Interpret the intersection of this plane with the x_1x_2-plane $z = 0$.

7. Find a linear functional $L: \mathbb{R}^3 \longrightarrow \mathbb{R}$ such that $LX_1 = 0$, $LX_2 = 0$, where $X_1 = (1, 1, 0)$ and $X_2 = (1, 0, 1)$. Is L uniquely determined? If it is, say why; if it is not, say what additional data might be given to determine it uniquely.

WHAT IS LINEAR PROGRAMMING?

We content ourselves with

A PRACTICAL EXAMPLE: You are a traveling snake-oil salesman, selling brand A at \$5 per gallon and brand B at \$3 per gallon. You ply your trade in a territory that craves snake oil, and so you can be sure of selling all you carry at the prices mentioned. Things are not quite so simple, however. Snake oil requires the magical ingredient M. More precisely, brand A requires 4 oz of ingredient M per 100 gal of oil, and brand B requires 3 oz per 100 gal. And you have only 12 oz of ingredient M. See Fig. 3.17.

FIGURE 3.17

If the scarcity of ingredient M were the only constraint, then it is easy to see that you would maximize gross income by producing and selling only brand A. There is one more fly in the ointment. Brand A weighs 15 lb per gal; brand B weighs one-third this much. And you are capable of transporting only 3,000 lb of snake oil. How much should be brand A and how much brand B?

THE APPROACH: We proceed as follows. Let x = amount of brand A in hundreds of gallons, y = amount of brand B in hundreds of gallons.

We wish to compute x and y so that the gross income is maximized. This leads to a linear functional

$$\text{Gross income} = L(x, y) = 500x + 300y$$

since, for example, brand A sells for \$500 per 100 gal.

Now we may not use more than 12 oz of ingredient M; that is

$$4x + 3y \leq 12.$$

Also brands A and B weigh 1,500 and 500 lb per 100 gal, respectively, so that we must satisfy

$$1{,}500x + 500y \leq 3{,}000,$$

which is the same as

$$3x + y \leq 6.$$

Hence the problem may be stated in mathematical terms thus: maximize $L(x, y) = 500x + 300y$ subject to the constraints

$$x \geq 0, \quad y \geq 0 \qquad \text{amounts nonnegative,}$$
$$4x + 3y \leq 12 \qquad \text{ingredient } M \text{ limited,}$$
$$3x + y \leq 6 \qquad \text{weight limited.}$$

A GRAPHICAL SOLUTION: It is not hard to see that the set $\mathscr{K}$ of points (x, y) satisfying all four inequalities above looks as shown in Fig. 3.18. This set is **convex** in that if (x_1, y_1) and (x_2, y_2) are two points of $\mathscr{K}$, then the straight-line segment connecting them is also in $\mathscr{K}$.

Now for certain numbers c, the straight line $L(x, y) = 500x + 300y = c$ actually intersects the set $\mathscr{K}$—for example, $c = 0$, $(x, y) = (0, 0)$. What we seek is the *largest* number c having this property and the point (or points) (x, y) of intersection. The lines $500x + 300y = c$ all have the same slope; they are parallel. Examination of the picture should convince you that the maximum value of $L(x, y)$ for $(x, y) \in \mathscr{K}$ occurs at the point $X^* = (x^*, y^*) = (\frac{6}{5}, \frac{12}{5})$, one of the "vertices" (corners) of the convex set $\mathscr{K}$. This means that you should produce 120 gal of brand A and 240 gal of brand B in order to maximize your total income, which will be \$1,320.

For the general linear-programming problem, we have a linear functional $L: \mathbb{R}^n \to \mathbb{R}$, like $L(x, y) = 500x + 300y$ above, and some convex set $\mathscr{K}$, not a subspace, of $\mathbb{R}^n$. This constraint set $\mathscr{K}$ of admissible values is usually defined by many inequalities (we had four in the example). The problem is to find X^* in $\mathscr{K}$ that maximizes (or minimizes) L, so that $LX^* \geq LX$ for all X in the set $\mathscr{K}$ of admissible values. An important

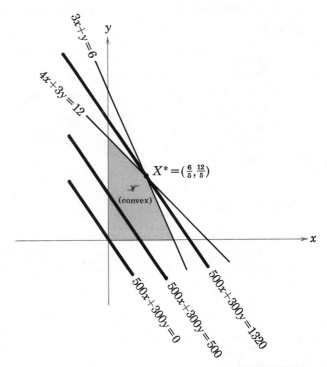

FIGURE 3.18

theorem asserts that the desired "best point" X^* is always one of the vertices (corners) of $\mathcal{K}$. Computers are essential in solving these problems.

REMARK: Note that we are maximizing a *linear* functional L here. Problems of maxima and minima for *nonlinear* functions are discussed in Chap. 6.

3.3 THE ALGEBRA OF LINEAR MAPS

3.3a Introduction

So far in this chapter we have considered linear maps singly, one at a time. In this section we learn how to combine them. We will study the sum of linear maps, multiplication of a linear map by a scalar, and, finally, the composition (a kind of multiplication) of two linear maps. From this we may conclude that the algebra of linear maps is at least as rich as that of vectors.

3.3b Addition of linear maps

Let L and M be linear maps from $\mathbb{R}^n$ into $\mathbb{R}^q$; we may write this L, M: $\mathbb{R}^n \to \mathbb{R}^q$. The sum of L and M, denoted $L + M$, is a function from $\mathbb{R}^n$ into $\mathbb{R}^q$ defined, for each $X \in \mathbb{R}^n$, by

$$(L + M)X = LX + MX.$$

This definition makes sense, because both LX and MX are in $\mathbb{R}^q$ and hence may be added by the usual addition of vectors.

EXAMPLE: Let $L, M: \mathbb{R}^3 \to \mathbb{R}$ be functionals given by $LX = 3x_1 - x_2 + 12x_3$, $MX = 16x_1 + x_2 + 17x_3$, where $X = (x_1, x_2, x_3)$. Then, using the definition,

$$(L + M)X = (3x_1 - x_2 + 12x_3) + (16x_1 + x_2 + 17x_3)$$
$$= 19x_1 + 29x_3.$$

Note here that $L + M$ is again linear. In a moment we will prove generally that the sum of linear maps is linear. You should be able to do this now, using the definition of linear map.

THEOREM 8: Let $L, M, P: \mathbb{R}^n \to \mathbb{R}^q$ be linear maps. Then:

1. The sum $L + M$ of two linear maps is also a linear map.
2. Associativity: $L + (M + P) = (L + M) + P$.
3. Commutativity: $L + M = M + L$.
4. Denoting by O the zero map that assigns to each $X \in \mathbb{R}^n$ the zero vector in $\mathbb{R}^q$, we have $L + O = L$.
5. Denoting by $-L$ the map that assigns to each $X \in \mathbb{R}^n$ the vector $-(LX)$ in $\mathbb{R}^q$, we have $L + (-L) = O$, the zero map.

Proof: 1. We must show that $L + M$ has the additivity and homogeneity properties. To verify additivity, note that

$$(L + M)(X + X') = L(X + X') + M(X + X') \quad \text{definition of} \; L + M$$

$$= LX + LX' + MX + MX' \quad \text{by additivity of } L \text{ and } M$$

$$= LX + MX + LX' + MX' \quad \text{commutativity in } \mathbb{R}^q$$

$$= (L + M)X + (L + M)X' \quad \text{definition of } L + M.$$

We leave the verification of homogeneity to you. It is entirely straight-forward.

Parts 2 to 5 of the theorem follow from the definition of the sum of linear maps and the analogous properties in $\mathbb{R}^q$. You should be able to check them all. ≪

3.3c Multiplication of a linear map by a scalar

Just as with vectors, it is possible to multiply a linear map by a scalar. Let $L: \mathbb{R}^n \to \mathbb{R}^q$ be linear and let $\alpha \in \mathbb{R}$. We define αL to be the function given, for all $X \in \mathbb{R}^n$, by

$$(\alpha L)X = \alpha(LX).$$

For example, if $L: \mathbb{R}^3 \to \mathbb{R}$ is given by $LX = x_1 - 2x_2 + 4x_3$ and $\alpha = -5$, then $(-5L)X = -5x_1 + 10x_2 - 20x_3$.

It is easy to verify the following theorem.

THEOREM 9: Let $L, M: \mathbb{R}^n \to \mathbb{R}^q$ be linear maps and α, β scalars. Then:

1. $\alpha L: \mathbb{R}^n \to \mathbb{R}^q$ is a linear map.
2. $(\alpha\beta)L = \alpha(\beta L)$.
3. $1L = L$.
4. $(\alpha + \beta)L = \alpha L + \beta L$.
5. $\alpha(L + M) = \alpha L + \alpha M$.

REMARK: We digress to point out the meaning of Theorems 8 and 9. If we let $\mathscr{L}$ denote the set of *all* linear maps $\mathbb{R}^n \to \mathbb{R}^q$, that is,

$$\mathscr{L} = \{L: \mathbb{R}^n \to \mathbb{R}^q \,|\, L \text{ is linear}\},$$

and if we define $L + M$ and αL as above, then *the set $\mathscr{L}$ is itself a vector space.* For Theorems 8 and 9 assure us that the requirements for associa-tivity, commutativity, zero, additive inverse, and so on, are all met. This is the first example we have encountered of a vector space that does not appear to be a Euclidean space $\mathbb{R}^n$ or a subspace of $\mathbb{R}^n$. The vector space $\mathscr{L}$ consists of functions, linear maps, and might therefore be called a "func-tion space." The study of vector spaces of functions, using generalizations of the geometric notions of length, norm, orthogonality, and inner product, is an important and fruitful branch of modern higher mathematics.

3.3d The composition of linear maps

The third algebraic operation we consider for linear maps is composition. In calculus, if $f(u) = e^u$ and $u(x) = 2x^2 + 1$, then the composition is $f(u(x)) = e^{2x^2+1}$. We now extend this notion to higher dimensions.

Let $L: \mathbb{R}^n \to \mathbb{R}^q$ and $M: \mathbb{R}^q \to \mathbb{R}^r$ be linear maps. We denote this $\mathbb{R}^n \xrightarrow{L} \mathbb{R}^q \xrightarrow{M} \mathbb{R}^r$. Note that if $X \in \mathbb{R}^n$, then $LX \in \mathbb{R}^q$, and $\mathbb{R}^q$, the target of the first map, is the domain of the second map M. Hence we may define the product, or **composition**, denoted ML, of L and M as

$$(ML)X = M(LX).$$

Thus $ML: \mathbb{R}^n \to \mathbb{R}^r$ is a function given by the rule: (1) to $X \in \mathbb{R}^n$, apply L, obtaining $LX \in \mathbb{R}^q$; (2) to LX now apply the map M, obtaining $M(LX)$ $\in \mathbb{R}^r$. We have $(ML)X = M(LX)$. See Fig. 3.19.

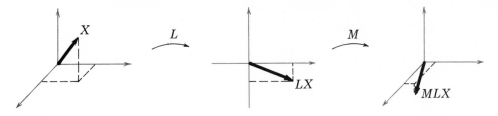

The composition of linear maps

FIGURE 3.19

Note that we also write simply MLX. Another useful notation is $L^2 = LL$, so $L^2X = L(LX)$; $L^3 = LL^2$, etc.

Two warnings: First, if ML is defined, it need not be true that LM is defined. Thus, if $L: \mathbb{R}^3 \to \mathbb{R}^2$ and $M: \mathbb{R}^2 \to \mathbb{R}^7$, then $ML: \mathbb{R}^3 \to \mathbb{R}^7$ is defined; on the other hand, $L(MX)$ is meaningless, because $MX \in \mathbb{R}^7$, and L operates on vectors in $\mathbb{R}^3$ only.

The second warning is this. Even if LM and ML are both defined—say that $L: \mathbb{R}^2 \to \mathbb{R}^2$ and $M: \mathbb{R}^2 \to \mathbb{R}^2$—then it need not be true that $LM = ML$; in fact, this is rather rare. In general, the operation of composing two linear maps is *not* commutative.

EXAMPLES: 1. Let $L: \mathbb{R}^2 \to \mathbb{R}^3$ be determined by

$$Le_1 = \bar{e}_1 - \bar{e}_2 + 5\bar{e}_3, \qquad Le_2 = 2\bar{e}_1 + \bar{e}_2 + \bar{e}_3,$$

where e_1, e_2 and $\bar{e}_1$, $\bar{e}_2$, $\bar{e}_3$ are the standard bases for $\mathbb{R}^2$ and $\mathbb{R}^3$ respectively. Also, let $M: \mathbb{R}^3 \to \mathbb{R}^3$ by

$$M\bar{e}_1 = \bar{e}_2, \qquad M\bar{e}_2 = \bar{e}_3, \qquad M\bar{e}_3 = \bar{e}_1.$$

Then $ML: \mathbb{R}^2 \to \mathbb{R}^3$ operates on e_1, e_2 as follows:

$$MLe_1 = M(Le_1) = M(\bar{e}_1 - \bar{e}_2 + 5\bar{e}_3) = M\bar{e}_1 - M\bar{e}_2 + 5M\bar{e}_3$$
$$= 5\bar{e}_1 + \bar{e}_2 - \bar{e}_3.$$

We leave it to you to compute MLe_2.

Note that LM is not defined.

2. Let $L, M : \mathbb{R}^2 \to \mathbb{R}^2$ be linear maps determined as follows:

$$Le_1 = e_2, \quad Le_2 = -e_1 \qquad \text{rotation } 90° \text{ counterclockwise,}$$

$$Me_1 = e_1, \quad Me_2 = -e_2 \qquad \text{reflection across horizontal axis.}$$

We first note that LM and ML are both defined. For ML, see Fig. 3.20. We see that

$$MLe_1 = -e_2, \qquad MLe_2 = -e_1.$$

On the other hand, for LM, see Fig. 3.21. Thus

$$LMe_1 = e_2, \qquad LMe_2 = e_1.$$

Hence, $LM \neq ML$; in fact, $LM = -ML$.

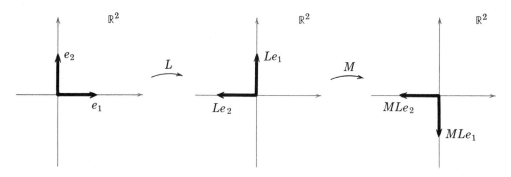

First rotation, then reflection

FIGURE 3.20

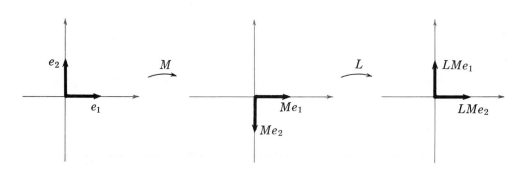

Reflection first, then rotation

FIGURE 3.21

In this same example we see that $L^2e_1 = L(Le_1) = Le_2 = -e_1$ and $L^2e_2 = -e_2$. Thus $L^2 = -I$ is rotation by $180°$. What is L^3?

3. Again let $L, M : \mathbb{R}^2 \longrightarrow \mathbb{R}^2$, but this time let $LX = 7X$ for all $X \in \mathbb{R}^2$, and let M be rotation through a fixed angle θ. Then LM and ML are both defined and, as you can verify by drawing pictures, $LM = ML$. The reason for commutativity in this case is the fact that the map L is very simple; it amounts to multiplication by the scalar 7. A formal proof that $LM = ML$ follows:

$$LMX = 7MX = M(7X) = M(LX) = MLX, \qquad \text{for all } X \in \mathbb{R}^2.$$

You may have noted in these examples that the composition ML, when defined, was linear. We now go on to verify this in general and collect some other properties of composition.

THEOREM 10: Let $\mathbb{R}^n \xrightarrow{L} \mathbb{R}^q \xrightarrow{M} \mathbb{R}^r \xrightarrow{P} \mathbb{R}^s$ be linear maps. Then:

1.   The composition $ML : \mathbb{R}^n \longrightarrow \mathbb{R}^r$ is a linear map.
2.   Associativity: $P(ML) = (PM)L$ as a map $\mathbb{R}^n \longrightarrow \mathbb{R}^s$.

The following distributive laws also hold:

3. $M(L_1 + L_2) = ML_1 + ML_2$, where $L_1, L_2 : \mathbb{R}^n \longrightarrow \mathbb{R}^q$ and $M : \mathbb{R}^q \longrightarrow \mathbb{R}^r$.
4. $(M_1 + M_2)L = M_1L + M_2L$, where $L : \mathbb{R}^n \longrightarrow \mathbb{R}^q$ and $M_1, M_2 : \mathbb{R}^q \longrightarrow \mathbb{R}^r$.

Further, if we denote by $I_q : \mathbb{R}^q \longrightarrow \mathbb{R}^q$ the identity map $IX = X$ for $X \in \mathbb{R}^q$, then:

5. $I_qL = L = LI_n$, where $L : \mathbb{R}^n \longrightarrow \mathbb{R}^q$.

Proof: 1. We must show that $ML(X_1 + X_2) = MLX_1 + MLX_2$ (additivity) and also that $ML(\alpha X) = \alpha MLX$ (homogeneity).

To prove additivity, note that $ML(X_1 + X_2)$ means $M(L(X_1 + X_2))$, which equals $M(LX_1 + LX_2) = MLX_1 + MLX_2$ by the separate additivity of L and M. Done.

We leave homogeneity to you.

2. Left to you. You should convince yourself that the composition of mappings, when defined, of course, is always associative.

3, 4, 5. Left to you. $\ll$

Exercises

Here are some simple calisthenics in the algebra of linear maps.

1.   Let $L, M : \mathbb{R}^2 \longrightarrow \mathbb{R}^2$ be linear maps determined by

$$Le_1 = e_1 + e_2, \qquad Le_2 = e_1 - e_2;$$
$$Me_1 = 2e_1, \qquad Me_2 = e_1 + 3e_2.$$

Compute the following in terms of e_1, e_2 (don't use coordinates or matrices):

(a) $(L + M)e_1$;   (g) $(ML)e_2$;

(b) $(L + M)e_2$;   (h) $(ML)X$, with $X = 2e_1 + e_2$;

(c) $(L + M)X$, where $X = 2e_1 + e_2$;   (i) $(LM)e_1$;

(d) $(4L)e_2$;   (j) $(LM)e_2$;

(e) $(4L - M)e_1$;   (k) $L^2 e_1$, that is, $L(Le_1)$.

(f) $(ML)e_1$;

2. (a) Write down the 2 by 2 matrices $[L]$, $[M]$ for the maps in Exercise 1.

　 (b) Since $L + M$ is a linear map $(L + M)\colon \mathbb{R}^2 \longrightarrow \mathbb{R}^2$, it has a 2 by 2 matrix $[L + M]$. Using $(L + M)e_1$, $(L + M)e_2$, write down this matrix.

　 (c) How is $[L + M]$ obtained from $[L]$ and $[M]$? We study this in the following section.

3. Let $L, M, P\colon \mathbb{R}^2 \longrightarrow \mathbb{R}^2$ be linear maps given by

$$Le_1 = e_2, \qquad Le_2 = -e_1 \qquad \text{rotation } 90° \text{ counterclockwise;}$$

$$Me_1 = -e_1, \qquad Me_2 = e_2 \qquad \text{flip across } x_2 \text{ axis;}$$

$$PX = -X \qquad\qquad\qquad \text{reflection across the origin.}$$

　 (a) Draw pictures describing the actions of the maps L, M, P and the compositions LM, ML, LP, PL, MP, PM.

　 (b) Which pairs of these maps commute?

4. Let L, M, P be as in Exercise 3. True or false:

(a) $L^2 = P$;   (e) $M^2 = I$;

(b) $P^2 = I$, the identity map;   (f) $M^3 = M$;

(c) $L^4 = I$;   (g) $MPM = P$;

(d) $L^5 = L$;   (h) $PMP = L$.

5. Invertibility of linear maps. The linear map $L\colon \mathbb{R}^n \longrightarrow \mathbb{R}^n$ is said to be *invertible* or *nonsingular* if and only if there exists a linear map $M\colon \mathbb{R}^n \longrightarrow \mathbb{R}^n$ such that $LM = I$ and $ML = I$; that is, $LMX = X$ and $MLX = X$ for all $X \in \mathbb{R}^n$. In this case, the map M is called the *inverse* of L and denoted L^{-1}. True or false (if true, give the easy proof):

(a) If L is invertible, then every equation $LX = Y$ has a unique solution given by $X = L^{-1}Y$.

(b) If L is invertible, then L^{-1} is invertible.

(c) If $LX = 5X$ for all X, then $L^{-1}X = \frac{1}{5}X$.

(d) If $L\colon \mathbb{R}^2 \longrightarrow \mathbb{R}^2$ rotates the plane 90° counterclockwise, then L^{-1} rotates the plane 90° clockwise.

(e) If $L\colon \mathbb{R}^2 \longrightarrow \mathbb{R}^2$ rotates the plane 90° counterclockwise, then L^{-1} rotates the plane 270° counterclockwise.

(f) The zero map, $LX = O$ for all X, is invertible.

(g) If L is invertible, then $L^{-1}O = O$.

(h) If $LX = O$ for some $X \neq O$, then L is not invertible.

(i) The identity map I is the only map $L\colon \mathbb{R}^2 \longrightarrow \mathbb{R}^2$ that is its own inverse, $L = L^{-1}$.

6. Given $L: \mathbb{R}^2 \to \mathbb{R}^2$, compute L^{-1}; that is, write $L^{-1}e_1$, $L^{-1}e_2$ in terms of e_1, e_2, for each of the following maps:

(a) $Le_1 = e_2$, $Le_2 = -e_1$ (familiar?);

(b) $Le_1 = e_1$, $Le_2 = e_1 + e_2$ (Note: $L^{-1}e_2 = \alpha e_1 + \beta e_2$; solve for α, β.)

7. (a) Given L as in Exercise 6(a), solve $LX = Y$, where $Y = 1{,}728e_1 - 1{,}984e_2$.

(b) Given L as in Exercise 6(b), solve $LX = Y$, Y as in Exercise 7(a).

8. Let $L: \mathbb{R}^3 \to \mathbb{R}^2$, $M: \mathbb{R}^2 \to \mathbb{R}^2$. Which of the following maps make sense?

(a) $2L + M$,   (e) LM,

(b) L^2,   (f) M^2L,

(c) M^2,   (g) $ML - 3M$.

(d) ML,

9. Let L and $M: \mathbb{R}^n \to \mathbb{R}^n$. Use the definition to prove $(L + M)^2 = L^2 + LM + ML + M^2$. Can this be simplified to $(L + M)^2 = L^2 + 2LM + M^2$? Why?

10. Let L and $M: \mathbb{R}^n \to \mathbb{R}^n$. Prove one of the following and give a counterexample to show that the other is false:

(a) $(LM)^2 = L^2 M^2$,   (b) $(LM)^2 = LMLM$.

11. Define L and $M: \mathbb{R}^2 \to \mathbb{R}^2$ as $L(x_1, x_2) = (0, x_1)$ and $M(x_1, x_2) = (x_1, 0)$:

(a) Compute LMX, MLX, L^2X, and M^2X to deduce that $LM = L$, $L^2 = 0$, $M^2 = M$, $ML = 0$.

(b) Use the results in Exercise 11(a) to express LML, MLM, $(LM)^2$, $(ML)^2$, M^{57}, $(L + M)^2$, $(2L + M)^2$, $(L - 2M)^3$ in the form $aL + bM$.

12. Let L, $M: \mathbb{R}^n \to \mathbb{R}^n$. For each of the following, give either a proof or a counterexample (see Exercises 4 and 11):

(a) $L^2 = 0 \Rightarrow L = 0$.

(b) $L^2 = L \Rightarrow L = I$ or $L = 0$.

(c) $L^2 = I \Rightarrow L = I$ or $L = -I$.

13. Complete the proof of Theorem 8.

14. Complete the proof of Theorem 10.

15. Let $L: \mathbb{R}^n \to \mathbb{R}^n$. Prove that $\mathcal{N}(L) \subset \mathcal{N}(L^2)$ and $\mathcal{I}(L^2) \subset \mathcal{I}(L)$.

16. Let $L: \mathbb{R}^n \to \mathbb{R}^n$. If for every $X \in \mathbb{R}^n$, there is a number k (depending on X) such that $L^k X = 0$, prove that there is a p such that $L^p = 0$.

17. Let $L: \mathbb{R}^n \to \mathbb{R}^n$. Prove that $\mathcal{I}(L) \subset \mathcal{N}(L) \Leftrightarrow L^2 = 0$.

GEOMETRY OF LINEAR MAPS

Let $L: \mathbb{R}^2 \to \mathbb{R}^2$ be a linear map. We know that L is determined by its effect on the standard basis e_1, e_2 of $\mathbb{R}^2$. Figure 3.22 illustrates the geometric effect of some special linear maps in terms of their matrix representation. In each case we have drawn the image of the unit square with corners at $(0, 0)$, $(1, 0)$, $(1, 1)$, $(0, 1)$. We hope that you will ponder them.

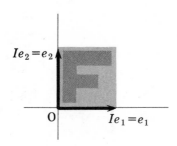

$$Ie_2 = e_2$$
$$O \qquad Ie_1 = e_1$$

(a) Identity $I = \begin{bmatrix} 1 & 0 \\ 0 & 1 \end{bmatrix}$

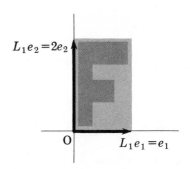

$$L_1 e_2 = 2e_2$$
$$O \qquad L_1 e_1 = e_1$$

(b) Stretching $L_1 = \begin{bmatrix} 1 & 0 \\ 0 & 2 \end{bmatrix}$

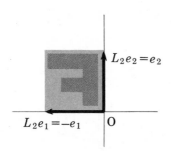

$$L_2 e_2 = e_2$$
$$L_2 e_1 = -e_1 \qquad O$$

(c) Reflection $L_2 = \begin{bmatrix} -1 & 0 \\ 0 & 1 \end{bmatrix}$

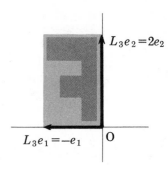

$$L_3 e_2 = 2e_2$$
$$L_3 e_1 = -e_1 \qquad O$$

(d) Stretching and reflection
$$L_3 = \begin{bmatrix} -1 & 0 \\ 0 & 2 \end{bmatrix}$$

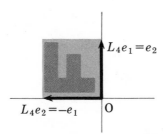

$$L_4 e_1 = e_2$$
$$L_4 e_2 = -e_1 \qquad O$$

(e) Rotation $L_4 = \begin{bmatrix} 0 & -1 \\ 1 & 0 \end{bmatrix}$

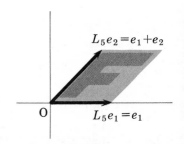

$$L_5 e_2 = e_1 + e_2$$
$$O \qquad L_5 e_1 = e_1$$

(f) Shear $L_5 = \begin{bmatrix} 1 & 1 \\ 0 & 1 \end{bmatrix}$

FIGURE 3.22

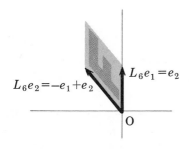

$L_6e_2=-e_1+e_2$ $L_6e_1=e_2$

O

(g) Shear and rotation $L_6=\begin{bmatrix}0&-1\\1&1\end{bmatrix}$

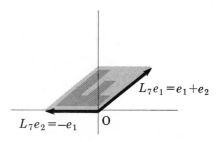

$L_7e_1=e_1+e_2$

$L_7e_2=-e_1$ O

(h) Rotation and shear $L_7=\begin{bmatrix}1&-1\\1&0\end{bmatrix}$

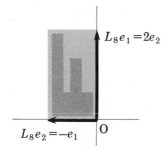

$L_8e_1=2e_2$

$L_8e_2=-e_1$ O

(i) Rotation and stretching
$$L_8=\begin{bmatrix}0&-1\\2&0\end{bmatrix}$$

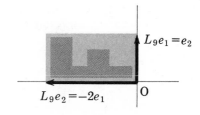

$L_9e_1=e_2$

$L_9e_2=-2e_1$ O

(j) Stretching and rotation
$$L_9=\begin{bmatrix}0&-2\\1&0\end{bmatrix}$$

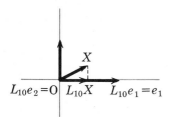

X

$L_{10}e_2=0$ $L_{10}X$ $L_{10}e_1=e_1$

(k) Projection $L_{10}=\begin{bmatrix}1&0\\0&0\end{bmatrix}$

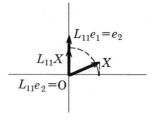

$L_{11}e_1=e_2$

$L_{11}X$ X

$L_{11}e_2=0$

(l) Projection and rotation $L_{11}=\begin{bmatrix}0&0\\1&0\end{bmatrix}$

FIGURE 3.22 (cont.)

119

3.4 MATRICES

3.4a Introduction

At the end of Sec. 1 of this chapter we used a rectangular array of numbers, a so-called matrix, as a shorthand representation of a linear map. In this section we elaborate on the relation between linear maps and matrices and then go on to transfer the algebraic operations for linear maps—that is, addition, multiplication by scalars, and composition—to matrices. Having done this, we may use matrices not only to represent maps, as a mere shorthand device, but also to carry out all numerical computations involving linear maps, a role for which the matrix is especially well suited.

3.4b The matrix of a linear map

Let us recall how we associate a matrix $[L]$ to a linear map L. Say that $L\colon \mathbb{R}^3 \to \mathbb{R}^2$ is given by

$$Le_1 = \alpha_1\bar{e}_1 + \alpha_2\bar{e}_2, \qquad Le_2 = \beta_1\bar{e}_1 + \beta_2\bar{e}_2, \qquad Le_3 = \gamma_1\bar{e}_1 + \gamma_2\bar{e}_2,$$

where e_1, e_2, e_3 and $\bar{e}_1$, $\bar{e}_2$ are the standard bases for $\mathbb{R}^3$ and $\mathbb{R}^2$ respectively. Then we associate to L the matrix

$$[L] = \begin{bmatrix} \alpha_1 & \beta_1 & \gamma_1 \\ \alpha_2 & \beta_2 & \gamma_2 \end{bmatrix}.$$

Note that the first (left-hand) column of $[L]$ gives information on Le_1, the second column on Le_2, the third column on Le_3.

REMARK: Though it may seem odd at first, we will become used to writing this information vertically. It will expedite computations later.

Now let us observe that we can reverse this process. Given a matrix, such as

$$\begin{bmatrix} 3 & 2 & -2 \\ -1 & 0 & 1 \end{bmatrix},$$

we can use it to define a linear map $L\colon \mathbb{R}^3 \to \mathbb{R}^2$; namely, define

$$Le_1 = 3\bar{e}_1 - \bar{e}_2, \qquad Le_2 = 2\bar{e}_1, \qquad Le_3 = -2\bar{e}_1 + \bar{e}_2.$$

Having defined the values of L on the basis vectors e_1, e_2, e_3, we know from Theorem 2 that there is precisely one linear map L from $\mathbb{R}^3$ into $\mathbb{R}^2$ that assumes these values at e_1, e_2, e_3.

We note further that if we take the map L just defined and write down $[L]$, we return to the array given just above.

In brief, we have set up a one-to-one correspondence between two sets: the set of all linear maps $L: \mathbb{R}^3 \to \mathbb{R}^2$ and the set of all matrices (rectangular arrays of numbers) having two horizontal rows and three vertical columns. To each map L there is precisely one matrix and vice versa.

3.4c   Matrix nomenclature

As indicated above, a **matrix** is any rectangular array of numbers, such as

$$A = \begin{bmatrix} \alpha_{11} & \alpha_{12} & \cdots & \alpha_{1n} \\ \alpha_{21} & \alpha_{22} & \cdots & \alpha_{2n} \\ \cdot & \cdot & & \cdot \\ \cdot & \cdot & & \cdot \\ \cdot & \cdot & & \cdot \\ \alpha_{q1} & \alpha_{q2} & \cdots & \alpha_{qn} \end{bmatrix}.$$

The numbers $\alpha_{11}, \alpha_{12}, \ldots$ are the **entries** or **elements** of the matrix A. It is natural to call α_{ij} the i, j**th entry**. The first index i in α_{ij} is called the **row index**, the second index j the **column index** (don't mix them up!). The ith **row** of the matrix A is

$$[\alpha_{i1} \quad \alpha_{i2} \quad \cdots \quad \alpha_{in}];$$

the jth **column** of A is

$$\begin{bmatrix} \alpha_{1j} \\ \alpha_{2j} \\ \cdot \\ \cdot \\ \cdot \\ \alpha_{qj} \end{bmatrix}.$$

Note that $i = 1, 2, \ldots, q$ and $j = 1, 2, \ldots, n$; A has q rows and n columns and is thereby called a q *by* n matrix. (Remember: rows *before* columns!) We denote the set of all q by n matrices $\mathcal{M}(q; n)$.

Note that a 1 by 1 matrix is the same as a real number. We may call a 1 by n matrix, say,

$$[\alpha_1 \quad \alpha_2 \quad \cdots \quad \alpha_n],$$

a **row vector,** and a q by 1 matrix, say,

$$\begin{bmatrix} \alpha_1 \\ \alpha_2 \\ \cdot \\ \cdot \\ \cdot \\ \alpha_q \end{bmatrix},$$

a **column vector.** These will be especially prominent in what follows.

We use capital letters $A, B, C, \ldots$ from the beginning of the alphabet to denote matrices. There is a further useful notation also. Let $A \in \mathscr{M}(q; n)$, and let the i, jth entry of A be α_{ij}, where it is understood that $i = 1, \ldots, q$ and $j = 1, \ldots, n$. Then it is customary to write $A = [\alpha_{ij}]$. This notation can save space and writing and causes no confusion, provided that we keep in mind that $1 \leq i \leq q$ and $1 \leq j \leq n$.

We remark that some authors would write (α_{ij}) instead of $[\alpha_{ij}]$ and therefore

$$\begin{pmatrix} 1 & 7 \\ -3 & 4 \end{pmatrix} \quad \text{instead of} \quad \begin{bmatrix} 1 & 7 \\ -3 & 4 \end{bmatrix}.$$

3.4d Addition of matrices

Our aim now is to use the correspondence between linear maps and matrices to define the three algebraic operations—addition, multiplication by scalars, multiplication—for matrices. We deal with addition first.

Let $L, M: \mathbb{R}^3 \to \mathbb{R}^2$ be linear maps. Suppose that

$$Le_1 = \alpha_{11}\bar{e}_1 + \alpha_{21}\bar{e}_2, \quad Le_2 = \alpha_{12}\bar{e}_1 + \alpha_{22}\bar{e}_2, \quad Le_3 = \alpha_{13}\bar{e}_1 + \alpha_{23}\bar{e}_2;$$

and

$$Me_1 = \beta_{11}\bar{e}_1 + \beta_{21}\bar{e}_2, \quad Me_2 = \beta_{12}\bar{e}_1 + \beta_{22}\bar{e}_2, \quad Me_3 = \beta_{13}\bar{e}_1 + \beta_{23}\bar{e}_2.$$

Thus we get the associated matrices

$$[L] = \begin{bmatrix} \alpha_{11} & \alpha_{12} & \alpha_{13} \\ \alpha_{21} & \alpha_{22} & \alpha_{23} \end{bmatrix}, \quad [M] = \begin{bmatrix} \beta_{11} & \beta_{12} & \beta_{13} \\ \beta_{21} & \beta_{22} & \beta_{23} \end{bmatrix}.$$

How should we define $[L] + [M]$? It is natural to do this as follows: Add the *maps* L and M, obtaining the map $L + M$, and then define $[L] + [M]$ to be the associated matrix $[L + M]$.

Adding the maps as in the preceding section, we obtain

$$(L + M)e_1 = Le_1 + Me_1$$
$$= (\alpha_{11}\bar{e}_1 + \alpha_{21}\bar{e}_1) + (\beta_{11}\bar{e}_1 + \beta_{21}\bar{e}_2)$$
$$= (\alpha_{11} + \beta_{11})\bar{e}_1 + (\alpha_{21} + \beta_{21})\bar{e}_2.$$

Likewise, we obtain

$$(L + M)e_2 = (\alpha_{12} + \beta_{12})\bar{e}_1 + (\alpha_{22} + \beta_{22})\bar{e}_2,$$
$$(L + M)e_3 = (\alpha_{13} + \beta_{13})\bar{e}_1 + (\alpha_{23} + \beta_{23})\bar{e}_2.$$

Hence we may immediately write down $[L + M]$:

$$[L + M] = \begin{bmatrix} \alpha_{11} + \beta_{11} & \alpha_{12} + \beta_{12} & \alpha_{13} + \beta_{13} \\ \alpha_{21} + \beta_{21} & \alpha_{22} + \beta_{22} & \alpha_{23} + \beta_{23} \end{bmatrix}.$$

Now we *define* this matrix to be $[L] + [M]$.

Thus, motivated by the correspondence between maps and matrices, let us make our definition in full generality.

Let A and B be matrices of the same shape, say $A = [\alpha_{ij}]$ and $B = [\beta_{ij}]$, with $i = 1, \ldots, q$ and $j = 1, \ldots, n$. Then we define their **sum**, denoted $A + B$, to be the q by n matrix

$$A + B = [\alpha_{ij} + \beta_{ij}].$$

Thus the i, jth entry of $A + B$ is the sum of the i, jth entries of A and B.

Note that if A and B are of different shapes, their sum is undefined.

EXAMPLES: Here are some examples of addition.

1. $A = \begin{bmatrix} 5 & -1 \\ 2 & 0 \end{bmatrix}$, $\qquad B = \begin{bmatrix} \frac{1}{2} & 4 \\ -3 & 3 \end{bmatrix}$, $\qquad A + B = \begin{bmatrix} \frac{11}{2} & 3 \\ -1 & 3 \end{bmatrix}$;

2. $A = \begin{bmatrix} 1 \\ 0 \\ -2 \end{bmatrix}$, $\qquad B = \begin{bmatrix} 7 \\ -2 \\ -2 \end{bmatrix}$, $\qquad A + B = \begin{bmatrix} 8 \\ -2 \\ -4 \end{bmatrix}$;

3. $\begin{bmatrix} 2 & 1 & 6 & -\frac{1}{2} \\ 3 & 0 & -1 & 0 \end{bmatrix} + \begin{bmatrix} 1 & 4 & 5 & 1 \\ 2 & 0 & -1 & -3 \end{bmatrix} = \begin{bmatrix} 3 & 5 & 11 & \frac{1}{2} \\ & \text{left to you} \end{bmatrix}.$

3.4e Multiplication of a matrix by a scalar

Let A be a q by n matrix and c any scalar. How shall we define cA? Again we are guided by the correspondence between maps and matrices. Let $A = [\alpha_{ij}]$, and suppose that $L: \mathbb{R}^n \longrightarrow \mathbb{R}^q$ is a linear map whose associated matrix is A; that is, $[L] = A$. This means that $Le_1 = \alpha_{11}\bar{e}_1 + \alpha_{21}\bar{e}_2 + \cdots + \alpha_{q1}\bar{e}_q$, and so on, for $Le_2, \ldots, Le_n$. Now we know that

$$(cL)e_1 = c(Le_1) = c(\alpha_{11}\bar{e}_1 + \alpha_{21}\bar{e}_2 + \cdots + \alpha_{q1}\bar{e}_q)$$
$$= c\alpha_{11}\bar{e}_1 + c\alpha_{21}\bar{e}_2 + \cdots + c\alpha_{q1}\bar{e}_q.$$

That is, the first column of $[L]$ is multiplied by c. The others are also.

Hence we define cA to be $[cL]$; that is,

$$cA = [c\alpha_{ij}].$$

EXAMPLES:

1. $A = \begin{bmatrix} 3 & 2 \\ -1 & 5 \end{bmatrix}$, $4A = 4\begin{bmatrix} 3 & 2 \\ -1 & 5 \end{bmatrix} = \begin{bmatrix} 12 & 8 \\ -4 & 20 \end{bmatrix}$.

2. If A is as above and $c = -1$, then

$$cA = -1A = \begin{bmatrix} -3 & -2 \\ 1 & -5 \end{bmatrix}.$$

3. If A is any matrix and if the scalar c equals zero, then $cA = 0$, the zero matrix; that is, $0A = 0$.

You should be expecting the following theorem, which does for matrices what Theorem 9 did for linear maps.

THEOREM 11: Let A, B be q by n matrices and α, β scalars. Then:

1. αA is a q by n matrix.
2. $(\alpha\beta)A = \alpha(\beta A)$.
3. $1A = A$.
4. $(\alpha + \beta)A = \alpha A + \beta A$.
5. $\alpha(A + B) = \alpha A + \alpha B$.

3.4f Multiplication of matrices

Thus far we have defined addition of matrices and multiplication of a matrix by a scalar, guided by the correspondence between matrices and linear maps. Now we ask, "How should we define a multiplication for matrices?"

We proceed as follows. Let $L: \mathbb{R}^n \longrightarrow \mathbb{R}^q$ and $M: \mathbb{R}^q \longrightarrow \mathbb{R}^r$ be linear maps. In the more succinct notation, we have $\mathbb{R}^n \overset{L}{\longrightarrow} \mathbb{R}^q \overset{M}{\longrightarrow} \mathbb{R}^r$. We know that the composition is a linear map $ML: \mathbb{R}^n \longrightarrow \mathbb{R}^r$. As usual, we have the associated matrices $[L] \in \mathcal{M}(q; n)$, $[M] \in \mathcal{M}(r; q)$, and $[ML] \in \mathcal{M}(r; n)$, and we define $[M][L]$ to be the matrix $[ML]$. We must obtain an expression for the entries of $[ML]$ in terms of the entries of $[M]$ and $[L]$. This will be our rule for multiplying.

First let us look at a special case $\mathbb{R}^3 \overset{L}{\longrightarrow} \mathbb{R}^2 \overset{M}{\longrightarrow} \mathbb{R}^2$. Then the matrices $[L]$, $[M]$ appear as follows:

$$[L] = \begin{bmatrix} \alpha_{11} & \alpha_{12} & \alpha_{13} \\ \alpha_{21} & \alpha_{22} & \alpha_{23} \end{bmatrix},   [M] = \begin{bmatrix} \beta_{11} & \beta_{12} \\ \beta_{21} & \beta_{22} \end{bmatrix}.$$

What is $[ML]$?

To answer this, we know that the first column of $[ML]$ must give the coefficients of MLe_1 when this vector is written as a linear combination of $\bar{e}_1$, $\bar{e}_2$. Now we have

$$
\begin{aligned}
MLe_1 &= M(Le_1) \\
&= M(\alpha_{11}\bar{e}_1 + \alpha_{21}\bar{e}_2) \\
&= \alpha_{11}M\bar{e}_1 + \alpha_{21}M\bar{e}_2 \\
&= \alpha_{11}(\beta_{11}\bar{e}_1 + \beta_{21}\bar{e}_2) + \alpha_{21}(\beta_{12}\bar{e}_1 + \beta_{22}\bar{e}_2) \\
&= (\beta_{11}\alpha_{11} + \beta_{12}\alpha_{21})\bar{e}_1 + (\beta_{21}\alpha_{11} + \beta_{22}\alpha_{21})\bar{e}_2.
\end{aligned}
$$

Hence, we have so far

$$
[ML] = \begin{bmatrix} \beta_{11}\alpha_{11} + \beta_{12}\alpha_{21} & ? & ? \\ \beta_{21}\alpha_{11} + \beta_{22}\alpha_{21} & ? & ? \end{bmatrix}.
$$

We obtain the second and third columns of $[ML]$ similarly, by computing MLe_2 and MLe_3 in terms of $\bar{e}_1$, $\bar{e}_2$. Finally, we get

$$
[ML] = \begin{bmatrix} \beta_{11}\alpha_{11} + \beta_{12}\alpha_{21} & \beta_{11}\alpha_{12} + \beta_{12}\alpha_{22} & \beta_{11}\alpha_{13} + \beta_{12}\alpha_{23} \\ \beta_{21}\alpha_{11} + \beta_{22}\alpha_{21} & \beta_{21}\alpha_{12} + \beta_{22}\alpha_{22} & \beta_{21}\alpha_{13} + \beta_{22}\alpha_{23} \end{bmatrix}.
$$

This is what the product $[M][L]$ should be. Hence, we make it our *definition*:

$$
[M][L] = [ML];
$$

that is,

$$
\begin{bmatrix} \beta_{11} & \beta_{12} \\ \beta_{21} & \beta_{22} \end{bmatrix} \begin{bmatrix} \alpha_{11} & \alpha_{12} & \alpha_{13} \\ \alpha_{21} & \alpha_{22} & \alpha_{23} \end{bmatrix}
$$
$$
= \begin{bmatrix} \beta_{11}\alpha_{11} + \beta_{12}\alpha_{21} & \beta_{11}\alpha_{12} + \beta_{12}\alpha_{22} & \beta_{11}\alpha_{13} + \beta_{12}\alpha_{23} \\ \beta_{21}\alpha_{11} + \beta_{22}\alpha_{21} & \beta_{21}\alpha_{12} + \beta_{22}\alpha_{22} & \beta_{21}\alpha_{13} + \beta_{22}\alpha_{23} \end{bmatrix}.
$$

Having this definition of $[M][L]$, we would like a means of remembering it, so that we need not repeat the entire process of computing MLe_1, MLe_2, MLe_3.

If we examine $[ML]$, we observe that its upper left-hand entry $\beta_{11}\alpha_{11} + \beta_{12}\alpha_{21}$ can be computed as follows: Take the first *row* of $[M]$ and first *column* of $[L]$, namely,

$$[\beta_{11} \quad \beta_{12}] \quad \text{and} \quad \begin{bmatrix} \alpha_{11} \\ \alpha_{21} \end{bmatrix},$$

and define their product thus:

$$[\beta_{11} \quad \beta_{12}]\begin{bmatrix} \alpha_{11} \\ \alpha_{21} \end{bmatrix} = \beta_{11}\alpha_{11} + \beta_{12}\alpha_{21}.$$

It may be helpful to view the right-hand side as the inner product of the row vector and the column vector. This yields the upper left-hand entry of $[ML]$. More generally, the i, jth term $\beta_{i1}\alpha_{1j} + \beta_{i2}\alpha_{2j}$ of $[ML]$ is obtained by taking the ith *row* of $[M]$ and the jth *column* of $[L]$ and forming the product (a number):

$$[\beta_{i1} \quad \beta_{i2}]\begin{bmatrix} \alpha_{1j} \\ \alpha_{2j} \end{bmatrix} = \beta_{i1}\alpha_{1j} + \beta_{i2}\alpha_{2j}.$$

Guided by this example, we now define multiplication of matrices in general.

DEFINITION: Let B be an r by q matrix and A a q by n matrix, so that B has as many columns as A has rows. Say that $B = [\beta_{hi}]$, $A = [\alpha_{ij}]$, with $h = 1, \ldots, r; i = 1, \ldots, q; j = 1, \ldots . n$. Then the **product** BA is the r by n matrix $[\gamma_{hj}]$, where

$$\gamma_{hj} = \beta_{h1}\alpha_{1j} + \beta_{h2}\alpha_{2j} + \cdots + \beta_{hq}\alpha_{qj}.$$

REMARKS: 1. We do not even consider forming the product BA unless the *number of columns of B equals the number of rows of A*.

2. The h, jth entry γ_{hj} of BA is obtained by taking the hth *row* of B and the jth *column* of A and forming the product (an inner product of vectors, if you will):

$$[\beta_{h1} \quad \beta_{h2} \quad \cdots \quad \beta_{hq}]\begin{bmatrix} \alpha_{1j} \\ \alpha_{2j} \\ \vdots \\ \alpha_{qj} \end{bmatrix} = \beta_{h1}\alpha_{1j} + \beta_{h2}\alpha_{2j} + \cdots + \beta_{hq}\alpha_{qj}.$$

This is γ_{hj}.

EXAMPLES:

1. $\begin{bmatrix} 5 & 2 \\ -1 & -3 \end{bmatrix}\begin{bmatrix} 2 & 0 & 1 \\ -1 & 4 & 2 \end{bmatrix} = \begin{bmatrix} 8 & 8 & 9 \\ 5 & -12 & -5 \end{bmatrix}.$

Note, for instance, that the 2,1st (lower left-hand) entry of the product is obtained from

$$[1 \quad -3]\begin{bmatrix} 2 \\ -1 \end{bmatrix} = 1 \cdot 2 + (-3) \cdot (-1) = 5.$$

2. $\begin{bmatrix} 1 & 2 \\ 3 & 4 \end{bmatrix}\begin{bmatrix} 4 & 2 \\ 3 & 1 \end{bmatrix} = \begin{bmatrix} 10 & 4 \\ 24 & 10 \end{bmatrix}.$

3. $\begin{bmatrix} 3 & 0 & 1 \\ 0 & 2 & -1 \\ 1 & 1 & 5 \end{bmatrix}\begin{bmatrix} 2 \\ -3 \\ 1 \end{bmatrix} = \begin{bmatrix} 7 \\ -7 \\ 4 \end{bmatrix}.$

4. $\begin{bmatrix} 3 & 0 & 1 \\ 0 & 2 & -1 \\ 1 & 1 & 5 \end{bmatrix}\begin{bmatrix} 2 & 0 \\ -3 & 0 \\ 1 & 1 \end{bmatrix} = \begin{bmatrix} 7 & 1 \\ -7 & -1 \\ 4 & 5 \end{bmatrix}.$

We suggest that you go back and perform some of these multiplications yourself (mentally, of course), keeping the right-hand side (product) covered and then comparing.

In a moment we will list the basic properties of matrix multiplication as Theorem 12. First, however, we designate as **identity matrix** the n by n (square) matrix $[I]$, where $I : \mathbb{R}^n \to \mathbb{R}^n$ is the identity map. It is customary to drop the brackets and denote the identity *matrix* by I.

What does this matrix look like? Since $IX = X$ for all $X \in \mathbb{R}^n$, we have $Ie_1 = e_1, \ldots, Ie_n = e_n$. Thus, when $n = 3$ and 4 respectively, we get the identity matrices

$$I = \begin{bmatrix} 1 & 0 & 0 \\ 0 & 1 & 0 \\ 0 & 0 & 1 \end{bmatrix}, \quad I = \begin{bmatrix} 1 & 0 & 0 & 0 \\ 0 & 1 & 0 & 0 \\ 0 & 0 & 1 & 0 \\ 0 & 0 & 0 & 1 \end{bmatrix}.$$

Similarly, for a general n the identity matrix I is readily seen to have 1's on the **main diagonal** (upper left to lower right) and 0's elsewhere.

We might expect that the matrix I plays a role in matrix multiplication similar to that played by the number 1 in multiplication of real numbers. It does.

Having this, we state the basic properties of matrix multiplication.

They may all be proved by recalling that each matrix determines a linear map and that multiplication of matrices parallels the composition of linear maps.

THEOREM 12:   Let A, B, C be matrices. Then:

     1. Associativity: $A(BC) = (AB)C$, provided that all products are defined.

     2. Left and right distributivity: if the products and sums are defined, we have

$$A(B + C) = AB + AC,$$
$$(A + B)C = AC + BC.$$

     3. Identity: $AI = A, IA = A$ provided that the products are defined.

REMARK: If A, B, C, I are n by n square matrices, then all products and sums are defined.

Proof: This is simple, although many people find it logically difficult at first. Let L, M, P be the maps corresponding to A, B, C, respectively, so that $[L] = A$, and so on. Then by *definition* of matrix multiplication, $AB = [L][M] = [LM]$, so that by Theorem 10 in the third equality, $A(BC) = [L][MP] = [L(MP)] = [(LM)P] = [LM][P] = (AB)C$. This proves the associativity. The rest are similar and are left to you. ≪

3.4g   Matrices as functions

Before going on, let us sum up what we have done in Secs. 3.3 and 3.4 to date:

1. In Sec. 3.3 we developed the algebra of linear maps: addition, multiplication by scalars, composition.

2. Then to each linear map we associated a matrix and vice versa.

3. Finally, using this correspondence, we transferred the algebraic operations from maps to matrices.

Our aim in all this is the study of linear maps L. These are, after all, functions. They "map" or "transform" or "assign" to a given vector X another vector LX. Consequently, we should expect matrices $[L]$ to operate on "vectors" also, to transform one "vector" into another. Indeed, this is the case, as we shall now see.

The key step is this: Just as we correspond to each map L a matrix $[L]$, we correspond to each vector X an object $[X]$ that combines with matrices in the proper manner. To be less vague, let $X = (x_1, \ldots, x_n)$

be any vector in $\mathbb{R}^n$. We correspond to X the **column vector** (an n by 1 matrix) $[X]$ given by

$$[X] = \begin{bmatrix} x_1 \\ x_2 \\ \cdot \\ \cdot \\ \cdot \\ x_n \end{bmatrix}.$$

Thus to $X = (1, 3, -2)$ in $\mathbb{R}^3$ or $Y = (0, 4)$ in $\mathbb{R}^2$ we correspond

$$[X] = \begin{bmatrix} 1 \\ 3 \\ -2 \end{bmatrix} \quad \text{and} \quad [Y] = \begin{bmatrix} 0 \\ 4 \end{bmatrix},$$

respectively.

We add column vectors (of the same size), just as in $\mathscr{M}(n; 1)$, and likewise for multiplication by scalars.

How does a matrix operate on a column vector? Happily, we have answered this already. Let us look at an example. Say that $L: \mathbb{R}^2 \to \mathbb{R}^3$ so that $[L] = [\alpha_{ij}]$, with $i = 1, 2, 3$ and $j = 1, 2$. Let $X = (x_1, x_2)$ in $\mathbb{R}^2$. Then

$$[L][X] = \begin{bmatrix} \alpha_{11} & \alpha_{12} \\ \alpha_{21} & \alpha_{22} \\ \alpha_{31} & \alpha_{32} \end{bmatrix} \begin{bmatrix} x_1 \\ x_2 \end{bmatrix} = \begin{bmatrix} \alpha_{11}x_1 + \alpha_{12}x_2 \\ \alpha_{21}x_1 + \alpha_{22}x_2 \\ \alpha_{31}x_1 + \alpha_{32}x_2 \end{bmatrix},$$

where we have multiplied a 3 by 2 matrix by a 2 by 1 column vector to obtain a 3 by 1 *column* vector (don't let its apparent width fool you; it is a vector). Given the matrix $[\alpha_{ij}]$ corresponding to L, this computation was automatic—just matrix multiplication.

On the other hand, let us compute LX directly. We have

$$\begin{aligned} LX &= L(x_1e_1 + x_2e_2) \\ &= x_1(\alpha_{11}\bar{e}_1 + \alpha_{21}\bar{e}_2 + \alpha_{31}\bar{e}_3) + x_2(\alpha_{12}\bar{e}_1 + \alpha_{22}\bar{e}_2 + \alpha_{32}\bar{e}_3) \\ &= (\alpha_{11}x_1 + \alpha_{12}x_2)\bar{e}_1 + (\alpha_{21}x_1 + \alpha_{22}x_2)\bar{e}_2 + (\alpha_{31}x_1 + \alpha_{32}x_2)\bar{e}_3 \\ &= (\alpha_{11}x_1 + \alpha_{12}x_2, \alpha_{21}x_1 + \alpha_{22}x_2, \alpha_{31}x_1 + \alpha_{32}x_2). \end{aligned}$$

This is a bit more tedious than matrix multiplication.

Now we can write the column vector corresponding to LX:

$$[LX] = \begin{bmatrix} \alpha_{11}x_1 + \alpha_{12}x_2 \\ \alpha_{21}x_1 + \alpha_{22}x_2 \\ \alpha_{31}x_1 + \alpha_{32}x_2 \end{bmatrix}.$$

But this is just the product $[L][X]$ obtained earlier; that is, $[LX] = [L][X]$.

MORAL: Once we have the matrix $[L]$, we can readily compute LX for various X, using $[LX] = [L][X]$.

Although our analysis was done in a special case, it is readily verified that it holds true in all dimensions. We state it as a theorem.

THEOREM 13:   Let $L: \mathbb{R}^n \to \mathbb{R}^q$ be a linear map. Let X be a vector in $\mathbb{R}^n$ and $[X]$ the corresponding column vector defined above. Then $[LX]$, the column vector corresponding to $LX \in \mathbb{R}^q$, is given by

$$[LX] = [L][X],$$

where the product on the right is matrix multiplication.

3.4h Some remarks of a general nature

We began this chapter by giving a qualitative definition of a linear map, that is, a function that enjoys certain properties with respect to vector addition and multiplication by scalars. We then showed, through examples, that linear maps were discernible in a wide variety of mathematical situations: linear equations, rotations of the plane, and so on. This variety may have been a bit bewildering.

It is important to realize that our subsequent work has unified this picture somewhat. More specifically, when we meet a linear map in life (from some geometric, numerical, or physical problem), it is usually presented in one of three ways, namely, (1) standard basis, (2) as a matrix, or (3) in coordinates. Let us review these in turn:

1. Using the standard bases. We have learned that $Le_1 = \alpha_{11}\bar{e}_1 + \cdots + \alpha_{q1}\bar{e}_q$, and so on, and that the coefficients α_{ij} determine L precisely. Different coefficients, different map. This is Theorem 2.

2. Using the matrix. We know that L is also determined by its matrix $[L] = [\alpha_{ij}]$. In this representation, the standard bases are implicit. This tidies up the notation, prompts the introduction of matrix multiplication, and thereby facilitates computing with L.

3. Using coordinates. We observe that *every* L may be written

$$LX = (\alpha_{11}x_1 + \cdots + \alpha_{1n}x_n, \ldots, \alpha_{q1}x_1 + \cdots + \alpha_{qn}x_n),$$

where $X = (x_1, \ldots, x_n)$. Thus every linear map looks like the "left-hand side" of a system of linear equations. This is the coordinate form of L.

EXAMPLE: Here is a quick way to get the coordinate representation for a linear map $L: \mathbb{R}^n \to \mathbb{R}^q$. Remember, we want LX in terms of $x_1, \ldots, x_n$.
   For definiteness, let $L: \mathbb{R}^2 \to \mathbb{R}^3$ be given by the matrix

$$[L] = \begin{bmatrix} 2 & 12 \\ -1 & 8 \\ 3 & -17 \end{bmatrix}.$$

Now let $X = (x_1, x_2)$, and compute $[LX]$ as $[L][X]$:

$$\begin{bmatrix} 2 & 12 \\ -1 & 8 \\ 3 & -17 \end{bmatrix} \begin{bmatrix} x_1 \\ x_2 \end{bmatrix} = \begin{bmatrix} 2x_1 + 12x_2 \\ -x_1 + 8x_2 \\ 3x_1 - 17x_2 \end{bmatrix}.$$

Hence $LX = (2x_1 + 12x_2, -x_1 + 8x_2, 3x_1 - 17x_2)$.

3.4i The inverse of a matrix

Before beginning, we remark that the material of this subsection and the next is not essential to the study of nonlinear maps beginning in Chap. 4.
   Now let us begin with the following.

DEFINITION: Let A be an n by n (square) matrix. Then A is **invertible,** or **nonsingular** if and only if there exists an n by n matrix C such that $AC = I$ and $CA = I$, where I is the n by n identity matrix. If C exists, it is called the **inverse of** A and denoted A^{-1} rather than C.

   For example, if

$$A = \begin{bmatrix} 3 & 1 \\ 5 & 2 \end{bmatrix},$$

then

$$A^{-1} = \begin{bmatrix} 2 & -1 \\ -5 & 3 \end{bmatrix}.$$

You should check that their product $AA^{-1} = A^{-1}A$ is

$$\begin{bmatrix} 1 & 0 \\ 0 & 1 \end{bmatrix},$$

the identity matrix. See Fig. 3.23.

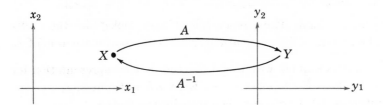

A^{-1} undoes the effect of A:

$$A^{-1}AX = X, \quad AA^{-1}Y = Y$$

FIGURE 3.23

NOTE: *Not* all square matrices are invertible. In fact you should convince yourself that the zero matrix is not invertible, just as the number 0 is not invertible in the sense of ordinary multiplication of numbers.

It is not hard to prove that if A^{-1} exists, then it is unique: *a matrix has at most one inverse.* For if both C_1 and C_2 are inverses of A, then we have

$$C_1 = C_1 I = C_1(AC_2) = (C_1 A)C_2 = IC_2 = C_2,$$

so that any two inverses are equal. Done.

Here are some natural questions about inverses.

Question 1. Given a square matrix A, how can we tell if it is invertible? Does A^{-1} exist?

Question 2. If A^{-1} exists, how can we compute it?

The knowing student distinguishes between *existence* questions, such as the first, and *construction* questions, of which the second is an example. One reason: The first kind of question is often much easier to answer. A second reason is that if the answer to the first question is *no*, then it would be a waste of time to try to compute A^{-1}. A third reason is that often a given situation or theoretical problem does not require the explicit matrix A^{-1} but only the assurance that A^{-1} exists.

Perhaps we should also raise the following.

Question 3. How are inverses used?

We discuss this in the following subsection on linear equations.

In answer to Question 1, we state that there is a complete but impractical theory, namely, determinants, that enables us to decide whether a given matrix is invertible. The basic statement here is that A is invertible if and only if its determinant (a certain number) is different from zero. The computation of the determinant of an n by n matrix A, although elementary, may be quite tedious, especially if n is large. In the 2 by 2 and 3 by 3 cases, however, it is often possible to decide whether A^{-1} exists by a quick inspection of the matrix A (and only here are determinants practical).

In answer to Question 2, we remark that the construction of A^{-1} may be effected in several ways, all tedious. The simplest methods reduce the problem to the solution of a system of linear equations, as with so many problems we have met. Here is one such method.

Let $A = [\alpha_{ij}]$ be n by n and invertible, and let

$$X = \begin{bmatrix} x_1 \\ x_2 \\ \cdot \\ \cdot \\ \cdot \\ x_n \end{bmatrix}, \qquad Y = \begin{bmatrix} y_1 \\ y_2 \\ \cdot \\ \cdot \\ \cdot \\ y_n \end{bmatrix},$$

where $x_1, \ldots, x_n, y_1, \ldots, y_n$ are independent variables. If $AX = Y$, then $A^{-1}AX = A^{-1}Y$; that is, $X = A^{-1}Y$. To find the matrix $A^{-1} = [\gamma_{ij}]$, write $AX = Y$ as

$$\alpha_{11}x_1 + \cdots + \alpha_{1n}x_n = y_1$$
$$\vdots \qquad\qquad \vdots \qquad \vdots$$
$$\alpha_{n1}x_1 + \cdots + \alpha_{nn}x_n = y_n$$

We know the α_{ij}'s; they are given. But also $X = A^{-1}Y$ has the form

$$x_1 = \gamma_{11}y_1 + \cdots + \gamma_{1n}y_n$$
$$\vdots \qquad\qquad \vdots \qquad \vdots$$
$$x_n = \gamma_{n1}y_1 + \cdots + \gamma_{nn}y_n$$

where we want to find the γ_{ij}'s, since they are the elements of A^{-1}. The method is now clear: Solve in the first system for the x's in terms of the y's. This gives each x_i as a linear combination of $y_1, \ldots, y_n$, that is, the second system of equations above. But *the coefficients in the second system just obtained are the γ_{ij}'s, the matrix entries of A^{-1} that we have been seeking.*

EXAMPLE: Compute A^{-1}, where

$$A = \begin{bmatrix} 3 & 5 \\ 1 & 3 \end{bmatrix}.$$

We write $AX = Y$ as

$$3x_1 + 5x_2 = y_1,$$
$$x_1 + 3x_2 = y_2.$$

Now we want to solve for the x's in terms of the y's by elimination as in high school algebra, to find

$$x_1 = \tfrac{3}{4}y_1 - \tfrac{5}{4}y_2,$$
$$x_2 = -\tfrac{1}{4}y_1 + \tfrac{3}{4}y_2.$$

By the discussion above, we conclude that

$$A^{-1} = \begin{bmatrix} \tfrac{3}{4} & -\tfrac{5}{4} \\ -\tfrac{1}{4} & \tfrac{3}{4} \end{bmatrix}.$$

How would you check this?

NOTE: We assumed from the outset that A^{-1} existed here. If it did not, we would not have obtained unique solutions for the x's in terms of the y's.

ANOTHER EXAMPLE: Compute A^{-1}, where

$$A = \begin{bmatrix} 1 & 0 & 2 \\ -1 & 3 & 0 \\ 0 & 3 & 1 \end{bmatrix}.$$

We form the equations for $AX = Y$, namely,

$$x_1 \qquad\quad + 2x_3 = y_1,$$
$$-x_1 + 3x_2 \qquad\quad = y_2,$$
$$3x_2 + \;\; x_3 = y_3.$$

The standard method of elimination yields, as you should be able to verify, $X = A^{-1}Y$ in the form

$$x_1 = -y_1 \;\; - 2y_2 + 2y_3,$$
$$x_2 = -\tfrac{1}{3}y_1 - \tfrac{1}{3}y_2 + \tfrac{2}{3}y_3,$$
$$x_3 = \;\;\;\; y_1 + \;\; y_2 - y_3.$$

Thus

$$A^{-1} = \begin{bmatrix} -1 & -2 & 2 \\ -\tfrac{1}{3} & -\tfrac{1}{3} & \tfrac{2}{3} \\ 1 & 1 & -1 \end{bmatrix}.$$

You might check here that $AA^{-1} = I$ and $A^{-1}A = I$.

A GEOMETRIC EXAMPLE: Let $L\colon \mathbb{R}^2 \to \mathbb{R}^2$ be the rotation through an angle θ, counterclockwise. Then clearly the inverse map, L^{-1}, is rotation by $-\theta$. For example, if $\theta = \pi/2$, then

$$[L] = \begin{bmatrix} 0 & -1 \\ 1 & 0 \end{bmatrix} \quad \text{and} \quad [L]^{-1} = \begin{bmatrix} 0 & 1 \\ -1 & 0 \end{bmatrix}.$$

MORE ON THE 2 BY 2 CASE: The following observations will transform you into a lightning inverter of 2 by 2 matrices.

Let

$$A = \begin{bmatrix} \alpha & \beta \\ \gamma & \delta \end{bmatrix}.$$

The **determinant** of A is the number

$$\Delta = \alpha\delta - \beta\gamma.$$

Then, as you may verify,

THEOREM 14: A 2 by 2 matrix A is invertible $\Leftrightarrow$ the determinant Δ is different from zero. Further, if $\Delta \neq 0$, then

$$A^{-1} = \frac{1}{\Delta} \begin{bmatrix} \delta & -\beta \\ -\gamma & \alpha \end{bmatrix} = \begin{bmatrix} \dfrac{\delta}{\Delta} & -\dfrac{\beta}{\Delta} \\ -\dfrac{\gamma}{\Delta} & \dfrac{\alpha}{\Delta} \end{bmatrix}.$$

For example, if

$$A = \begin{bmatrix} 3 & 5 \\ 1 & 3 \end{bmatrix},$$

then $\Delta = 4$, and so

$$A^{-1} = \begin{bmatrix} \frac{3}{4} & -\frac{5}{4} \\ -\frac{1}{4} & \frac{3}{4} \end{bmatrix},$$

as we saw in the example above.

REMARK: The formula for A^{-1} given here is the 2 by 2 version of *Cramer's Rule*, a more complicated version of which applies in the n by n case. For large matrices, however, the most efficient way of computing inverses is to solve the associated system of linear equations by eliminating one variable at a time, as we have done. This is what computers do.

3.4j Linear equations, matrices, and mappings

This section, like the previous one, will not be utilized in our study of nonlinear maps.

Suppose that we are to solve a system of q equations in n unknowns:

$$\alpha_{11}x_1 + \cdots + \alpha_{1n}x_n = y_1$$
$$\vdots \qquad\qquad \vdots \qquad \vdots \;;$$
$$\alpha_{q1}x_1 + \cdots + \alpha_{qn}x_n = y_q$$

that is, the α's and y's are prescribed, and the x's are to be found. We may rewrite this system in matrix notation as

$$AX = Y,$$

where

$$A = \begin{bmatrix} \alpha_{11} & \cdots & \alpha_{1n} \\ \cdot & & \cdot \\ \cdot & & \cdot \\ \cdot & & \cdot \\ \alpha_{q1} & \cdots & \alpha_{qn} \end{bmatrix}, \qquad X = \begin{bmatrix} x_1 \\ \cdot \\ \cdot \\ \cdot \\ x_n \end{bmatrix}, \qquad Y = \begin{bmatrix} y_1 \\ \cdot \\ \cdot \\ \cdot \\ y_q \end{bmatrix}$$

This both saves writing and leads to a valuable insight, as we are about to see.

Some natural questions now arise.

Question 1.   Does this system have a solution $x_1, \ldots, x_n$?

Question 2.   If so, how many?

Question 3.   Compute one.

Question 4.   Compute all.

A *no* answer to the first question will, of course, save us from worrying about the remaining questions.

We have dealt with Question 3 several times in this book, every time we have solved an explicit system such as

$$x_1 + x_2 + x_3 = 0 \quad \text{or} \quad \begin{matrix} 2x_1 - x_2 = 1 \\ x_1 + x_2 = 4 \end{matrix}.$$

One way of dealing with the second system above would be to compute the inverse of the matrix of coefficients, but there are even more simple-minded ways.

Our aim now is to describe a way of thinking about the other questions. We have alluded to it before. It is this: *The matrix A of the system $AX = Y$ determines a linear map $L\colon \mathbb{R}^n \to \mathbb{R}^q$. The system has a solution X if and only if Y is in $\mathscr{I}(L)$, the image subspace of $\mathbb{R}^q$.*

NOTE: Here we do not distinguish between the column vector Y and the vector in $\mathbb{R}^q$ with the same coordinates. We could, if we chose to, use the same letter A for both matrix and linear map, although this may cause confusion in more advanced theories.

Thus when confronted with a system $AX = Y$, we visualize some linear mapping L sending all of $\mathbb{R}^n$ onto a subspace $\mathscr{I}(L)$ (proper or improper) of $\mathbb{R}^q$. Since we know of cases where $\mathscr{I}(L)$ is a proper subspace of $\mathbb{R}^q$, we are rid of the childish hope that *all* systems $AX = Y$ have a solution. We are able to visualize maps that remind us that a given system may have no solutions at all for a fixed Y, or just one solution, or many (actually an infinite number).

Rather than work out a fully detailed theory of linear equations, let us content ourselves with visualizing two situations of frequent occurence.

FIRST SITUATION: $AX = 0, n > q$.

This is the homogeneous case of $Y = 0$ with more unknowns than equations. The trivial solution is $X = 0$; that is, $x_1 = 0, \ldots, x_n = 0$. The usual question is "Is there a nontrivial solution X?"

There is, provided that $n > q$. To see this, think of L, with matrix $[L] = A$, mapping big $\mathbb{R}^n$ into little $\mathbb{R}^q$. Things are squashed. Some vector Y in $\mathbb{R}^q$ is the image of more than one vector in $\mathbb{R}^n$; that is, $LX_1 = LX_2 = Y$, with $X_1 \neq X_2$. Then $X = X_1 - X_2$ is mapped by L to 0, since L is linear. Thus any scalar multiple αX is also mapped to 0. This is the picture behind the following restatement of the theorem in Chap. 1, Appendix 1.

THEOREM 15: A system of homogeneous linear equations with more unknowns than equations has a nontrivial solution.

To solve such a system, say

$$x_1 + x_2 - x_3 = 0,$$
$$x_1 - 5x_2 + 8x_3 = 0,$$

we could, of course, begin by eliminating x_1 in the second equation. The informal picture here is something like Fig. 3.24.

Question. What does $n > q$ imply for the null space, $\mathscr{N}(L)$?

SECOND SITUATION: $AX = Y, n = q$.

Here the matrix A determines a map $L: \mathbb{R}^n \to \mathbb{R}^n$, and we may visualize the special case of $n = 3$. What happens? We forget Y for a moment and concentrate on the image subspace $\mathscr{I}(L)$. Inside 3-space, $\mathscr{I}(L)$ must be either all of $\mathbb{R}^3$, or a plane through the origin, or a straight line through

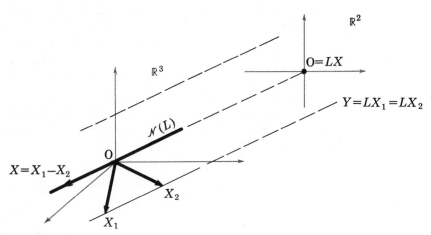

FIGURE 3.24

the origin, or {O} itself. (Can you cook up matrices that yield each of these cases?)

Let us interpret the case of $\mathscr{I}(L) = \mathbb{R}^3$; that is, L maps $\mathbb{R}^3$ **onto** $\mathbb{R}^3$. Thus for every Y there is an X such that $LX = Y$, so that the original system of equations does have a solution. But we can say more. If L is to map $\mathbb{R}^3$ *onto* $\mathbb{R}^3$, then, roughly speaking, it cannot waste vectors X by sending too many of them to the one vector Y; that is, L is **one-to-one**, or **injective**, in the sense that $X_1 \neq X_2$ implies $LX_1 \neq LX_2$. Thus the system $AX = Y$ has a *unique* solution X for each Y. In particular the homogeneous equation $AX = O$ has only the trivial solution $X = O$ in this case. Again, all this depends strongly on the fact that $n = q$.

You might meditate on the following theorem, most of which should be believable.

THEOREM 16: Let $L: \mathbb{R}^n \longrightarrow \mathbb{R}^n$ be linear. Then the following are equivalent (if one happens, then so do all the others):

1. L is one-to-one.
2. L is onto.
3. The homogeneous system $AX = O$, where $A = [L]$, has only the trivial solution $X = O$.
4. For every Y, the system $AX = Y$, where $A = [L]$, has a unique solution X.
5. A^{-1} exists.
6. The determinant of A is nonzero.

MORAL: There are many ways of characterizing the same phenomenon.

Exercises

These give you practice in carrying out the basic algebraic operations on matrices —a skill that, like the multiplication table learned in grade school, will be taken for granted from now on.

1.   Write down the 2 by 3 matrix $A = [\alpha_{ij}]$ whose entries are $\alpha_{11} = 2$, $\alpha_{21} = -3$, $\alpha_{12} = 0$, $\alpha_{22} = -1$, $\alpha_{13} = 4$, $\alpha_{23} = 144$.

2.   Perform the additions indicated:

(a) $\begin{bmatrix} 2 & 0 & 4 \\ -3 & -1 & 144 \end{bmatrix} + \begin{bmatrix} 1 & 1 & 2 \\ 0 & 1 & 2 \end{bmatrix}$,

(c) $\begin{bmatrix} 1 \\ 4 \\ 6 \\ 1 \end{bmatrix} + \begin{bmatrix} 0 \\ -4 \\ -6 \\ 0 \end{bmatrix}$,

(b) $\begin{bmatrix} 1 & 3 \\ 2 & 0 \end{bmatrix} + \begin{bmatrix} -1 & 2 \\ 3 & 1 \end{bmatrix}$,

(d) $\begin{bmatrix} 1 & 0 & -1 \\ 3 & 1 & -1 \\ 2 & 2 & 1 \end{bmatrix} + \begin{bmatrix} 4 & 0 & 0 \\ 0 & 4 & 0 \\ 0 & 0 & 4 \end{bmatrix}$.

3.   Perform the operations indicated (note: *linear combinations* of matrices):

(a) $\begin{bmatrix} 1 & 2 \\ 2 & 1 \end{bmatrix} + 4 \begin{bmatrix} 0 & -1 \\ 1 & 1 \end{bmatrix}$,

(c) $7 \begin{bmatrix} 3 & 4 \\ 1 & -1 \\ 1 & 0 \end{bmatrix}$,

(b) $2 \begin{bmatrix} 1 \\ 0 \end{bmatrix} - 7 \begin{bmatrix} 0 \\ 1 \end{bmatrix}$,

(d) $2 \begin{bmatrix} 1 \\ 1 \\ 0 \end{bmatrix} + 4 \begin{bmatrix} -1 \\ 0 \\ -1 \end{bmatrix} - \begin{bmatrix} -2 \\ 2 \\ 4 \end{bmatrix}$.

4.   Perform the following matrix multiplications:

(a) $\begin{bmatrix} 2 & 1 \\ 0 & 3 \end{bmatrix} \begin{bmatrix} -2 & 1 \\ 1 & 3 \end{bmatrix}$,

(f) $\begin{bmatrix} 2 & 1 \\ 0 & 3 \end{bmatrix} \begin{bmatrix} 1 & 0 & 1 \\ 0 & 1 & 1 \end{bmatrix}$,

(b) $\begin{bmatrix} 1 & 0 \\ 0 & 1 \end{bmatrix} \begin{bmatrix} 5 & 6 \\ 7 & 8 \end{bmatrix}$,

(g) $\begin{bmatrix} 1 & 2 & 0 \\ 3 & 1 & 2 \\ -1 & 0 & 2 \end{bmatrix} \begin{bmatrix} -1 & 1 & 1 \\ 1 & -1 & 1 \\ 1 & -1 & -1 \end{bmatrix}$,

(c) $\begin{bmatrix} 2 & 1 \\ 0 & 3 \end{bmatrix} \begin{bmatrix} 1 \\ 0 \end{bmatrix}$,

(h) $\begin{bmatrix} 1 & 2 & 0 \\ 3 & 1 & 2 \\ -1 & 0 & 2 \end{bmatrix} \begin{bmatrix} 2 & -1 \\ 1 & 1 \\ -1 & 1 \end{bmatrix}$,

(d) $\begin{bmatrix} -2 & 1 \\ 4 & -2 \end{bmatrix} \begin{bmatrix} 1 & 3 \\ 2 & 6 \end{bmatrix}$,

(i) $\begin{bmatrix} 1 & 2 & 0 \\ 3 & 1 & 2 \\ -1 & 0 & 2 \end{bmatrix} \begin{bmatrix} 5 & 0 & 0 \\ 0 & 5 & 0 \\ 0 & 0 & 5 \end{bmatrix}$,

(e) $\begin{bmatrix} 2 & 1 \\ 0 & 3 \end{bmatrix} \begin{bmatrix} 0 \\ 1 \end{bmatrix}$,

(j) $\begin{bmatrix} -1 & 1 & 1 \\ 1 & -1 & 1 \\ 1 & -1 & -1 \end{bmatrix} \begin{bmatrix} 0 \\ 1 \\ 0 \end{bmatrix}$.

5. Compute numbers $\alpha, \beta, \gamma, \delta$ (not all zero) so that the following equations hold:

(a) $\begin{bmatrix} 2 & 1 \\ 0 & 1 \end{bmatrix}\begin{bmatrix} \alpha & \beta \\ \gamma & \delta \end{bmatrix} = \begin{bmatrix} 1 & 0 \\ 0 & 1 \end{bmatrix},$ (b) $\begin{bmatrix} 3 & -1 \\ -6 & 2 \end{bmatrix}\begin{bmatrix} \alpha & \beta \\ \gamma & \delta \end{bmatrix} = \begin{bmatrix} 0 & 0 \\ 0 & 0 \end{bmatrix}.$

(Hint: Solve linear equations.)

6. The matrix of a linear map. Let $L \colon \mathbb{R}^2 \longrightarrow \mathbb{R}^2$ be the linear map determined by

$$Le_1 = -2e_1 + e_2, \qquad Le_2 = e_1 + 3e_2.$$

(a) What is the matrix $[L]$?

(b) What (column) vectors are associated with the basis vectors e_1, e_2, that is, $[e_1], [e_2]$?

(c) What is the relation between the image vector Le_1 and the first column of the matrix $[L]$?

(d) What is the product of the matrix $[L]$ and the column vector $[e_1]$?

(e) Given $X = (4, 5) = 4e_1 + 5e_2$, what is the column vector $[X]$?

(f) Compute LX in two ways: directly, using knowledge of Le_1, Le_2; and by matrix multiplication. Which is more efficient? (Hopefully your answers agree.)

7. Composition of maps and multiplication of matrices. Let L be as above, and let $M \colon \mathbb{R}^2 \longrightarrow \mathbb{R}^2$ be linear also, determined by

$$Me_1 = 2e_1, \qquad Me_2 = e_1 + 3e_2.$$

(a) What is the matrix $[M]$?

(b) Find $[ML]$. Here ML is composition, $(ML)X = M(LX)$.

(c) What is the relation between the image vector MLe_1 and the first column of the matrix $[ML]$? What about MLe_2?

8. Different prescriptions of a linear map. Suppose that $L \colon \mathbb{R}^2 \longrightarrow \mathbb{R}^2$ is linear and we are given its matrix only (2 by 2, of course):

$$[L] = \begin{bmatrix} -2 & 1 \\ 1 & 3 \end{bmatrix}.$$

We know that L may just as well be prescribed in other ways, any of which we are likely to encounter:

(a) What are Le_1, Le_2, as explicit linear combinations of e_1, e_2?

(b) Why do the two image vectors Le_1, Le_2 determine *all* image vectors LX, for all X? (This is the essence of linearity.)

(c) Let $X = (x_1, x_2)$ be a typical vector in the domain $\mathbb{R}^2$. Write the typical image LX in terms of coordinates; that is, $LX = Y = (y_1, y_2)$, where y_1 and y_2 are each linear expressions $y_1 = \alpha x_1 + \beta x_2$, $y_2 = \gamma x_1 + \delta x_2$. Compute $\alpha, \beta, \gamma, \delta$.

9. Inverses:

(a) What is the inverse of the 1 by 1 matrix 5, or if you insist, [5]?

(b) Solve $5x = 12$.

(c) Which 1 by 1 matrices do not have inverses? Note: Most do.

(d) What is the inverse of the 2 by 2 matrix

$$A = \begin{bmatrix} 5 & 0 \\ 0 & 6 \end{bmatrix}?$$

(e) Solve for X in $AX = Y$, where

$$X = \begin{bmatrix} x_1 \\ x_2 \end{bmatrix}, \qquad Y = \begin{bmatrix} 12 \\ -13 \end{bmatrix}.$$

(Hint: Attack both sides of the equation with A^{-1}, keeping in mind that $AA^{-1} = A^{-1}A = I$.)

(f) Why *must* X be unique in Exercise 9(e)?

(g) Now find X such that $AX = O$, where O is

$$\begin{bmatrix} 0 \\ 0 \end{bmatrix}.$$

Is X unique?

10.   (a) Compute inverses for the following matrices:

$$A = \begin{bmatrix} -2 & 1 \\ 0 & 1 \end{bmatrix}, \qquad B = \begin{bmatrix} -2 & 1 \\ 1 & 1 \end{bmatrix}, \qquad C = \begin{bmatrix} -2 & 0 \\ 0 & 1 \end{bmatrix}, \qquad D = \begin{bmatrix} -2 & 1 \\ 1 & 3 \end{bmatrix},$$

(b) Solve for x_1, x_2 in terms of y_1, y_2 in the system (see B above)

$$-2x_1 + x_2 = y_1,$$
$$x_1 + x_2 = y_2.$$

(c) Likewise in the system

$$-2x_1 + x_2 = y_1,$$
$$x_1 + 3x_2 = y_2.$$

11.   The existence of a matrix inverse. Let A be an n by n (square) matrix, and let X, X_1, X_2, Y be n by 1 column vectors. True or false:

(a) If A^{-1} exists and Y is given, then there exists precisely one X such that $AX = Y$.

(b) If A^{-1} exists and $AX = O$, then $X = O$.

(c) If $AX_1 = Y$ and also $AX_2 = Y$ for $X_1 \neq X_2$, then there exists a nonzero X such that $AX = O$.

(d) If $AX = O$ for some $X \neq O$, then A does not have an inverse.

(e) If $AX_1 = Y$ and also $AX_2 = Y$ with $X_1 \neq X_2$, then A does not have an inverse.

(f) If A does not have an inverse, then A is the zero matrix.

(g) There is a nonzero column vector X such that $AX = O$, where

$$A = \begin{bmatrix} 0 & 1 \\ 0 & 2 \end{bmatrix}.$$

(h) The matrix A in Exercise 11(g) has an inverse.

(i) The matrix

$$\begin{bmatrix} 1 & 1 \\ 1 & 1 \end{bmatrix}$$

has an inverse.

12. Let

$$A = \begin{bmatrix} 1 & -2 \\ 2 & 0 \\ 3 & 1 \end{bmatrix}, \qquad B = \begin{bmatrix} -1 & 0 & -2 \\ 2 & 1 & 0 \end{bmatrix}, \qquad C = \begin{bmatrix} 3 & 0 & 2 \\ 1 & 4 & -1 \\ 0 & -2 & 0 \end{bmatrix}.$$

Compute all the following products that make sense: $AB, BA, AC, CA, BC,$ $CB, A^2, B^2, C^2, ABC, CAB.$

13. Compute A^{-1} if

(a) $A = \begin{bmatrix} 1 & 2 & 0 \\ 0 & 1 & -1 \\ 0 & 0 & 1 \end{bmatrix},$ (b) $A = \begin{bmatrix} 2 & 0 & 3 \\ 0 & 1 & 0 \\ 1 & 0 & 2 \end{bmatrix},$ (c) $A = \begin{bmatrix} 1 & 1 & 0 & 0 \\ 0 & 1 & 1 & 0 \\ 0 & 0 & 1 & 1 \\ 0 & 0 & 0 & 1 \end{bmatrix}.$

14. If

$$A = \begin{bmatrix} 1 & 0 \\ 0 & 0 \end{bmatrix} \quad \text{and} \quad B = \begin{bmatrix} 0 & 0 \\ 1 & 0 \end{bmatrix},$$

express A^2, B^2, AB, BA, ABA, BAB, $(AB)^2$, $(BA)^2$, A^5, $(A + 2B)^2$, $(3A - B)^2$ in the form $aA + bB$.

15. If A and B are n by n matrices with A invertible, prove that
$$(ABA^{-1})^k = AB^kA^{-1}, \quad k = 1, 2, \ldots.$$

16. Let

$$A = \begin{bmatrix} \frac{1}{2} & -\frac{\sqrt{3}}{2} \\ \frac{\sqrt{3}}{2} & \frac{1}{2} \end{bmatrix} \quad \text{and} \quad B = \begin{bmatrix} 5 & \sqrt{3} \\ \sqrt{3} & 3 \end{bmatrix}.$$

Compute ABA^{-1} and $(ABA^{-1})^{99}$. Use this to compute B^{99}.

17. Given two 2 by 2 matrices A, B such that $AB = 0$, which of the following statements are *always* true? Proof or counterexample:

(a) $A = 0.$
(b) $B = 0.$
(c) Either $A = 0$ or $B = 0.$
(d) A is not invertible.
(e) If A^{-1} exists, then $B = 0.$
(f) $BA = 0.$
(g) If $A \neq 0$ and $B \neq 0$, then neither A nor B is invertible.

18. Matrix equations. Often one meets the following kind of problem: Given matrices A, B, find C so that $AC = B$, assuming that A, B, C are n by n and A is invertible. To solve this, we multiply both sides of $AC = B$ on the *left* by A^{-1}, to give $A^{-1}AC = A^{-1}B$; that is, $C = A^{-1}B$. Use this method to solve the following for C. Here

$$A = \begin{bmatrix} 2 & 3 \\ 1 & 2 \end{bmatrix}, \qquad B = \begin{bmatrix} 1 & -1 \\ 1 & 0 \end{bmatrix}.$$

(a) $AC = B,$
(b) $CA = B,$
(c) $BC = 0,$

(d) $CA = A,$
(e) $ACA = I,$
(f) $ACB = A + B.$

19. Find a 2 by 2 matrix A such that $A^2 = -I$.

20. If a square matrix A satisfies $A^2 - 2A = I$, find A^{-1} in terms of A. (Suggestion: By inspection find B such that $AB = BA = I$.)

21. If a square matrix A satisfies $A^4 = 0$, verify that $(I - A)^{-1} = I + A + A^2 + A^3$. Find $(I + A)^{-1}$ too.

22. Complete the proof of Theorem 12.

23. Let $L, M: \mathbb{R}^n \longrightarrow \mathbb{R}^n$ be invertible maps. Prove:
$$(LM)^{-1} = M^{-1}L^{-1}.$$

ALGEBRA OF LINEAR EQUATIONS

As we have seen, the problem of solving simultaneous linear equations occurs repeatedly throughout mathematics and its applications. The most general system of k equations in n variables is

$$a_{11}x_1 + a_{12}x_2 + \cdots + a_{1n}x_n = b_1$$
$$a_{21}x_1 + a_{22}x_2 + \cdots + a_{2n}x_n = b_2$$
$$\vdots$$
$$a_{k1}x_1 + a_{k2}x_2 + \cdots + a_{kn}x_n = b_k$$

where the coefficients a_{11}, and so on, and the right sides $b_1, b_2, \ldots$ are prescribed. The same system of equations with $b_1 = b_2 = \cdots = b_k = 0$ is called the **homogeneous** system associated with the original **inhomogeneous** system, that is, some $b_j \neq 0$.

Three questions already raised in the last section:

1. Are there any solutions?
2. If so, how many solutions are there?
3. Can you actually find all the solutions?

We shall restrict ourselves to the first two questions. Of course, questions 1 and 2 are merely preliminary, but in many cases, the answers to them would be sufficient, the actual solutions being of secondary interest. Moreover, if the answer to question 1 is *no*, then one cannot proceed further. The actual computation of solutions, if they exist, is often best left to electronic computers.

The main theorem on linear equations is

THEOREM: Given a system of k equations in n variables, then:

1. If there are more variables than equations $(n > k)$, then the homo-

geneous system always has a solution other than the trivial solution $x_1 = x_2 = x_n = 0$.

2. If there is the same number of equations as variables $(n = k)$, then for any inhomogeneous system there exists a unique solution $\Leftrightarrow$ the only solution of the homogeneous system is the trivial solution $x_1 = x_2 = \cdots = x_n = 0$.

3. If there are fewer variables than equations $(n < k)$, then there is at least one set of numbers $b_1, \ldots, b_k$ such that the corresponding inhomogeneous equations has no solutions.

REMARK: Notice that part 1 asserts that under certain conditions a solution *exists*. On the other hand, part 3 asserts that under certain conditions a solution does *not exist*. Part 2 is an assertion concerning both *existence* and *uniqueness*. It may be restated that, for $n = k$, we have existence $\Leftrightarrow$ uniqueness. (Compare with Theorem 16, parts **3** and **4**).

Proof: The key is to observe that the given system of equations defines a linear map $L: \mathbb{R}^n \to \mathbb{R}^k$, $LX = B$, where B is the vector $B = (b_1, \ldots, b_k)$ and L is the matrix $[a_{ij}]$. Now:

1. The solutions of the homogeneous equation $LX = O$ are precisely the elements of $\mathcal{N}(L)$.

2. There is a solution of $LX = B$ if and only if B is in the image of L. By the Dimension Theorem,

$$n = \dim \mathcal{D}(L) = \dim \mathcal{I}(L) + \dim \mathcal{N}(L). \tag{*}$$

Also $\mathcal{I}(L) \subset \mathbb{R}^k$, so that $\dim \mathcal{I}(L) \leq k$:

1. If $n > k$, then $\dim \mathcal{N}(L) = n - \dim \mathcal{I}(L) \geq n - k > 0$. Therefore the homogeneous system has a nontrivial solution $LX = O$.

2. Let $n = k$.

($\Leftarrow$)   If the only solution of the homogeneous equation is the trivial solution $X = O$, then $\mathcal{N}(L) = \{O\}$. Thus from Eq. (*) $\dim \mathcal{I}(L) = n$. Consequently $\mathcal{I}(L)$ must be all of $\mathbb{R}^n$. This means that, given any $B \in \mathbb{R}^n$, there is some X such that $LX = B$. Moreover, this X is unique, for if there were another X' satisfying $LX' = B$, let $Y = X - X'$. Then $LY = LX - LX' = B - B = O$; that is, $Y \in \mathcal{N}(L) = \{O\}$, so that $X = X'$.

($\Rightarrow$)   If you can always solve $LX = B$ for any $B \in \mathbb{R}^n$, then $\dim \mathcal{I}(L) = n$. Therefore from Eq. (*) $\dim \mathcal{N}(L) = 0$, which implies that $\mathcal{N}(L) = \{O\}$.

3. Let $n < k$. From Eq. (*) $\dim \mathcal{I}(L) \leq n < k$, so that $\mathcal{I}(L)$ is not

all of $\mathbb{R}^k$. This is another way of saying that there are vectors $B \in \mathbb{R}^k$ such that the equation $LX = B$ has no solution. Done. $\ll$

WARNING: Even if $n = k$, it may happen that a certain system of equations has no solution. An example is

$$x_1 - x_2 = 3,$$
$$-2x_1 + 2x_2 = 5,$$

for if (x_1, x_2) satisfies the first equation, then $-2x_1 + 2x_2 = -2(x_1 - x_2)$ $= -6 \neq 5$. The homogeneous equations have the nontrivial solution $x_1 = 1$, $x_2 = 1$.

As a complement to the result above, we have a result also mentioned in Sec. 3.1.

THEOREM: The general solution of $LX = Y$ is of the form $X = X_0 + X_1$, where X_1 is a particular solution of $LX_1 = Y$ and X_0 is any solution of the homogeneous equation $LX_0 = 0$. In other words, the solution space of $LX = Y$ is the affine subspace parallel to $\mathcal{N}(L)$ that contains a particular solution X_1.

Proof: Let X be any solution of $LX = Y$ and let $Z = X - X_1$. Then $LZ = LX - LX_1 = Y - Y = 0$; that is, Z is a solution of the homogeneous equation. Since $X = Z + X_1$, this is the assertion of the theorem with Z replaced by the letter X_0. $\ll$

Exercises

1. For the following systems of equations find bases for $\mathcal{N}(L)$, and compute $\dim \mathcal{N}(L)$ and $\dim \mathcal{I}(L)$:

    (a) $2x_1 - x_2 = b_1,$
        $x_1 + x_2 = b_2;$

    (c) $2x_1 - x_2 = b_1,$
        $-2x_1 + x_2 = b_2,$
        $2x_1 - x_2 = b_3.$

    (b) $2x_1 - x_2 + x_3 = b_1,$
        $-2x_1 + x_2 - x_3 = b_2;$

2. Say that you have k equations in n variables, which we write as $LX = B$. True or false? If false, give a counterexample:

    (a) If $n = k$, there is always *at most one* solution of $LX = B$.
    (b) If $n > k$, you can always solve $LX = B$.
    (c) If $n = k$, you can always solve $LX = B$.
    (d) If $n < k$, you can never solve $LX = B$.
    (e) If $n > k$, you can never solve $LX = B$.
    (f) If $n < k$, the only solution of $LX = 0$ is $X = 0$.

3.5 AFFINE MAPS

3.5a Introduction

These are interesting for several reasons: They are almost, but not quite, linear. They are built from linear maps. Their graphs are affine subspaces (compare planes in $\mathbb{R}^3$), and their images are also affine subspaces. (Recall that an affine subspace is obtained by translating an ordinary subspace away from the origin in $\mathbb{R}^n$).

DEFINITION: A function $T: \mathbb{R}^n \longrightarrow \mathbb{R}^q$ is **affine** if and only if there is a fixed $Y_0 \in \mathbb{R}^q$ and linear map $L: \mathbb{R}^n \longrightarrow \mathbb{R}^q$ such that, for all $X \in \mathbb{R}^n$,

$$T(X) = Y_0 + LX.$$

Thus, an affine map is accomplished by first operating on X with a linear map L and then translating LX in $\mathbb{R}^q$ by the vector Y_0.

Note that the affine map is linear if and only if the vector Y_0 is O.

Examples of Affine Maps: 1. In the case of $n = q = 1$, the vector X is a real number x, and an affine map is of the form $T(x) = b + ax$, where b and a are fixed scalars. The graph $y = T(x)$ is a straight line.

2. Fix $Y_0 \in \mathbb{R}^n$, and define $T(X) = Y_0 + X$. Here $T: \mathbb{R}^n \longrightarrow \mathbb{R}^n$ "translates" every vector X by the fixed vector Y_0. In this special case $L = I$, the identity map.

3. Suppose that $T: \mathbb{R}^2 \longrightarrow \mathbb{R}^2$ is the affine map given by $T(X) = Y_0 + LX$, where $Y_0 = (6, -7)$ and $LX = (x_1 - x_2, 2x_1 + x_2)$. Then T is given in coordinate form by

$$T(X) = (6 + x_1 - x_2, -7 + 2x_1 + x_2).$$

Note that each coordinate of $T(X)$ has a constant term—for example, 6 —and a linear term—for example, $x_1 - x_2$. Is this true for a general affine map $T: \mathbb{R}^n \longrightarrow \mathbb{R}^q$?

3.5b A property of affine maps

A basic property of affine maps is conveyed by Figure 3.25. Here $\mathscr{A}$ is an affine subspace (a plane) in $\mathbb{R}^3$, and T is an affine map $T: \mathbb{R}^3 \longrightarrow \mathbb{R}^3$. Figure 3.25 indicates that the image of the set $\mathscr{A}$ under the map T, namely,

$$T(\mathscr{A}) = \{Y \in \mathbb{R}^3 \mid Y = T(X), X \in \mathscr{A}\},$$

is also an affine subspace of the target space $\mathbb{R}^3$. In fact, $T(\mathscr{A})$ is here

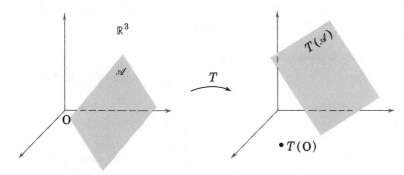

FIGURE 3.25

pictured to be a plane also, just as $\mathscr{A}$ was. Thus affine maps preserve affine subspaces. To be precise, we state

THEOREM 17: Let $T \colon \mathbb{R}^n \longrightarrow \mathbb{R}^q$ be an affine map and $\mathscr{A}$ an affine subspace of $\mathbb{R}^n$. Then the set $T(\mathscr{A})$ defined by

$$T(\mathscr{A}) = \{Y \in \mathbb{R}^q \mid Y = T(X),\ X \in \mathscr{A}\}$$

is an affine subspace of $\mathbb{R}^q$. In other words, the image $\mathscr{I}(T)$ of $\mathscr{A}$ under the map T is an affine subspace of $\mathbb{R}^q$.

Proof: Let $T(X) = Y_0 + LX$, with Y_0 fixed in $\mathbb{R}^q$ and $L \colon \mathbb{R}^n \longrightarrow \mathbb{R}^q$ linear. Since $\mathscr{A}$ is affine, $\mathscr{A} = \mathscr{S} + Z$, where Z is a fixed vector in $\mathscr{A}$ and $\mathscr{S}$ a vector subspace of $\mathbb{R}^n$. Thus if $X \in \mathscr{A}$, we know that $X = X' + Z$, with $X' \in \mathscr{S}$. Applying the map T gives $T(X) = T(X' + Z) = Y_0 + LZ + LX'$. It follows that $T(\mathscr{A})$ consists of all vectors in $\mathbb{R}^q$ of the form $(Y_0 + LZ) + LX'$, with $X' \in \mathscr{S}$. Note that $Y_0 + LZ$ is a constant vector; it does not depend on X.

Now since $\mathscr{S}$ is a subspace and L is linear, the set of all LX' with $X' \in \mathscr{S}$ is again a subspace (see Theorem 3). Let us denote this subspace $L(\mathscr{S})$. Thus

$$T(\mathscr{A}) = (Y_0 + LZ) + L(\mathscr{S}),$$

and so $T(\mathscr{A})$ is an affine subspace of $\mathbb{R}^q$. $\ll$

FOOD FOR THOUGHT: You might reflect on how much more cluttered the proof of Theorem 17 would be if, instead of using the coordinate-free vector notation, all the maps and subspaces were specified using the coordinates $x_1, \ldots, x_n$ and $y_1, \ldots, y_q$ and systems of linear equations in these coordinates. We would be lost in clouds of linear equations.

IMPORTANT: Later on we will see examples of maps that "bend" or "fold" planes. They will not be affine maps, of course.

3.5c   The graph of an affine map

This will be used later when we study nonlinear functions using tangent planes to their graphs. Let $T: \mathbb{R}^2 \to \mathbb{R}$ be an affine function. It is easy to see that such a map has the form

$$T(X) = \alpha + \beta x_1 + \gamma x_2,$$

where x_1, x_2 are the coordinates of $\mathbb{R}^2$ and α, β, γ are fixed scalars. Consider the **graph** of $y = T(X)$, that is, the set of all points (x_1, x_2, y) in $\mathbb{R}^3$ with $y = \alpha + \beta x_1 + \gamma x_2$. This set is precisely the set of solutions to

$$\beta x_1 + \gamma x_2 - y = -\alpha$$

and hence, by Theorem 6 in Chap. 1, is a plane in $\mathbb{R}^3$.

For example, let $T(X) = 3 - x_1 - 2x_2$. Then $T(O) = 3$, $T(e_1) = 2$, $T(e_2) = 1$, and so the graph $y = T(X)$ is the plane through the three points $(0, 0, 3)$, $(1, 0, 2)$, $(0, 1, 1)$ in $\mathbb{R}^3$. See Fig. 3.26.

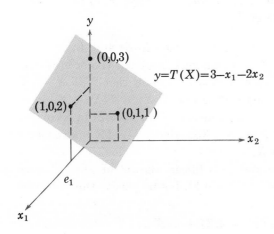

The graph of an affine map $T: \mathbb{R}^2 \to \mathbb{R}$

FIGURE 3.26

Exercises

These give you practice in recognizing, constructing, and picturing affine maps. Their graphs play the same role for us that the tangent line did in one-variable calculus.

1. Which of the following maps T are affine?
$$T: \mathbb{R} \longrightarrow \mathbb{R}$$

(a) $T(x) = 1 + x^2$,   (c) $T(x) = \frac{1}{2}x$;

(b) $T(x) = 1 + 2x$,

$$T: \mathbb{R}^2 \longrightarrow \mathbb{R}, \; X = (x_1, x_2)$$

(d) $T(X) = x_1^2 + x_2^2$,   (f) $T(X) = 1 + x_1 + 3x_2$;

(e) $T(X) = x_1 - x_2$,

$$T: \mathbb{R}^2 \longrightarrow \mathbb{R}^2$$

(g) $T(X) = X$ (identity map),   (i) $T(X) = Y_0 + X$, Y_0 fixed,

(h) $T(X) = \langle X, X \rangle X$,   (j) $T(X) = (x_1, 2 + x_1 + x_2)$;

$$T: \mathbb{R} \longrightarrow \mathbb{R}^2 \qquad \text{the vector variable in the domain}$$
$$\mathbb{R} \text{ is a scalar } t$$

(k) $T(t) = (t, t^2)$,   (m) $T(t) = (t, t)$,

(l) $T(t) = (1 - t, 2 + 3t)$,   (n) $T(t) = (0, 4)$ (constant).

2. Let $T: \mathbb{R}^2 \longrightarrow \mathbb{R}$ be the affine map $T(X) = 7 - x_1 + 4x_2$:

(a) Compute $T(e_1)$, where $e_1 = (1, 0)$, as usual.

(b) Compute $T(e_2)$.

(c) Compute $T(O)$.

(d) If we write this same map in the form $T(X) = y_0 + LX$, then what is the number y_0?

(e) What is the coordinate expression for the linear map L?

(f) What is the matrix $[L]$?

3. Planes and affine maps:

(a) Let x_1, x_2, y be coordinates in $\mathbb{R}^3$. The plane given by $3x_1 - x_2 + 6y = 7$ is the graph of an affine map $T: \mathbb{R}^2 \longrightarrow \mathbb{R}$, $y = T(X)$. Which map?

(b) Now let x_1, x_2, z be coordinates in $\mathbb{R}^3$, as we shall denote them in Chap. 5. Same question for the plane $x_1 + x_2 - 2z = 1$; what is $z = T(X)$, which has this plane as its graph?

4. (a) Write down an affine map (coordinate form $T(X) = \alpha + \beta x_1 + \gamma x_2$) such that $T(O) = 7$, $T(e_1) = -1$, $T(e_2) = 4$. Is this affine map unique?

(b) Compute the equation of the plane in $\mathbb{R}^3$ (coordinates x_1, x_2, z) through the three points $(0, 0, 7)$, $(1, 0, -1)$, $(0, 1, 4)$.

(c) Write down an affine map $T: \mathbb{R}^2 \longrightarrow \mathbb{R}$, $T(X) = \alpha + \beta x_1 + \gamma x_2$, such that $T(1, 1) = 10$, $T(1, -1) = 2$, $T(-1, 1) = 12$.

5. True or false:

(a) Every affine map is linear.

(b) Every constant map is affine.

(c) If $T: \mathbb{R}^2 \longrightarrow \mathbb{R}$ is affine, then it is uniquely determined by the two values $T(e_1)$, $T(e_2)$.

(d) If $T: \mathbb{R}^2 \longrightarrow \mathbb{R}$ is affine, then its graph in $\mathbb{R}^3$ might be a straight line.

(e) The graph of every affine map $T: \mathbb{R}^2 \longrightarrow \mathbb{R}$ contains the origin of $\mathbb{R}^3$.

(f) If $T: \mathbb{R}^2 \longrightarrow \mathbb{R}$ is affine, then $T(X) = T(O) + T(e_1)x_1 + T(e_2)x_2$.

(g) The map $T: \mathbb{R}^2 \longrightarrow \mathbb{R}$ given by $T(X) = 3 + 4(x_1 - 1) + (x_2 + 1)$ is affine.

(h) If $T(X) = \alpha + \beta x_1 + \gamma x_2$, then the graph of $z = T(X)$ in $\mathbb{R}^3$ is horizontal $\Leftrightarrow \beta = \gamma = 0$.

(i) The graphs of the maps $T(X) = \alpha + \beta x_1 + \gamma x_2$ and $T^*(X) = \alpha^* + \beta^* x_1 + \gamma^* x_2$ are parallel planes in $\mathbb{R}^3 \Leftrightarrow \beta = \beta^*$ and $\gamma = \gamma^*$.

(j) If T is affine, then $T(-X) = -T(X)$.

(k) Every affine map is a linear functional.

(l) Every linear map is affine.

6. Let $T: \mathbb{R}^n \longrightarrow \mathbb{R}^q$ be affine. Prove that T is linear $\Leftrightarrow T(O) = O$.

7. (a) Let $T: \mathbb{R}^2 \longrightarrow \mathbb{R}^3$ be affine, and say that $T(O) = (1, -2, 3)$, $T(e_1) = (0, 1, 1)$ $T(e_2) = (1, -1, 0)$. If $X = (x_1, x_2)$, find $T(X)$. In particular, find $T(1, 2)$.

(b) Let $T: \mathbb{R}^n \longrightarrow \mathbb{R}^q$ be affine, and say that $T(O) = Y_0$, $T(e_j) = Y_j$, $j = 1, \ldots, n$. If $X = (x_1, \ldots, x_n)$, show that

$$TX = Y_0 + x_1(Y_1 - Y_0) + \cdots + x_n(Y_n - Y_0).$$

This shows that an affine map is uniquely determined if one knows the images of O and of the standard basis vectors.

8. (a) Let $T: \mathbb{R}^n \longrightarrow \mathbb{R}^q$ be a given map, and let $\alpha + \beta = 1$. Prove that T is affine $\Rightarrow$ for any $X, Y \in \mathbb{R}^n$, we have $T(\alpha X + \beta Y) = \alpha T(X) + \beta T(Y)$.

(b) Prove the converse, $\Leftarrow$, of Exercise 8(a).

9. If $T: \mathbb{R}^n \longrightarrow \mathbb{R}^q$ is an affine map, we define its **zero set** (a substitute for the null space) as the set of all $X \in \mathbb{R}^n$ whose image is zero, $TX = O$. Prove that the zero set of an affine map is an affine subspace of $\mathbb{R}^n$.

10. Let $T(X) = Y_0 + LX$ be an affine map. If $TX_0 = Y_1$, prove that $T(X) = Y_1 + L(X - X_0)$.

11. Let $T(X) = Y_0 + LX$ be an affine map from $\mathbb{R}^n$ to $\mathbb{R}^n$. Prove that T is invertible, that is, there is a map S such that for all $X, Y, S(TX) = X$ and $T(SY) = Y$, so that $S = T^{-1}$, $\Leftrightarrow$ the linear map L is invertible. (Suggestion: Find a formula for T^{-1} involving Y_0 and L^{-1} by thinking of T geometrically, as a linear map is followed by a translation.)

4.
From Linear
to
Nonlinear

4.0 INTRODUCTION

Now we begin calculus proper. This chapter is preparatory; we shall make some basic definitions and ask some questions.

In ordinary calculus we considered a function $y = f(x)$, with $x \in \mathbb{R}$. Such a function assigns to each real x another real number y. Thus we may write

$$f: \mathbb{R} \longrightarrow \mathbb{R}.$$

What are the simplest such scalar-valued functions of a scalar variable? Linear functions, of course,

$$f(x) = ax$$

and the closely related affine functions

$$g(x) = ax + b.$$

Note that a linear function is a special kind of affine function, with $b = 0$. Some authors would call both these functions "linear," because the graphs $y = f(x)$ and $y = g(x)$ are straight lines in each case (if $b \neq 0$, the graph of the affine function $g(x)$ does not pass through the origin.) See Fig. 4.1.

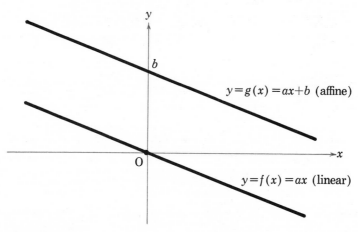

FIGURE 4.1

At an early age we enlarged our vocabulary of functions to a more general class that includes as examples

$$f_1(x) = x^2 + 1, \qquad f_2(x) = \sin x, \qquad f_3(x) = \sqrt{x}, \qquad f_4(x) = \frac{1}{x}.$$

None of these functions is linear. Recall, for example, that $\sin(x_1 + x_2)$

$\neq \sin x_1 + \sin x_2$. Each maps some subset of the reals into the reals. The subset of the reals for which they are defined is called their *domain*, or *domain of definition*, and denoted $\mathscr{D}(f)$. Thus

$$\mathscr{D}(f_1) = \mathbb{R}, \qquad\qquad \mathscr{D}(f_2) = \mathbb{R},$$
$$\mathscr{D}(f_3) = \{x \in \mathbb{R} \mid x \geq 0\}, \qquad \mathscr{D}(f_4) = \{x \in \mathbb{R} \mid x \neq 0\}.$$

Now on to differential calculus. The main problem was this: *Given a function, such as $y = f(x) = \sqrt{x}$ in Fig. 4.2, and given a point $x_0 \in \mathscr{D}(f)$, define and construct the tangent line to the graph at the point x_0.* This tangent line is the graph of an affine function $y = T(x)$, and when x is close to x_0, $T(x)$ is close to $f(x)$; that is, $T(x)$ approximates $f(x)$. Later on we shall speak of the tangent line $y = T(x)$ as the **"best affine approximation"** to $f(x)$ near the point x_0.

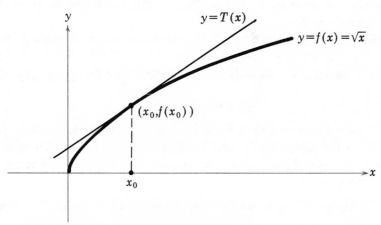

The tangent line at x_0

FIGURE 4.2

In the sections to come, we parallel the study above. Now we introduce nonlinear functions

$$F\colon \mathbb{R}^n \longrightarrow \mathbb{R}^q,$$

with $n \geq 1$ and $q \geq 1$. Writing $Y = F(X)$ with $X \in \mathbb{R}^n$, $Y \in \mathbb{R}^q$, we say that F is a "vector-valued" function of a vector. Then, in order to understand the behavior of $F(X)$ for X near some fixed vector X_0, we construct a "best affine approximation"

$$T\colon \mathbb{R}^n \longrightarrow \mathbb{R}^q$$

such that $T(X_0)$ actually equals $F(X_0)$ and, moreover, $T(X)$ is close to $F(X)$ when X is close to X_0. This function $Y = T(X)$ plays the same role in vector calculus as that of the tangent line in ordinary calculus. The third section of this chapter consists of an informal discussion of the best affine approximation.

We remind you of the four-stage process we outlined in the introduction to Chap. 1, namely, (1) vector algebra, (2) vector geometry, (3) linear functions, (4) nonlinear functions. The first three stages have been accomplished. The present chapter is best viewed as a final preparation for the last and most important stage.

4.1 NONLINEAR FUNCTIONS

This section is devoted to the definition and examples of general functions from $\mathbb{R}^n$ to $\mathbb{R}^q$.

A *function* or *map* F from $\mathbb{R}^n$ to $\mathbb{R}^q$ is a rule that assigns to each X in a certain subset $\mathscr{D}(F)$ of $\mathbb{R}^n$ a vector $Y = F(X)$ in $\mathbb{R}^q$. The subset $\mathscr{D}(F)$ is called the **domain** of F. We write—inaccurately—but for typographical convenience

$$F \colon \mathbb{R}^n \longrightarrow \mathbb{R}^q,$$

although it should be noted that F need not be defined for all vectors in $\mathbb{R}^n$ but only for the subset $\mathscr{D}(F)$, which may be smaller than $\mathbb{R}^n$. In order to specify F, we should first state what the domain, $\mathscr{D}(F)$, is if it is not clear from the context.

Also, if $\mathscr{S}$ is any subset of $\mathscr{D}(F)$, then the **image** of $\mathscr{S}$ under F is the set

$$F(\mathscr{S}) = \{F(X) \in \mathbb{R}^q \mid X \in \mathscr{S}\}.$$

EXAMPLES OF FUNCTIONS: 1. Any linear map $L \colon \mathbb{R}^n \longrightarrow \mathbb{R}^q$. We have said a good deal about these already. In this case, $\mathscr{D}(L) = \mathbb{R}^n$.

2. We recall that an *affine* map $T \colon \mathbb{R}^n \longrightarrow \mathbb{R}^q$ is any function of the form

$$T(X) = Y_0 + LX,$$

where Y_0 is a fixed vector in $\mathbb{R}^q$ and L is a linear map

$$L \colon \mathbb{R}^n \longrightarrow \mathbb{R}^q.$$

Thus an affine map is accomplished by first operating with a linear map L and then translating by a vector Y_0. Note that $\mathscr{D}(T) = \mathbb{R}^n$ for affine maps T.

3. Let $f\colon \mathbb{R}^2 \to \mathbb{R}$ be given by $f(X) = x_1^2 + x_2^2$, where $X = (x_1, x_2)$. Here the domain, $\mathscr{D}(f)$, is all of $\mathbb{R}^2$. For instance, if $X = (2, -8)$, then $f(X) = 68$.

There is an interesting picture associated with this function. If we write $z = f(X) = x_1^2 + x_2^2$, then, by erecting a vertical z-axis above the plane of (x_1, x_2), we may graph the function as shown in Fig. 4.3. The surface obtained in this case is a **paraboloid of revolution**; it could also be obtained by graphing $z = x_2^2$ in the (x_2, z) plane and then rotating the graph about the z-axis.

4. Let us define a function $f\colon \mathbb{R}^3 \to \mathbb{R}$ as follows. Let $\mathscr{D}(f)$ be some room (a subset of $\mathbb{R}^3$), and, for a point X in this room, define $f(X)$ to be the temperature, in degrees centigrade, say, at that point. See Fig. 4.4. As X varies throughout the room, the *temperature* $f(X)$ is likely to vary also: cooler near a window, warmer toward the ceiling, and so on.

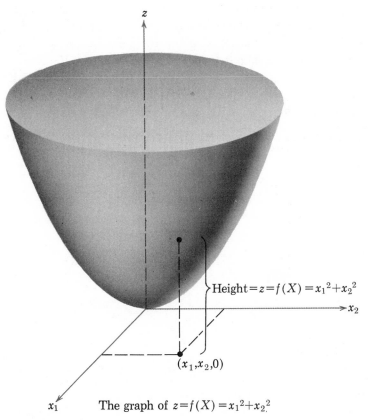

Height $= z = f(X) = x_1{}^2 + x_2{}^2$

$(x_1, x_2, 0)$

The graph of $z = f(X) = x_1{}^2 + x_2{}^2$

FIGURE 4.3

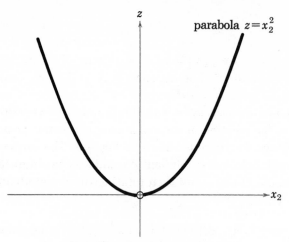

parabola $z = x_2^2$

Revolving parabola about
z—axis gives graph of $z = f(X) = x_1{}^2 + x_2{}^2$

FIGURE 4.4

Note that we do not furnish a formula, in terms of the space variables x_1, x_2, x_3, say, for the temperature function $f(X)$. In fact, a substantial problem in mathematical physics consists in trying to compute a formula for the temperature as a function of X, given the general theory of heat flow and specific information about your room.

Note also that we cannot draw a graph of $z = f(X)$, as we did in the previous example. Since $X = (x_1, x_2, x_3)$, *four* dimensions would be required to depict points of the form $(x_1, x_2, x_3, f(X))$, that is, (x_1, x_2, x_3, z).

NOTE: It is also possible to speak of the temperature in a flat two-dimensional heated plate. This leads to a function $f \colon \mathbb{R}^2 \to \mathbb{R}$.

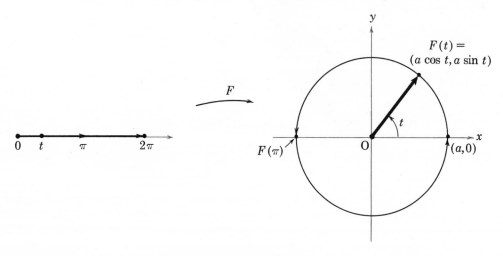

FIGURE 4.5 $F(t)$ traces out the circle

5. Now we discuss a nonlinear function $F: \mathbb{R} \to \mathbb{R}^2$ that may be familiar. Let $\mathscr{D}(F)$ be the interval $0 \le t < 2\pi$. Let $a > 0$ be fixed, and define

$$F(t) = (a \cos t, a \sin t).$$

Note that as t varies from $t = 0$ to $t = 2\pi$, the point (arrowhead) $F(t)$ traverses the circle of radius a centered at the origin of $\mathbb{R}^2$ in a counterclockwise direction. See Fig. 4.5.

One way to verify that $F(t)$ traces out the circle of radius a is to write $F(t) = (x, y)$, so that $x = a \cos t$, $y = a \sin t$. Then $x^2 + y^2 = a^2$, and since this equation defines the circle, $F(t)$ must be on the circle.

In this example, we may interpret the number t as the angle, in radians, between the positive x-axis and the vector $F(t)$ whose arrowhead traces out the circle.

Language: we have "parametrized" the circle of radius a by the map $F(t)$; the variable t(angle) is the **parameter**. Obtaining the circle $x^2 + y^2 = a^2$ as the image of the interval $0 \le t < 2\pi$ emphasizes its one-dimensionality; the circle is an interval that has been "curved" by the function F.

6. Having obtained the well-known circle as the image of a nonlinear function, we now mention that even more exotic constructions are possible. We are going to define a function $G: \mathbb{R}^2 \to \mathbb{R}^3$ that maps a rectangular piece of the plane $\mathbb{R}^2$ onto a torus embedded in $\mathbb{R}^3$. See Fig. 4.6.

Important: the torus, although doughnut-shaped, is hollow—a *surface*, not a solid.

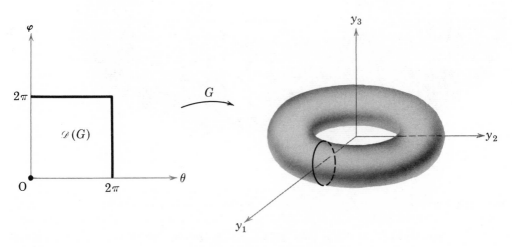

G maps the square onto the torus.

FIGURE 4.6

To give a formula for the function G, it is helpful to denote a typical point of $\mathbb{R}^2$ as (θ, φ), leading to a horizontal θ-axis and vertical φ-axis, as pictured in Fig. 4.6. Likewise, we denote a typical point of $\mathbb{R}^3$ by (y_1, y_2, y_3). Let $\mathscr{D}(G)$ be the square in $\mathbb{R}^2$ determined by $0 \leq \theta < 2\pi$, $0 \leq \varphi < 2\pi$. Then $G(\theta, \varphi) = Y = (y_1, y_2, y_3)$, where

$$y_1 = (a + b \cos \varphi) \cos \theta,$$
$$y_2 = (a + b \cos \varphi) \sin \theta,$$
$$y_3 = b \sin \varphi,$$

and a, b are scalars satisfying $0 < b < a$.

A more detailed picture is Fig. 4.7.

We strongly recommend that you take the time to convince yourself that the image of the square under the mapping G is indeed a torus.

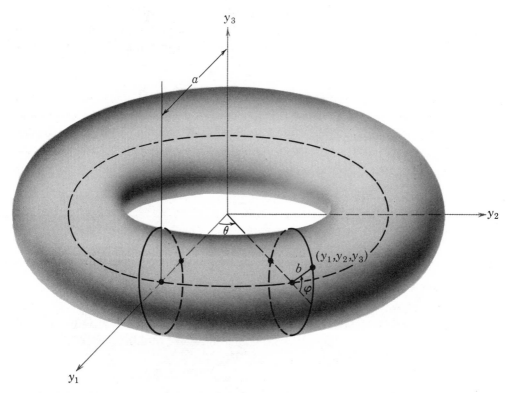

FIGURE 4.7

REMARKS: 1. Note that this torus is *not* the graph of a scalar-valued function $y_3 = f(y_1, y_2)$ defined for points of the $y_1 y_2$-plane.

In fact, there are *two* values of y_3 for each y_1, y_2, one above and one below the horizontal ($y_3 = 0$) plane. Both values give a point on the torus. We might be able to consider the torus as the union of the graph of *two* functions, $f(y_1, y_2) \geq 0$ giving the top half and $-f(y_1, y_2) \leq 0$ giving the bottom half, but that is another story. *Embedding* a piece of the $\theta\varphi$ plane into $\mathbb{R}^3$, as done here by the function G, is another, more general way of producing surfaces. A surface that is the *graph* of a scalar-valued function $y_3 = f(y_1, y_2)$ clearly has at most *one* point in $y_1y_2y_3$-space above each point in the horizontal plane $y_3 = 0$.

2. The last two functions we have seen are examples of nonlinear vector-valued functions, since to each point in a domain they assign a point (vector) in $\mathbb{R}^q$ (with $q = 2$ in the case of the circle map and $q = 3$ in the case of the torus.)

We will see more of these when we discuss curves in Chap. 7 and general vector-valued functions in Chap. 8.

ON COORDINATE FORM: Let us point out something you may have noticed in the examples. If we have a function $F: \mathbb{R}^n \to \mathbb{R}^q$ and if $q \geq 2$, then $F(X)$ is a *vector* with q coordinates. We might write

$$F(X) = (y_1, \ldots, y_q).$$

As X varies, each of $y_1, \ldots, y_q$ may vary; that is, each y_j is a scalar function of X, $y_j = f_j(X)$. Hence we may write

$$F(X) = (f_1(X), \ldots, f_q(X)).$$

We specify F by specifying the q coordinate functions $f_1, \ldots, f_q$.

Thus, in Example 5, the circle map $F(t) = (a \cos t, a \sin t)$ could be given as $F = (f_1, f_2)$, where

$$f_1(t) = a \cos t, \qquad f_2(t) = a \sin t.$$

ON THE PICTURES WE DRAW: You should observe that the kind of picture we draw depends on the function involved:

1. If $f: \mathbb{R}^2 \to \mathbb{R}$, then we may sketch the **graph** of $z = f(X)$ by erecting a vertical z-axis above the plane of vectors X, obtaining $\mathbb{R}^3$. This will be the picture of a surface. See Fig. 4.8.

2. If $F: \mathbb{R} \to \mathbb{R}^2$ (or $F: \mathbb{R} \to \mathbb{R}^3$), giving a curve as the image, then we draw a picture of this curve in the plane, its target space. This is the **image picture**. See Fig. 4.9. Note that this is *not* a graph in the strict sense. The *graph* of $F: \mathbb{R} \to \mathbb{R}^n$ would be the set of all points $(t, F(t))$ with $t \in \mathscr{D}(F)$; this is a subset of $\mathbb{R} \times \mathbb{R}^n = \mathbb{R}^{n+1}$.

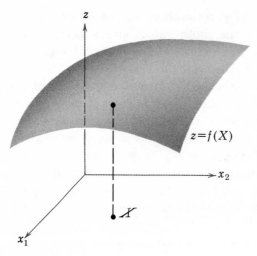

FIGURE 4.8

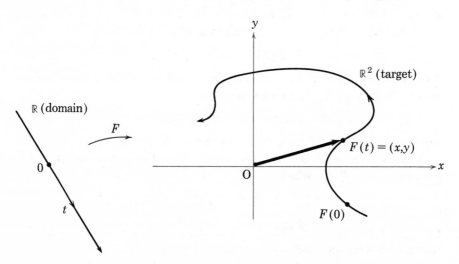

FIGURE 4.9

Note that, as in Examples 5 (the circle) and 6 (the torus), we also drew a picture of the domain of the function as well as of the image. Perhaps, then, a better name would be the **domain-image** picture.

SUMMARY: This is a rich variety of examples. Remember that *all* linear functions were lumped together as Example 1. These alone are indispensible in numerous mathematical and physical problems. But they are not enough; as the examples show, life can be nonlinear, and that's what makes it interesting.

In making a detailed study of nonlinear functions, it is helpful to consider them in smaller classes, as follows:

1. $f\colon \mathbb{R}^n \to \mathbb{R}$. These are scalar-valued functions of several variables $x_1, \ldots, x_n$, or, more compactly, of the vector X.

2. $F\colon \mathbb{R} \to \mathbb{R}^q$. These are vector-valued functions of a scalar and give curves in space, parametrized by a real variable. We discussed the circle in Example 5.

3. $F\colon \mathbb{R}^n \to \mathbb{R}^q$, with $n \geq 1$, $q \geq 1$. These are the vector-valued functions of a vector. An important special case to be studied later is $n = q$, when we interpret $F(X)$ as a **vectorfield** in $\mathbb{R}^n$.

Later, we devote separate chapters to each of these types. Each type of function generates its own problems. For example, we are interested, just as in one-variable calculus, in finding the maxima and minima of scalar-valued functions $f\colon \mathbb{R}^n \to \mathbb{R}$, such as the temperature function $f(X)$ = temperature at X. On the other hand, a study of curves in the plane leads directly to questions of motion along a curved path, velocity of motion, distance traveled.

So much for the differences. It is very important for you to realize, however, that we approach this great variety of nonlinear functions with one and the same approach. Given a function $Y = F(X)$ defined at a point X_0, we construct another, simpler function $Y = T(X)$, the "best affine approximation to F at X_0." This is just what we did in ordinary calculus by constructing the tangent line to the graph of $y = f(x)$. The properties of T hold the key to those of F.

Consequently, we devote the rest of this chapter to matters of general relevance, most notably a discussion of the best affine approximation. We begin this in the following section by speaking of distance, limits, and continuity in $\mathbb{R}^n$.

Exercises

These provide some familiarity with a few nonlinear functions and their graphs.

1.   Is each of the following linear, affine, or neither? Note that most of these maps are given in terms of coordinates:

(a) $f\colon \mathbb{R} \to \mathbb{R}$, $f(x) = 1 + 5x + \cos x$;

(b) $f\colon \mathbb{R}^2 \to \mathbb{R}$, $X = (x_1, x_2)$, $f(X) = 1 + x_1 + x_2$;

(c) $g\colon \mathbb{R}^2 \to \mathbb{R}$, $X = (x_1, x_2)$, $g(X) = x_1 x_2$;

(d) $F: \mathbb{R}^2 \longrightarrow \mathbb{R}^2$, $X = (x_1, x_2)$, $F(X) = (x_1 x_2, 0)$;
(e) $G: \mathbb{R}^2 \longrightarrow \mathbb{R}^2$, $X = (x_1, x_2)$, $G(X) = (1 + x_1 + x_2, \ 2 - x_1)$;
(f) $F: \mathbb{R}^2 \longrightarrow \mathbb{R}^3$, $X = (x_1, x_2)$, $F(X) = (0, x_1, x_1 x_2)$;
(g) $f: \mathbb{R}^2 \longrightarrow \mathbb{R}$, $X = (x_1, x_2)$, $f(X) = x_1$;
(h) $f: \mathbb{R}^3 \longrightarrow \mathbb{R}$, $f(X) = \langle X, X \rangle$;
(i) $g: \mathbb{R}^2 \longrightarrow \mathbb{R}$, $X = (x, y)$, $g(X) = e^{xy}$;
(j) $F: \mathbb{R} \longrightarrow \mathbb{R}^2$, $F(t) = (t, t^2)$;
(k) $G: \mathbb{R} \longrightarrow \mathbb{R}^2$, $G(t) = (t, t)$;
(l) $F: \mathbb{R} \longrightarrow \mathbb{R}^2$, $F(t) = (1 - t, 3 + 5t)$.

2. (a) Sketch graphs (surfaces) of the following nonlinear functions in $x_1 x_2 z$-space; here $X = (x_1, x_2)$: $z = f(X) = x_1^2 + x_2^2$, $z = g(X) = -x_1^2 - x_2^2$, $z = h(X) = 1 - x_1^2 - x_2^2$.
    (b) Locate in your sketch all points $(X, 0) = (x_1, x_2, 0)$ such that $h(X) = 1$. Do the same for $h(X) = 0$, $h(X) = -3$.

3. Sketch a graph of $z = f(X) = x_1 x_2$. To do so, it may help to locate all points $(X, 0) = (x_1, x_2, 0)$ in $x_1 x_2 z$-space such that $f(X) = 0$ and likewise for $f(X) = 1, f(X) = -1$, and so on. These sets of points should look familiar (compare $x_1 x_2 = 1$). They are called **level curves** of the function f.

4. Suppose that temperature in the unit square
    $$\mathscr{S} = \{(x_1, x_2) \mid 0 \le x_1 \le 1, \ 0 \le x_2 \le 1\}$$
    at the point X is given by $f(X) = 1 - x_1^2 - x_2^2$. At what points in $\mathscr{S}$ is the temperature zero? What points of $\mathscr{S}$ are hottest; coolest? (No advanced technique required; just look. We'll discuss other maximum problems in Chap. 6.)

4.2 LIMITS AND CONTINUITY

4.2a Open sets in $\mathbb{R}^n$

We single out certain kinds of subsets of $\mathbb{R}^n$ as being especially useful. The *ball of radius r centered at X_0* is the set

$$B(X_0; r) = \{X \in \mathbb{R}^n \mid \|X - X_0\| < r\}.$$

Here, of course, $r > 0$. Note that $B(X_0; r)$ does not contain those points whose distance from X_0 is either exactly r or greater than r. In the case of $n = 2$, $B(X_0; r)$ consists of all points inside but not on the circle of radius r drawn centered at X_0. It is a *disk*. See Fig. 4.10. Also, if $\mathbb{R}^n = \mathbb{R}$, the real line, then an open ball is just an interval with end points deleted. $B(X_0; r)$ is also called the *open ball*, to contrast it with the *closed ball*:

$$\bar{B}(X_0; r) = \{X \in \mathbb{R}^n \mid \|X - X_0\| \le r\}.$$

Note that $B(X_0; r)$ is a proper subset of $\bar{B}(X_0; r)$.

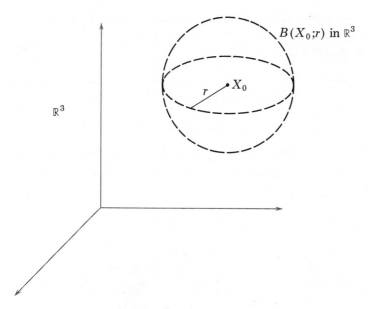

$B(X_0;r)$ in $\mathbb{R}^3$

$\mathbb{R}^3$

r X_0

An open ball in $\mathbb{R}^3$

FIGURE 4.10

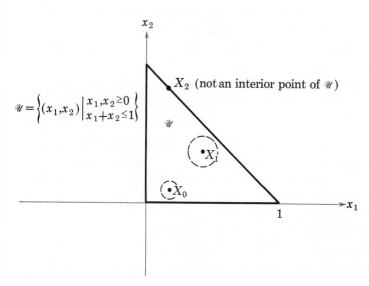

x_2

$\mathcal{U} = \left\{ (x_1,x_2) \,\middle|\, \begin{array}{l} x_1,x_2 \geq 0 \\ x_1+x_2 \leq 1 \end{array} \right\}$

X_2 (not an interior point of $\mathcal{U}$)

$\mathcal{U}$

$\cdot X_1$

$\cdot X_0$

1

x_1

X_0, X_1 are interior points of $\mathcal{U}$

FIGURE 4.11

DEFINITION: The point X of a subset $\mathcal{U}$ of $\mathbb{R}^n$ is an **interior point** of $\mathcal{U}$ if and only if there is an open ball $B(X; r)$ about X which fits inside $\mathcal{U}$; that is, $B(X; r) \subset \mathcal{U}$. See Fig. 4.11.

Intuitively, an interior point of $\mathcal{U}$ is "well inside" $\mathcal{U}$, since it is at the center of a ball contained entirely inside $\mathcal{U}$.

DEFINITION: The subset $\mathcal{U}$ of $\mathbb{R}^n$ is **open** if and only if every point of $\mathcal{U}$ is an interior point of $\mathcal{U}$.

EXAMPLES: 1. The entire space $\mathbb{R}^n$ is open, as is the empty set (having *no* points, *all* its points are interior).

2. Every open ball $B(X_0; r)$ is an open set.

3. In $\mathbb{R}^2$, the following are among the open sets:
$$\mathcal{U}_1 = \{X \in \mathbb{R}^2 \,|\, X \neq O\}, \text{ the "punctured plane"};$$
$$\mathcal{U}_2 = \{X \in \mathbb{R}^2 \,|\, X = (x_1, x_2) \text{ and } x_2 > 0\}, \text{ the upper half-plane};$$
$$\mathcal{U}_3 = \{X \in \mathbb{R}^2 \,|\, X = (x_1, x_2) \text{ } not \text{ on the curve } x_2 = x_1^2\}.$$
You should be able to find many more open sets.

4. The set $\{X \in \mathbb{R}^2 \,|\, X = (x_1, 0)\}$, the x_1-axis, is *not* open.

In case you are wondering, a subset $\mathcal{K}$ of $\mathbb{R}^n$ is *closed* if and only if its complement in $\mathbb{R}^n$, that is, the set of points *not* in $\mathcal{K}$, is open.

Some closed sets in $\mathbb{R}^n$ are $\mathbb{R}^n$ itself, since the empty set is open; the empty set, since $\mathbb{R}^n$ is open; any closed ball $\bar{B}(X_0; r)$; any set consisting of a single point; any set consisting of a finite number of points; any vector subspace not equal to the entire space $\mathbb{R}^n$. You might try proving these, especially the last.

REMARKS: 1. The entire space $\mathbb{R}^n$ and the empty set are both open and closed. No other subsets of $\mathbb{R}^n$ have this property (fortunately).

2. Unlike doors, subsets need not be open or closed. You might try finding a subset of $\mathbb{R}^2$ that is neither open nor closed. There are very many. One is $\{|x_1| \leq 1, |x_2| < 1\}$.

We need one additional concept concerning open sets, the idea that the set consists of only one "piece," not several pieces. The word used for sets consisting of only one "piece" is *connected*. Before formulating a definition, we look at several examples to decide, on the grounds of intuition, if such sets should be called connected. In Fig. 4.12, the set B has two "holes" in it, and the set C consists of two separate "pieces." Incidentally, in some sense, we feel that the set A is more connected than B. In more advanced

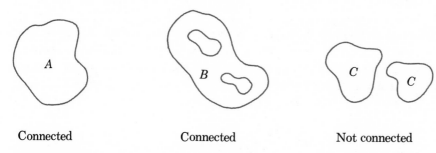

Connected   Connected   Not connected

FIGURE 4.12

work it is useful to distinguish various degrees of connectivity, but we do not need it here.

The official definition of connectivity for open sets is: An open set in $\mathbb{R}^n$ is *connected* if, given any two points in the set, there is a path consisting of straight-line segments joining those two points, and this path lies entirely in the given set. Presumably this definition meets the test of your intuition.

A set that is both open and connected is called a *domain*. See Fig. 4.13.

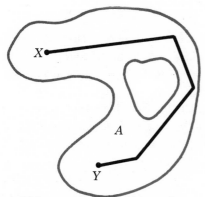

This open set A is connected

FIGURE 4.13

4.2b   Limits in $\mathbb{R}^n$

Let $X_1, X_2, \ldots$ be points in $\mathbb{R}^n$. We want to define what it means for this sequence of points to converge to the point X in $\mathbb{R}^n$. The standard notation for "X_k converges to X as k becomes large" is

$$\lim_{k \to \infty} X_k = X,$$

which is often shortened to $X_k \to X$.

The geometric idea is quite simple: Eventually, as k gets large, the points X_k should get close to the point X. See Fig. 4.14. More precisely, given the radius $\epsilon > 0$, there is a number N such that if $k > N$, then the points X_k are "trapped" in the open ball of radius ϵ about X. Now X_k is in this ball means that its distance from X is less than ϵ, $\|X_k - X\| < \epsilon$. Thus

$$k > N \Rightarrow \|X_k - X\| < \epsilon.$$

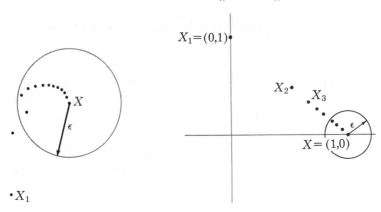

FIGURE 4.14   FIGURE 4.15

EXAMPLE: Let $X_k = (1 - 1/k, 1/k)$ in $\mathbb{R}^2$. We claim that $X_k \to X = (1, 0)$ as $k \to \infty$. Thus, given any radius $\epsilon > 0$, we must find an integer N such that $k > N \Rightarrow \|X_k - X\| < \epsilon$. See Fig. 4.15. To be specific, say that we are given $\epsilon = 1/100$. Now $X_k - X = (-1/k, 1/k)$, so that $\|X_k - X\| = \sqrt{2}/k$. A number N must be found so that if $k > N$, then $\sqrt{2}/k < 1/100$. It is evident that any integer N larger than $100\sqrt{2}$ will do, say $N = 200$. More generally, given any $\epsilon > 0$, we can let N be any integer larger than $\sqrt{2}/\epsilon$. This proves that $X_k \to X$ in our example.

For most uses of this book, an intuitive understanding of the concept of limit suffices. Advanced calculus courses study the concept of limit more thoroughly.

4.2c Continuity

Let $F: \mathbb{R}^n \to \mathbb{R}^q$ and let $X_0 \in \mathbb{R}^n$. Then we say that F is *continuous at* X_0 if and only if:

1. $X_0 \in \mathcal{D}(F)$, so that $F(X_0) \in \mathbb{R}^q$ is defined.
2. $\lim\limits_{X \to X_0} F(X) = F(X_0)$, $X \in \mathcal{D}(F)$.

If F is continuous at each point of its domain, then we say that F is *continuous*. Almost all the functions we encounter in this book are continuous.

Requirement 2 above reads "The limit of $F(X)$, as X approaches X_0 in $\mathscr{D}(F)$, equals $F(X_0)$." What does this mean? Intuitively, it means that whenever X varies in $\mathscr{D}(F)$ along some route so that the distance $\|X - X_0\|$ between X and X_0 tends to zero, then the distance $\|F(X) - F(X_0)\|$ should tend to zero also.

If you wish a more precise statement, you will find one in the following subsection.

EXAMPLES: 1. The function $f \colon \mathbb{R} \to \mathbb{R}$ defined by

$$f(x) = \begin{cases} 1 & 0 < x \leq 1 \\ 0 & \text{otherwise} \end{cases}$$

is defined for $x \in \mathbb{R}$ and is continuous at each x except at the two points $x_0 = 0$, $x_1 = 1$. At each of these points the graph of the function "jumps." See Fig. 4.16.

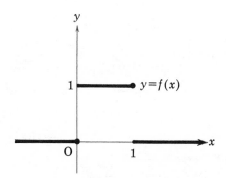

Discontinuities at $x_0 = 0$, $x_1 = 1$

FIGURE 4.16

2. The function $f \colon \mathbb{R}^2 \to \mathbb{R}$ given by

$$f(X) = \begin{cases} 1 & \|X\| \leq 1 \\ 0 & \text{otherwise} \end{cases}$$

is discontinuous at each point $\|X\| = 1$ of the unit circle and continuous elsewhere. See Fig. 4.17. To check discontinuity at X_1, where $\|X_1\| = 1$, let X approach X_1, where X is outside the unit disk, $\|X\| > 1$. Then

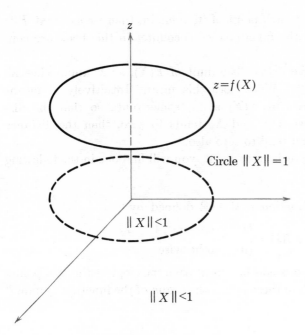

Discontinuities at $\|X\|=1$

FIGURE 4.17

$f(X) = 0$, and so $\lim_{X \to X_1} f(X) = 0$, which is different from $f(X_1) = 1$. Thus f is not continuous at X_1.

Here are some elementary facts about continuous functions.

THEOREM 1: Let $F, G: \mathbb{R}^n \longrightarrow \mathbb{R}^q$ both be continuous at $X_0 \in \mathbb{R}^n$. Then:
1. $\alpha F + \beta G$ is continuous at X_0, for all scalars α, β.
2. The scalar-valued function $\langle F, G \rangle$ is continuous at X_0.

REMARKS: 1. The function $\langle F, G \rangle(X) = \langle F(X), G(X) \rangle$, so that if $F = (f_1, \ldots, f_q)$ and $G = (g_1, \ldots, g_q)$, then $\langle F, G \rangle(X) = f_1(X)g_1(X) + \cdots + f_q(X)g_q(X)$.

2. In particular, if $q = 1$, then $\langle F, G \rangle$ is the usual multiplication of two scalar-valued functions, $\langle F, G \rangle(X) = F(X)G(X)$.

3. We do not speak of F/G because $G(X)$ is a vector (if $q \geq 2$) and *we cannot divide by vectors.*

4. Another fact you might consider is this: The function $F: \mathbb{R}^n \longrightarrow \mathbb{R}^q$, which in coordinate form is $F(X) = (f_1(X), \ldots, f_q(X))$, is continuous if and only if each scalar-valued coordinate function $f_j(X)$ is continuous.

5. Let $f\colon \mathbb{R}^n \to \mathbb{R}$ be a given function, say for $n = 2$, $z = f(x, y)$. It is tempting to believe that if f is continuous in x and y separately, that is, continuous in x for each fixed y, and vice versa, then f is continuous (in both variables together). This is false, unfortunately. An example illustrating this is given in Exercise 5.

4.2d Continuity revisited

For those who wish it, we now use the notion of open ball in $\mathbb{R}^n$ to make the standard, technically precise definition of continuity at a point. We say that $F\colon \mathbb{R}^n \to \mathbb{R}^q$ is *continuous at* X_0 if and only if:

1. $F(X_0)$ is defined.
2. Given any "target" ball $B(F(X_0); \epsilon)$, with $\epsilon > 0$, no matter how small, then we can find a "confidence" ball $B(X_0; \delta)$, with $\delta > 0$ —where the size of δ depends on ϵ, the point X_0, and the function F—that has the property that if X is any point in $B(X_0; \delta)$ and if $F(X)$ is defined, then we may be confident that $F(X)$ is in the target $B(F(X_0); \epsilon)$.

In other words, F must map the entire confidence ball $B(X_0; \delta)$ into the target ball. No matter how small a target $B(F(X_0); \epsilon)$ we are confronted with, we must be able to find such a confidence ball if F is to be continuous at X_0. See Fig. 4.18.

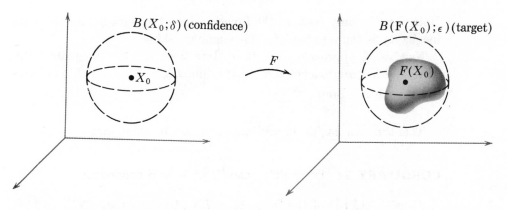

F maps confidence ball inside given target ball.

FIGURE 4.18

4.2e   Linear maps are continuous

To prove this, we use the defining properties (additivity and homogeneity) of linear functions, as well as the triangle inequality in Euclidean space.

THEOREM 2:   Every linear map $L: \mathbb{R}^n \to \mathbb{R}^q$ is continuous.

Proof: Let X_0 be any point of $\mathbb{R}^n$. We must show that $\lim_{X \to X_0} LX = LX_0$; that is, $\|LX - LX_0\|$ approaches zero as $\|X - X_0\|$ approaches zero.

Using the standard basis $e_1, \ldots, e_n$ of $\mathbb{R}^n$, let

$$X = x_1 e_1 + \cdots + x_n e_n, \qquad X_0 = c_1 e_1 + \cdots + c_n e_n.$$

Now we have

$$LX - LX_0 = L(X - X_0) \qquad\qquad\qquad \text{by linearity}$$

$$= (x_1 - c_1) L e_1 + \cdots + (x_n - c_n) L e_n \qquad \text{by linearity,}$$

so that

$$\|LX - LX_0\| \le \|(x_1 - c_1) L e_1\| + \cdots + \|(x_n - c_n) L e_n\|$$

$$= |x_1 - c_1| \, \|L e_1\| + \cdots + |x_n - c_n| \, \|L e_n\| \qquad \text{triangle inequality,}$$

since $\|\alpha Y\| = |\alpha| \, \|Y\|$ for all vectors Y and scalars α.

Now note that $|x_j - c_j| \le \|X - X_0\|$, and choose some positive number M that is greater than any of the n numbers $\|L e_1\|, \ldots, \|L e_n\|$. We may now conclude that

$$\|LX - LX_0\| \le nM \|X - X_0\|.$$

Note that the only trace of the function L on the right-hand side of this inequality is the number M. This simplifies things.

Now as X approaches X_0, it is clear from the last inequality that $\|LX - LX_0\|$ approaches zero; that is, $\lim_{X \to X_0} LX = LX_0$, and L is continuous at each X_0. Done. $\ll$

This theorem has an immediate extension to affine maps.

COROLLARY 3:   Every affine map $T: \mathbb{R}^n \to \mathbb{R}^q$ is continuous.

Proof: $T(X)$ is of the form $Y_0 + LX$. Observe that $T(X) - T(X_0)$ $= LX - LX_0$. Since we proved $\lim_{X \to X_0} LX = LX_0$, we conclude that $\lim_{X \to X_0} T(X) = T(X_0)$. Hence T is continuous at each X_0. Done. $\ll$

Exercises

Some elementary investigations into continuity.

1.   Polynomials in x_1, x_2. Given that $f_1(x_1, x_2) = x_1$ and $f_2(x_1, x_2) = x_2$ are continuous functions from $\mathbb{R}^2$ into $\mathbb{R}$ (in fact, they are linear):

    (a) Which results in this section allow you to conclude the continuity of the function $f(X) = x_1 x_2$ (or $x_1^2 x_2$ or, in general, $c x_1^i x_2^j$ for scalar c and positive integers i, j)?

    (b) Which results in this section allow you to conclude that any polynomial, say $g(X) = x_1 - 7x_2 + x_1 x_2 + x_1^2 x_2^3$, is continuous?

    (c) Let $X_0 = (2, 1)$. What is $\lim\limits_{X \to X_0} g(X)$, where $g(X)$ is the polynomial in Exercise 1(b)?

2.   Let $L: \mathbb{R}^3 \longrightarrow \mathbb{R}^3$, $LX = 7X$. Given $\epsilon = \frac{1}{10}$, find a number $\delta > 0$ such that L maps the confidence ball $B(O, \delta)$ into the ball $B(O; \epsilon)$ in $\mathbb{R}^3$.

3.   (a) Which theorems in this section imply that $F: \mathbb{R}^n \longrightarrow \mathbb{R}$ given by $F(X) = \langle X, X \rangle$ is continuous?

    (b) Same question for $G(X) = \langle X, LX \rangle$, where $L: \mathbb{R}^n \longrightarrow \mathbb{R}^n$ is linear. (The maps F and G here are so-called quadratic forms. They generalize $f(x) = ax^2$.)

4.   Compute $\lim\limits_{X \to X_0} F(X)$, where:

    (a) $X_0 = (1, 2)$, $F: \mathbb{R}^2 \longrightarrow \mathbb{R}$, $F(X) = \langle X, X \rangle$;

    (b) $X_0 = (2, 1)$, $F: \mathbb{R}^2 \longrightarrow \mathbb{R}^2$, $F(X) = (x_1 + x_2, x_1 x_2)$;

    (c) $X_0 = (0, 0)$, $F: \mathbb{R}^2 \longrightarrow \mathbb{R}^3$, $F(X) = (1, e^{x_1}, e^{x_2})$.

5.   This is an example of a function that is continuous in each variable separately but not in both variables together. Let

$$f(X) = f(x, y) = \begin{cases} \dfrac{xy}{x^2 + y^2}, & X = (x, y) \neq 0, \\ 0, & X = 0. \end{cases}$$

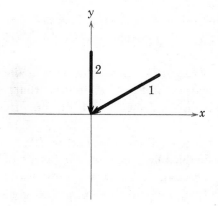

FIGURE 4.19

Since f is the quotient of two continuous functions, it is continuous (in each variable separately *and* in both variables together) except possibly at the origin, where the denominator is zero.

    (a) Show that $f(x, 0)$ is a continuous function of x and that $f(0, y)$ is a continuous function of y (this is clear).

    (b) Show that $f(X)$ is *not* continuous at the origin by computing $\lim f(X)$ as $X \to 0$ along the two paths 1 and 2 (see Fig. 4.19) and showing that

$$\lim_{\substack{X \to 0 \\ \text{path 1}}} f(X) \neq \lim_{\substack{X \to 0 \\ \text{path 2}}} f(X).$$

4.3 THE BEST AFFINE APPROXIMATION: A PREVIEW

4.3a What is differential calculus all about?

The chief triumph of differential calculus of one real variable, due to Barrow, Newton, Leibniz, and their contemporaries, is illustrated in Fig. 4.20. Given a function $f(x)$, it is possible to define, compute, and interpret the tangent line to the graph $y = f(x)$ above each point x_0. This is true provided that the function $f(x)$ is "differentiable" at x_0. Recall that this means that the limit

$$f'(x_0) = \lim_{x \to x_0} \frac{f(x) - f(x_0)}{x - x_0}$$

or, same thing, rewriting x as $x_0 + h$,

$$f'(x_0) = \lim_{h \to 0} \frac{f(x_0 + h) - f(x_0)}{h}$$

exists as a finite number, the **first derivative of** $f(x)$ at x_0.

If $f(x)$ is differentiable at x_0, then the tangent line there is itself the graph of a function, the affine function

$$y = T(x) = f(x_0) + f'(x_0)(x - x_0).$$

To convince yourself that this formula for $y = T(x)$ does give the tangent line to $y = f(x)$ at x_0, you should check three things:

1. The graph $y = T(x)$ is a straight line (affine function) in the xy-plane.
2. The line $y = T(x)$ goes through $(x_0, f(x_0))$, the point of tangency.
3. The line $y = T(x)$ has the correct slope $f'(x_0)$.

If these three criteria are satisfied, then $y = T(x)$ clearly deserves to be called the tangent line to $y = f(x)$ at x_0. *We urge you to verify these criteria; they will appear again.*

Note that most of the calculus involved in obtaining the tangent line centers on the definition and computation of $f'(x_0)$. It follows that a certain understanding of limits is necessary to obtain and interpret the tangent line.

Let us continue. Having the tangent line $y = T(x) = f(x_0) + f'(x_0)(x - x_0)$, it is often helpful to view the affine function $T(x)$ as an *approximation* to $f(x)$ when x is near x_0. This can best be seen geometrically. See Fig. 4.21. Near the point $(x_0, f(x_0))$, the straight line $y = T(x)$ gives a close, snug fit to the curve $y = f(x)$. Hence we are approximating a small

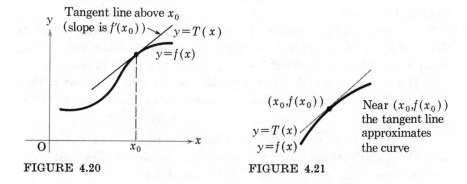

FIGURE 4.20   FIGURE 4.21

arc of the curved graph $y = f(x)$ by a small segment of the straight line $y = T(x)$.

Numerically, it follows that $f(x)$ and $T(x)$ are close, provided that the variable x is close to x_0. In fact, you can show directly from the definition of $f'(x_0)$ that

$$\lim_{x \to x_0} \frac{f(x) - T(x)}{x - x_0} = 0.$$

This means that the numerator $f(x) - T(x)$ must approach 0 even more rapidly than x approaches x_0. Hence the actual value $f(x)$ and the approximate value $T(x)$ are very close when x is close to x_0.

These remarks justify our calling $T(x)$ an *affine approximation* to $f(x)$ at x_0. Why is it the *best* affine approximation at x_0?

The answer is this: If $T_1(x)$ is any other affine function than $T(x)$ $= f(x_0) + f'(x_0)(x - x_0)$, then

$$\lim_{x \to x_0} \frac{f(x) - T_1(x)}{x - x_0} \neq 0.$$

The proof of this is not at all difficult (see the exercises). The meaning is that the difference $f(x) - T_1(x)$ either does not approach 0 at all as x approaches x_0, in which case $f(x_0) \neq T_1(x_0)$, or else $f(x) - T_1(x)$ approaches 0 *slowly* compared with $x - x_0$, in which case the slope of $y = T_1(x)$ is not that of $y = f(x)$ at x_0. In either case, $T_1(x)$ does not give so good an approximation to $f(x)$ as does the *best* approximation $T(x)$.

So much for the philosophy of the tangent line. Let us now recall some of its uses:

1. In seeking the local maxima and minima of $f(x)$, we know that we

need only consider those points x_0 at which the tangent line is horizontal, its slope $f'(x_0) = 0$.

2. In computing the values of fairly complicated functions, such as $f(x) = \sqrt{x}$, it often suffices to compute an approximate value $T(x_0)$ where $y = T(x)$ is the tangent line to $y = f(x)$ at a point x_0 near x.

3. If $y = f(x)$ is thought of as a position on the y-axis at time x, then the slope $f'(x_0)$ of the tangent line is the velocity or rate of change at the instant x_0.

Thus, an accurate answer to the question at the head of this subsection, "What is differential calculus all about?" would be "The best affine approximation" or, more simply, "The tangent line."

4.3b What comes now?

The answer to this is contained in the three pictures following:

1. In Fig. 4.22, we have the graph of the function $f: \mathbb{R}^2 \to \mathbb{R}$ given by $f(X) = x_1^2 + x_2^2$, where $X = (x_1, x_2)$. This graph is a surface in $x_1 x_2 z$-space. Above the point X_0, we have pictured the tangent plane to this surface. This tangent plane is the graph of some affine function $z = T(X)$, where $T: \mathbb{R}^2 \to \mathbb{R}$. We discussed these in Chap. 3. The question now is "Which affine function?"

We see that a different point X_0 would require a different tangent plane; hence the affine function depends on the point X_0. We ask, "In what way?" It is clear at the moment only that $T(X_0) = f(X_0)$.

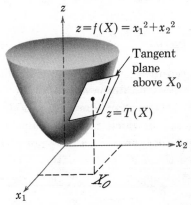

FIGURE 4.22

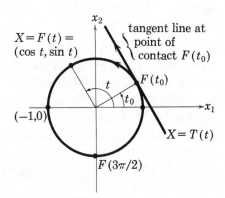

FIGURE 4.23

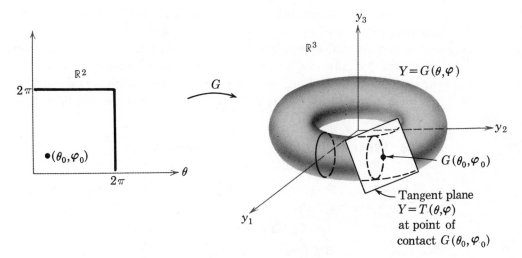

FIGURE 4.24

This tangent plane is, of course, the graph of the *best affine approxima-tion* to $z = f(X)$ at the point X_0. We obtain it in Chap. 5.

2. In Fig. 4.23, we have the unit circle portrayed as the image of the map $F: \mathbb{R} \to \mathbb{R}^2$, where $F(t) = (x_1, x_2)$, with

$$x_1 = \cos t \qquad x_2 = \sin t.$$

At a typical point $F(t_0)$ on the circle we have drawn the tangent line. This, too, is the image of a map, an affine map $T: \mathbb{R} \to \mathbb{R}^2$, $X = (x_1, x_2)$ $= T(t)$. We ask $x_1 = ?$, $x_2 = ?$ in terms of t.

Again, T is the *best affine approximation*, or *tangent map*, to F at the point t_0 of the domain $\mathbb{R}$. We make a precise definition and learn to compute it when we study curves in Chap. 7.

3. In Fig. 4.24, we have pictured the map $G: \mathbb{R}^2 \to \mathbb{R}^3$ whose image is a torus in $\mathbb{R}^3$. The formulae for $(y_1, y_2, y_3) = G(\theta, \varphi)$ were given earlier.

We have also pictured the image of the *best affine approximation*, or *tangent map*, to G at the point (θ_0, φ_0). This is an affine map $T: \mathbb{R}^2 \to \mathbb{R}^3$, $Y = T(\theta, \varphi)$. Since T is affine, its image is a plane, in this case the tangent plane to the torus at the point of contact $Y_0 = G(\theta_0, \varphi_0)$. We learn how to define and compute the affine map T in Chap. 8.

REMARKS: 1. The three examples above illustrate the three main kinds of functions we are considering: scalar-valued functions of a vector, vector-valued functions of a scalar (curves in space), and general vector-valued functions of a vector.

2. Note the common theme in the three examples. Given a nonlinear function and a point in its domain—X_0 or t_0 or (θ_0, φ_0)—we introduce an affine function that approximates the nonlinear one near the given point. Geometrically, the graph (or the image) of the approximating affine function gives a snug fit to the graph (or the image) of the nonlinear function near the point of contact, that is, $(X_0, f(X_0))$ in the graph, or $f(X_0)$ in the image.

4.3c What does the best affine approximation look like?

We recall that the tangent line to $y = f(x)$ at x_0 (ordinary one-variable calculus) is the graph of

$$y = T(x) = f(x_0) + f'(x_0)(x - x_0).$$

For this gives a line through $(x_0, f(x_0))$, and it has the "correct" slope $f'(x_0)$.

Now suppose that we have a nonlinear function $F: \mathbb{R}^n \to \mathbb{R}^q$, where n or q may equal 1, and suppose further that $Y = T(X)$ is the best affine approximation to $F(X)$ at a point X_0 in $\mathbb{R}^n$. What must this affine map look like in terms of F?

Since T is affine, we know that it has the form

$$T(X) = Y_1 + LX,$$

where Y_1 is a fixed vector in $\mathbb{R}^q$ and $L: \mathbb{R}^n \to \mathbb{R}^q$ is linear. Now let us examine Y_1 and L. Since T approximates F near X_0, we have certainly $T(X_0) = F(X_0)$. It follows that

$$Y_1 = T(X_0) - LX_0 = F(X_0) - LX_0.$$

Hence, using the linearity of L, we have

$$\begin{aligned} T(X) &= (F(X_0) - LX_0) + LX \\ &= F(X_0) + L(X - X_0). \end{aligned}$$

In the chapters to come, we define the **total derivative**, or **first derivative,** of F at X_0 to be this linear map L; that is, $F'(X_0) = L$, a linear map (*not* a number). Hence, in beautiful accordance with the familiar one-variable case of

$$y = T(x) = f(x_0) + f'(x_0)(x - x_0),$$

we have

$$Y = T(X) = F(X_0) + F'(X_0)(X - X_0).$$

Now, of course, the right-hand term denotes the operation of the linear map $L = F'(X_0)$ on the vector $X - X_0$.

Our task in the following chapters is to define and compute this total derivative $F'(X_0)$. This is one reason that we have studied linear maps. The remarks above indicate that, having obtained the total derivative and the best affine approximation, we are in a good position to study nonlinear vector functions by exploiting their similarity with the familiar one-variable calculus.

Exercises

These increase your intuitive understanding of the best affine approximation. For the moment we rely on pictures and memories of the one-variable case.

1. One variable. Let $f(x) = 1 - x^2$. Sketch the graph $y = f(x)$ and compute the slope $f'(x_0)$ and best affine approximation $y = T(x) = f(x_0) + f'(x_0)(x - x_0)$ at the following points x_0:

    (a) $x_0 = 0$,   (c) $x_0 = -2$,
    (b) $x_0 = 1$,   (d) $x_0 = 10$.

    Note: You obtain an entirely different function $T(x)$ for each x_0. Sketch them as tangent lines to $y = f(x)$.

2. (a) Sketch the surface $z = f(X) = 1 - x_1^2 - x_2^2$ in $x_1 x_2 z$-space.
    (b) On the same sketch, draw in the tangent plane to this surface at the point $(X_0, f(X_0))$, where $X_0 = (0, 0)$. Compare $(X_0, f(X_0)) = (0, 0, 1)$.
    (c) Likewise for $X_0 = (1, 0)$.
    (d) What affine function $T(X)$ has as its graph the plane sketched in Exercise 2(b)?
    (e) Which of the following is the correct expression for the affine map whose graph is the tangent plane sketched in Exercise 2(c)?

    i. $z = T(X) = 1 - x_1$,
    ii. $z = T(X) = -2x_1 - 2x_2$,
    iii. $z = T(X) = -2(x_1 - 1)$,
    iv. $z = T(X) = 1 - 2(x_1 - 1)$.

    (f) What is $f(X_1)$, where $X_1 = (1, 1)$?
    (g) One of the following planes contacts the surface $z = f(X)$ at the point $(X_1, f(X_1))$, where $X_1 = (1, 1)$ as above. Which one?

    i. $z = 2 - 2(x_1 - 1) - 2(x_2 - 1)$,
    ii. $z = -1 - 2(x_1 - 1) - 2(x_2 - 1)$,
    iii. $z = -2(x_1 - 1) - 2(x_2 - 1)$,
    iv. $z = 1 - 2(x_1 - 1) - 2(x_2 - 1)$.

3. (a) What is the equation of the tangent line to the graph of $y = f(x) = 2 + 3x$ above the point $x_0 = 0$?
    (b) What do you suspect is the equation of the tangent plane to the graph of $z = f(X) = 2 + 3x_1 - x_2$ above $X_0 = (0, 0)$?
    (c) Conjecture a general theorem on the tangent plane to the graph of an affine function. (It's not very surprising.)

4.   (a) Using the obvious substitution

$$x_1 = (x_1 - 5) + 5, \qquad x_2 = (x_2 - 3) + 3,$$

rewrite $z = T(X) = 2 + 2x_1 - x_2$ in the form $z = T(X) = \alpha + \beta(x_1 - 5) + \gamma(x_2 - 3)$.

    (b) What is the meaning of α (in terms of the function T) in the representation above?

    (c) Write the polynomial $z = f(X) = 3 + 2x_1 - x_2 + x_1^2 + 2x_2^2$ in powers of $x_1 + 1$, $x_2 - 1/4$. (In fact, you should obtain the form $z = f(X) = \alpha + \beta(x_1 + 1)^2 + \gamma(x_2 - 1/4)^2$ for certain α, β, γ. Note no linear terms.)

    (d) What is the equation of the tangent plane to the surface $z = f(X)$ above the point $(-1, 1/4)$? Use Exercise 4(c) (and compare Exercise 2).

5.   Let $X_0 = (5, 3)$ and $T(X) = 9 + 2(x_1 - 5) - (x_2 - 3)$ (compare Exercise 4).

    (a) If $T(X) = y_0 + L(X - X_0)$, what are y_0 and $[L]$?

    (b) What is $T(X_0)$?

    (c) If we rewrite $T(X)$ as $y_1 + LX$, what is y_1?

    (d) Are the linear maps L in Exercises 5(a) and (c) the same linear map?

    (e) Does a straight line in the ordinary plane have the same slope at each of its points?

6.   (a) Let $f: \mathbb{R} \longrightarrow \mathbb{R}$, $f(x) = x^2$. Compute the tangent line $T(x) = f(x_0) + f'(x_0)(x - x_0)$ when $x_0 = 1$. Sketch.

    (b) Now let $f: \mathbb{R}^2 \longrightarrow \mathbb{R}$, $f(X) = f(x_1, x_2) = x_2^2$. Note no dependence on x_1. Sketch the surface $z = f(X)$.

    (c) True or false: The tangent plane to $z = f(X)$ above $(0, 1)$ (compare Exercise 6(b)) is the same plane as that above $(x_1, 1)$ for any x_1.

    (d) Compute $z = T(X)$, the tangent plane to $z = f(X)$ at $X_0 = (0, 1)$; that is, $z = T(X) = \alpha + \beta x_1 + \gamma(x_2 - 1)$. Compute α, β, γ. (Actually we do not solve this kind of problem until the following chapter. But this particular example should yield to a well-drawn sketch and Exercise 6(a).)

7.   Prove that, for f differentiable at x_0,

$$\lim_{x \to x_0} \frac{f(x) - \{b + a(x - x_0)\}}{x - x_0} = 0$$

implies $y = b + a(x - x_0)$ is the tangent line to $y = f(x)$ at x_0; that is, $b = f(x_0)$, $a = f'(x_0)$. (Hint: First show $b = f(x_0)$. Then discover the quotient $\dfrac{f(x) - f(x_0)}{x - x_0}$).

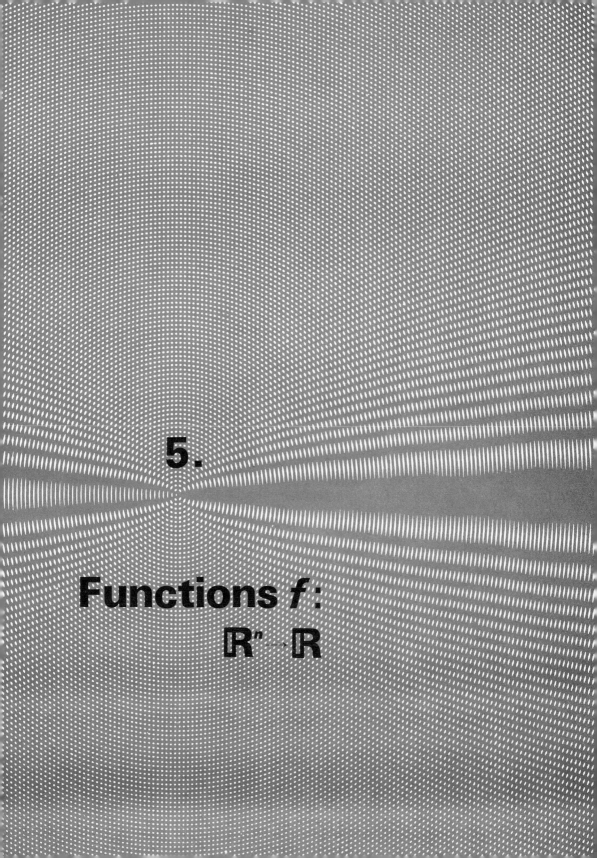

5.

Functions f: $\mathbb{R}^n \to \mathbb{R}$

5.0 INTRODUCTION

In this chapter we concentrate on the differential calculus of a function $f \colon \mathbb{R}^n \to \mathbb{R}$, a so-called scalar-valued function of a vector variable.

We recall that the notation $f \colon \mathbb{R}^n \to \mathbb{R}$ means that to every vector X in a subset $\mathscr{D}(f)$ (the **domain** of f) in $\mathbb{R}^n$ the function f assigns a real number $f(X)$. In most of our examples $\mathscr{D}(f) = \mathbb{R}^n$; the function f is defined for *all* $X \in \mathbb{R}^n$.

Since $X = (x_1, \ldots, x_n)$, one often sees $f(X)$ written $f(x_1, \ldots, x_n)$. Thus f is a function of n (scalar) variables, whence the name "multivariable calculus." Also in two or three dimensions, long custom has established the notation $f(x, y)$ instead of $f(x_1, x_2)$, and $f(x, y, z)$ instead of $f(x_1, x_2, x_3)$. We use these now and again.

Before giving some examples, we remark that formal rigorous proofs of assertions in calculus, of one and several variables, are considerably more difficult than those in linear algebra (the difficulty in the linear algebra we have done lies in its abstraction). Since the new concepts to be presented are already complicated, we often either sketch a difficult proof or eliminate it entirely. The interested reader—we hope that there are many—should look up the deleted proofs in any standard advanced calculus text.

EXAMPLES: 1. *Temperature in a room.* We have seen this before. Let $\mathscr{D}(f)$ denote some room, considered as a subset of a 3-space, and let $f(X)$ be the temperature at the point X. This assigns a real number to each point in the room and thereby defines a function $f \colon \mathscr{D}(f) \to \mathbb{R}$ or, as we shall write, $f \colon \mathbb{R}^3 \to \mathbb{R}$.

2. *The height function of a surface.* We saw this also in the preceding chapter. Let $f \colon \mathbb{R}^2 \to \mathbb{R}$ be given by $f(X) = x_1^2 + x_2^2$, where $X = (x_1, x_2)$. See Fig. 5.1. The graph of $z = f(X)$ in $x_1 x_2 z$-space $\mathbb{R}^3$ is a paraboloid of revolution, and the number $f(X)$ tells us the *height* of this surface above the point $X = (x_1, x_2)$ in the horizontal plane given by $z = 0$.

NOTE: Since functions from $\mathbb{R}^2$ into $\mathbb{R}$ are readily graphed and visualized as well as important in applications, we consider them frequently.

3. *Linear functions.* We recall from Chap. 3 that scalar-valued functions $L \colon \mathbb{R}^n \to \mathbb{R}$ are called functionals. We know that every linear functional is of the form

$$LX = \alpha_1 x_1 + \cdots + \alpha_n x_n,$$

where the α's are fixed coefficients and $X = (x_1, \ldots, x_n)$ is the vector of

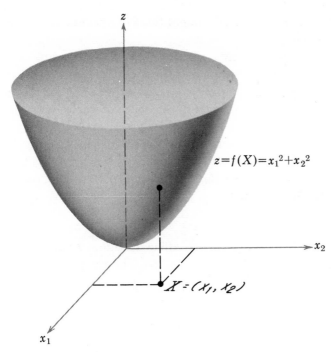

$$z = f(X) = x_1^2 + x_2^2$$

$$X = (x_1, x_2)$$

FIGURE 5.1

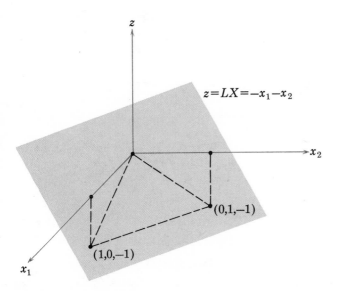

$$z = LX = -x_1 - x_2$$

$(0,1,-1)$

$(1,0,-1)$

The graph of a linear map

FIGURE 5.2

variables. This was done in Chap. 3. (We might think of a general $f: \mathbb{R}^n \to \mathbb{R}$ as being a *nonlinear* functional.)

If $n = 2$, we may write $z = LX = L(x_1, x_2)$ and draw a graph. (Of course, every function *has* a graph, but we cannot hope to draw it unless the dimensions are low.) For instance, if $z = L(x_1, x_2) = -x_1 - x_2$, then the graph is the plane through the origin in the space of triples (x_1, x_2, z) shown in Fig. 5.2. Since $L(1, 0) = -1$, $L(0, 1) = -1$, and $L(0, 0) = 0$, the graph $z = LX$ is the plane determined by the three points $(1, 0, -1)$, $(0, 1, -1)$, and the origin $(0, 0, 0)$.

Incidentally, this graph is, of course, the two-dimensional subspace of $\mathbb{R}^3$ determined by the linear equation $z - L(x_1, x_2) = 0$; that is, $z + x_1 + x_2 = 0$.

4. *Affine functions.* An example of an affine function $T: \mathbb{R}^2 \to \mathbb{R}$ is given by $T(X) = 3 - x_1 - x_2$, where $X = (x_1, x_2)$. The graph of $z = T(X)$

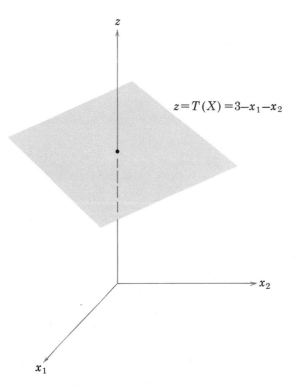

$$z = T(X) = 3 - x_1 - x_2$$

The graph of an affine map

FIGURE 5.3

is the plane in Example 3 translated up the z-axis a distance of three units, since $T(O) = 3$. See Fig. 5.3.

This plane (affine subspace) is determined by the equation $z - T(x_1, x_2) = 0$, which may assume the form $x_1 + x_2 + z = 3$ or $x_1 + x_2 + z - 3 = 0$.

5. *The function* $z = f(x, y) = y^2 - x^2$. The graph of this function is a very fancy surface—a **hyperbolic paraboloid**. See Fig. 5.4. We refer to it often in this chapter.

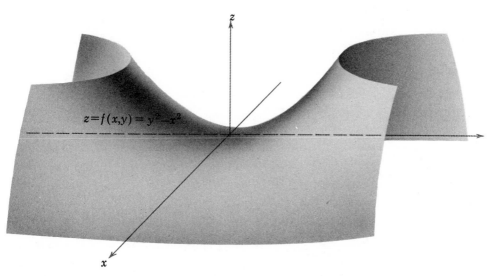

FIGURE 5.4

If this surface is cut by a plane parallel to the xy-plane, say $z = c$, where c is some constant, then the intersection is the curve $c = y^2 - x^2$. See Fig. 5.5. If $c > 0$, then this curve is a hyperbola that opens about the y-axis in the plane of all points (x, y, c). If $c < 0$, the curve determined by $c = y^2 - x^2$ is a hyperbola opening about the x-axis. When $c = 0$, we obtain two straight lines $y = \pm x$.

What happens if we intersect the surface $z = y^2 - x^2$ with planes parallel to the xz-plane?

The intersection of the hyperbolic paraboloid with the plane $y = c$ is a parabola that opens downward in this plane. For it is given by

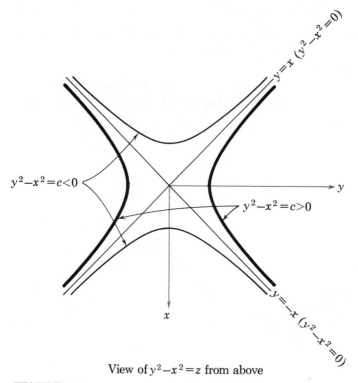

View of $y^2-x^2=z$ from above

FIGURE 5.5

$z = -x^2 + c^2$, which is clearly a parabola for which z decreases as x increases. See Fig. 5.6.

On the other hand, if we cut our surface with a plane $x = c$ parallel to the yz-plane, we obtain a parabola opening upward.

This surface is rightly called a *saddle*, and the origin $O = (0, 0, 0)$ a *saddle point*. Note that a particle standing at the origin can remain at rest there, or move on the surface in one direction and go up, or move on the surface in another direction and go down. We shall encounter saddle points again.

AN OUTLINE OF WHAT FOLLOWS: We recall from Chap. 4 that we wish to construct the tangent plane to a surface given by $z = f(X)$, $X = (x_1, x_2)$. This tangent plane is the graph of the best affine approximation to $f(X)$ at a point X_0. See Fig. 5.7. We approach this problem in easy stages, as follows:

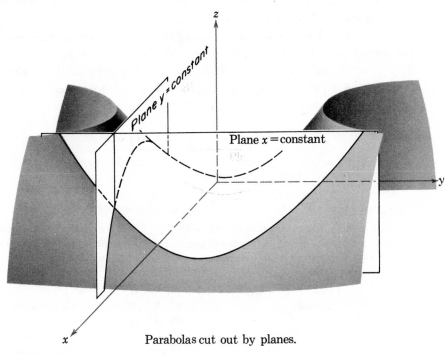

Plane y = constant

Plane x = constant

x Parabolas cut out by planes.

FIGURE 5.6

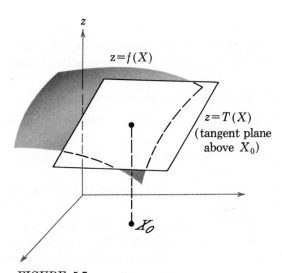

$z = f(X)$

$z = T(X)$
(tangent plane
above X_0)

X_0

FIGURE 5.7

1. The directional derivative $\nabla_e f$. This measures a certain "rate of change" of the function f in the direction of a vector e.

2. The partial derivatives $\partial f/\partial x_1, \ldots, \partial f/\partial x_n$. These are special cases of the directional derivative. They are readily computed using calculus.

3. The total derivative $f'(X_0)$. It is this which yields the best affine approximation $T(X) = f(X_0) + f'(X_0)(X - X_0)$. The total derivative is computed, as we shall see, from the various partial derivatives.

Exercises

1. Sketch the graphs of the following functions regarded as the equations of surfaces $z = f(X)$ in $\mathbb{R}^3$:

(a) $f(X) = 2 - x_1 - 2x_2$,

(b) $f(X) = y - x + 1$,

(c) $f(X) = (x - 1)^2 + y^2 + 2$,

(d) $f(X) = 1 + x^2 - y^2$.

2. Let $z = f(X)$ be the temperature at a point $X \in \mathbb{R}^2$. For each of the following, draw a sketch of $\mathbb{R}^2$ indicating the places where the temperature z is 0, 1, -1, and 4 (thus, if $f(x, y) = x^2 - y$, then the set where $z = 1$ is the parabola $x^2 - y = 1$ in the xy-plane):

(a) $f(x_1, x_2) = 3 + x_1 - 2x_2$,

(b) $f(X) = x_1^2 + x_2^2$,

(c) $f(x, y) = xy - 1$,

(d) $f(x, y) = y^2 - x^2$,

(e) $f(x_1, x_2) = x_1 - x_2^2$,

(f) $f(x, y) = x^2 - 3$.

5.1 THE DIRECTIONAL DERIVATIVE

5.1a Introduction

Think of $f: \mathbb{R}^2 \to \mathbb{R}$ as the temperature function, so that $f(X)$ is the temperature at the point X in the plane. We ask, somewhat vaguely, what is the rate of change of the temperature as one moves from a fixed point X_0? Now the temperature changes at different rates, depending on the direction in which one moves. So we refine our question and ask:

1. What is the rate of change of the temperature f as one moves from a given point X_0 in the direction of a given vector e?

2. Given the point X_0, in what direction e should one move so that the temperature increases (or decreases) most rapidly?

The first question is answered here, the second later on in this chapter (see the "Heat-seeking bug" in Sec. 5.4.)

Say that you are at a fixed point X_0 and intend to move in the direction of the unit vector e toward $X_0 + e$. See Fig. 5.8. Now all points on the straight line joining X_0 to $X_0 + e$ are of the form $X_0 + se$, where s is a real number (for $s = 0$ we are at X_0 and for $s = 1$ we are at $X_0 + e$).

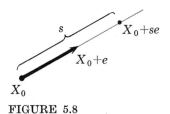

FIGURE 5.8

The difference between the temperature at X_0 and at $X_0 + se$ is $f(X_0 + se) - f(X_0)$. Thus, just as in elementary calculus, we are led to define the **directional derivative** of f at X_0 in the direction of the *unit* vector e, written $\nabla_e f(X_0)$, as the limit of a difference quotient,

$$\nabla_e f(X_0) = \lim_{s \to 0} \frac{f(X_0 + se) - f(X_0)}{s},$$

provided, of course, that the limit does exist. The symbol ∇ is usually called "del."

Note that the directional derivative is a number and that this number depends on the direction e. Although our discussion has been in terms of $\mathbb{R}^2$, there is absolutely no change for the general case $f: \mathbb{R}^n \to \mathbb{R}$. An example should help.

5.1b Computation of $\nabla_e f(X_0)$

Let $f: \mathbb{R}^2 \to \mathbb{R}$ be the function $f(X) = f(x_1, x_2) = x_1^2 + x_2^2$. Compute the directional derivative at $X_0 = (2, 1)$ in the direction of the vector $Z = (3, -4)$.

To solve this, we first find a unit vector e in the direction of Z. It is $e = (3/5, -4/5)$. Then $X_0 + se = (2 + 3s/5, 1 - 4s/5)$, so that

$$f(X_0 + se) = \left(2 + \frac{3s}{5}\right)^2 + \left(1 - \frac{4s}{5}\right)^2$$

$$= 5 + \frac{4s}{5} + s^2.$$

Since $f(X_0) = 5$, we find that $f(X_0 + se) - f(X_0) = 4s/5 + s^2$. Thus

$$\nabla_e f(X_0) = \lim_{s \to 0} \frac{f(X_0 + se) - f(X_0)}{s}$$

$$= \lim_{s \to 0} \frac{4s/5 + s^2}{s}$$

$$= \lim_{s \to 0} \left(\frac{4}{5} + s\right) = \frac{4}{5}.$$

In particular, because $\nabla_e f(X_0) = 4/5$ is positive, we conclude that if one starts at X_0 and moves in the direction of e, then the function f, which may be the temperature at X_0, increases. If temperature $f(X)$ is given in degrees and distance is measured in feet, then the rate of increase at X_0 is $4/5°$ per ft.

This "brute force" computation consists of two parts:

1. Compute the difference quotient explicitly as a function of s.
2. Take the limit as s tends to zero.

Both of these parts may be complicated for complicated functions. Later on, we give a much simpler procedure, so simple that the example above becomes a mental computation.

Right now, we present an alternative view of the procedure above. Although this will somewhat clean up the computation, the reason we present it is for later theoretical purposes. Let e be as above and define $\varphi(s) = f(X_0 + se)$, so that, as s varies, $\varphi(s)$ is the temperature at $X_0 + se$. Now by the definition of the ordinary derivative,

$$\varphi'(0) = \frac{d\varphi}{ds}\bigg|_{s=0} = \lim_{s\to 0} \frac{\varphi(s) - \varphi(0)}{s}.$$

But $\varphi(0) = f(X_0)$. Thus $\varphi(s) - \varphi(0) = f(X_0 + se) - f(X_0)$. Consequently, we find that

$$\varphi'(0) = \lim_{s\to 0} \frac{f(X_0 + se) - f(X_0)}{s},$$

that is,

$$\varphi'(0) = \nabla_e f(X_0).$$

In other words, the directional derivative of f in the direction of the unit vector e is the ordinary derivative of φ evaluated at $s = 0$.

EXAMPLE: The same one as before. There we found that

$$\varphi(s) = f(X_0 + se) = \left(2 + \frac{3s}{5}\right)^2 + \left(1 - \frac{4s}{5}\right)^2.$$

Thus

$$\varphi'(s) = 2\left(2 + \frac{3s}{5}\right)\frac{3}{5} + 2\left(1 - \frac{4s}{5}\right)\left(-\frac{4}{5}\right).$$

Letting $s = 0$, we have

$$\nabla_e f(X_0) = \varphi'(0) = \frac{12}{5} - \frac{8}{5} = \frac{4}{5},$$

just as found earlier. This is easier than the previous method but not as easy as a method to be presented later.

5.1c A geometric interpretation

Let us relate the directional derivative of f to the graph $z = f(X)$. Here we are supposing that X is in $\mathbb{R}^2$. The graph $z = f(X)$ is a surface in

$\mathbb{R}^3$. Given a point X_0 and unit vector e as before, we construct the line of all points $X_0 + se$ in the x_1x_2-plane. Having this, let us construct a vertical plane (parallel to the z-axis) that passes through the x_1x_2-plane along the line $X_0 + se$. See Fig. 5.9.

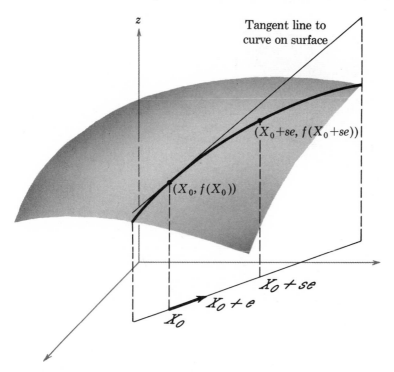

FIGURE 5.9

What is the intersection of this vertical plane with the graph $z = f(X)$? Clearly, it is the set of points on the graph that are directly *above* the points $X_0 + se$ of the line. Hence, this intersection is a curve that looks like the graph of the function $f(X_0 + se)$, the function we called $\varphi(s)$ above.

This gives us an interpretation of the number $\nabla_e f(X_0)$. Since $\nabla_e f(X_0) = \varphi'(0)$, it is the *slope* of the curve obtained by intersecting the graph $z = f(X)$ with the vertical plane through the line $X_0 + se$, computed at X_0, that is, $s = 0$.

This interpretation as slope of a curve underscores the fact that the directional derivative is very much a *one-dimensional* notion. You should convince yourself—it isn't difficult—that $\nabla_e f(X_0)$ depends on the direction e and is, in general, different for different directions e.

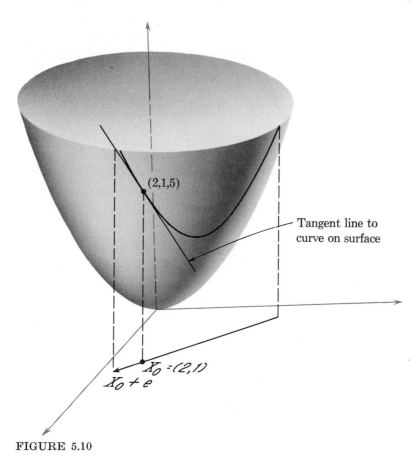

$(2,1,5)$

Tangent line to
curve on surface

$X_0 = (2,1)$

$X_0 + e$

FIGURE 5.10

EXAMPLE: Again the same example as before, $f(x_1, x_2) = x_1^2 + x_2^2$. The graph of this is the surface $z = x_1^2 + x_2^2$, a paraboloid. At $X_0 = (2, 1)$, the height of this surface is $f(X_0) = 5$. The intersection of the vertical plane along the line $X_0 + se$ is a curve—in fact, it is a parabola—whose slope above X_0 is $\nabla_e f(X_0) = 4/5$. As you can see from Fig. 5.10, the slope is positive, showing that the function f is increasing if one moves from X_0 in the direction of the unit vector e.

5.1d   The derivative along a vector

For a fixed point X_0 and *unit* vector e, we have defined the directional derivative of f at X_0 in the direction e to be

$$\nabla_e f(X_0) = \lim_{s \to 0} \frac{f(X_0 + se) - f(X_0)}{s}.$$

Examining the expression on the right-hand side, we see that there is nothing to prevent us from replacing the unit vector e in the difference quotient by any vector Z whatever.

Hence, if $f: \mathbb{R}^n \to \mathbb{R}$, X_0 is an interior point of $\mathscr{D}(f)$ and $Z \in \mathbb{R}^n$, we define the **derivative of f along the vector Z at the point** X_0 to be the number

$$\nabla_Z f(X_0) = \lim_{t \to 0} \frac{f(X_0 + tZ) - f(X_0)}{t},$$

provided that the limit exists.

REMARKS: 1. If Z happens to be a unit vector, then $\nabla_Z f(X_0)$ defined here *is* the directional derivative.

2. It will be useful for theoretical purposes to have the notion of derivative along any vector Z, not just a unit vector.

EXAMPLE: Compute $\nabla_Z f(X_0)$, where $f: \mathbb{R}^3 \to \mathbb{R}$ is given by $f(X) = x_1 x_2 + x_3$, with $X_0 = (1, 1, 1)$, $Z = (2, -3, 1)$.

Solution: We compute from the definition as usual. We have $X_0 + tZ = (1, 1, 1) + t(2, -3, 1) = (1 + 2t, 1 - 3t, 1 + t)$, whence we obtain

$$f(X_0 + tZ) = (1 + 2t)(1 - 3t) + (1 + t)$$
$$= 2 - 6t^2.$$

It follows that the difference quotient

$$\frac{f(X_0 + tZ) - f(X_0)}{t} = -6t,$$

whence $\nabla_Z f(X_0) = 0$, taking the limit as t approaches 0. Done.

You may check for yourself that the derivative $\nabla_{Z_1}(X_0)$, where $Z_1 = (2, -2, 2)$, is *different* from zero.

Here is a question. What is the relation between $\nabla_Z f(X_0)$ and $\nabla_e f(X_0)$, where e is the unit vector pointing in the same direction as Z; that is, $Z = \|Z\| e$?

To answer this, note that

$$\nabla_Z f(X_0) = \lim_{t \to 0} \frac{f(X_0 + t\|Z\|e) - f(X_0)}{t}$$

$$= \|Z\| \lim_{t \to 0} \frac{f(X_0 + t\|Z\|e) - f(X_0)}{t\|Z\|}$$

$$= \|Z\| \nabla_e f(X_0).$$

This is seen by using $s = t\|Z\|$ in the second line above and recalling the definition of $\nabla_e f(X_0)$. Note that s approaches 0 as t approaches 0. The point of all this is the fact that, just as there is no major difference between the unit vector e and the vector $Z = \|Z\|e$ (simply a scalar multiple), there is no major difference between the derivatives $\nabla_e f(X_0)$ and $\nabla_Z f(X_0)$.

Exercises

1. Compute the directional derivatives of each of the following; you may use any of the methods in this section:

   (a) $f(x_1, x_2) = 1 - 2x_1 + 3x_2$ at $(2, -1)$ in the direction $(3, 4)$;
   (b) $q(x, y) = 7 + x^2 - 2y^2$ at $(1, 1)$ in the direction $(1, -1)$;
   (c) $h(x, y) = (1 + x^2 + y^2)^{-1}$ at $(1, 2)$ in the direction $(1, 0)$;
   (d) $f(r, s) = 1 - 3r + rs$ at $(0, 6)$ in the direction $(3, -4)$;
   (e) $k(x, y) = x \sin y$ at $(1, \pi)$ in the direction $(0, 1)$;
   (f) $\varphi(x_1, x_2, x_3) = x_1^2 + 2x_2^2 + 3x_3^2$ at $(1, 0, -1)$ in the direction $(2, 2, 1)$;
   (g) $u(x, y, z) = xyz$ at $(1, 0, 3)$ in the direction $(3, 2, -6)$;
   (h) $p(x, y, z) = xye^{y+z}$ at $(1, 3, -3)$ in the direction $(4, -2, 4)$.

2. (a) Let $z = f(x, y) = x^2 + y^2$ be the temperature at a point $(x, y) \in \mathbb{R}^2$. Draw a sketch indicating the points in $\mathbb{R}^2$ where $z = 0$, $z = 1$, $z = 2$, and $z = 4$.
   (b) Compute the directional derivative of f at $(1, 1)$ in the eight directions $(1, 0)$, $(1, 1)$, $(0, 1)$, $(-1, 1)$, $(-1, 0)$, $(-1, -1)$, $(0, -1)$, and $(1, -1)$.
   (c) In which of these eight directions is the temperature increasing most rapidly; decreasing most rapidly; not changing?
   (d) How might one have deduced the result in Exercise 2(c) from the sketch in Exercise 2(a)?

3. Let $z = f(x, y) = x^2 + y^2$ be the equation of a surface in $\mathbb{R}^3$:

   (a) Draw a sketch of this surface, and indicate on it the places where $z = 0$, $z = 1$, $z = 2$, and $z = 4$.
   (b) Interpret the directional derivatives in Exercise 2(b) geometrically in terms of this surface. In particular, show how the results in Exercise 2(c) are geometrically evident.

4. Repeat all the parts in Exercise 2 for the function $f(x, y) = 4 - 2x - y$.

5. Repeat Exercise 3 for the function $f(x, y) = 4 - 2x - y$.

6. (a) What do you guess is the relation between

$$\nabla_e f(X) \quad \text{and} \quad \nabla_{-e} f(X)?$$

   (b) Prove it.

7. Let $f, g: \mathbb{R}^n \to \mathbb{R}$ be given functions, $Z, \overline{W} \in \mathbb{R}^n$ given vectors, and c a scalar. Show (assuming that all limits exist) that:

   (a) $\nabla_Z (f + g)(X_0) = \nabla_Z f(X_0) + \nabla_Z g(X_0)$,
   (b) $\nabla_Z cf(X_0) = c\nabla_Z f(X_0)$,

(c) $\nabla_Z(fg)(X_0) = f(X_0)\nabla_Z g(X_0) + g(X_0)\nabla_Z f(X_0)$,

(d) $\nabla_{cZ}f(X_0) = c\nabla_Z f(X_0)$.

5.2 THE PARTIAL DERIVATIVES

5.2a Introduction

In this section we define and compute the partial derivatives of a function $f\colon \mathbb{R}^n \to \mathbb{R}$. The definition is straightforward; the partial derivatives are special cases of the directional derivative we studied in the preceding section. They have two advantages, however:

1. They are readily computed by the methods of ordinary calculus (no need to take limits explicitly).

2. They may be used, as we shall see later, to compute the directional derivatives and the total derivative, avoiding the taking of limits in these derivatives as well.

5.2b Definition and computation

Let $f\colon \mathbb{R}^n \to \mathbb{R}$ be the function, and let X_0 be an interior point of the domain $\mathscr{D}(f)$. As usual, we denote a typical point $X \in \mathbb{R}^n$ by $X = (x_1, \ldots, x_n)$. Recall the standard basis vectors $e_1, \ldots, e_n$ of $\mathbb{R}^n$; they are given by

$$e_1 = (1, 0, \ldots, 0), \qquad e_2 = (0, 1, 0, \ldots, 0), \qquad \ldots, \qquad e_n = (0, \ldots, 0, 1).$$

Clearly each of these is a unit vector. Hence we may use any of them to define a directional derivative.

For instance, let us form the directional derivative $\nabla_{e_1}f(X_0)$. This is commonly called the **partial derivative** of f **with respect to** x_1 at the point X_0 and denoted $\partial f/\partial x_1\,(X_0)$. It is a real number, provided that $\nabla_{e_1}f(X_0)$ exists.

More generally, we define the partial derivative of f with respect to x_j at the point X_0 by

$$\frac{\partial f}{\partial x_j}(X_0) = \nabla_{e_j}f(X_0) \qquad j = 1, \cdots, n.$$

Hence at each interior point X_0 of its domain the function $f\colon \mathbb{R}^n \to \mathbb{R}$ has n distinct partial derivatives.

We mentioned above that the partial derivatives are readily computed using ordinary calculus. Let us see why this is so. We have, for instance, when $X_0 = (c_1, c_2, \ldots, c_n)$,

$$\frac{\partial f}{\partial x_1}(X_0) = \nabla_{e_1} f(X_0)$$

$$= \lim_{s \to 0} \frac{f(X_0 + se_1) - f(X_0)}{s}$$

$$= \lim_{s \to 0} \frac{f(c_1 + s, c_2, \ldots, c_n) - f(c_1, c_2, \ldots, c_n)}{s}.$$

For note that $X_0 + se_1 = (c_1, c_2, \ldots, c_n) + s(1, 0, \ldots, 0) = (c_1 + s, c_2, \ldots, c_n)$.

Now let us examine the last limit. Suppose that we form the function $f(x_1, c_2, \ldots, c_n)$ of the single variable x_1; each of the other coordinates slots has been filled by a fixed number $c_2, \ldots, c_n$. Examine the difference quotient from the limit above, namely,

$$\frac{f(c_1 + s, c_2, \ldots, c_n) - f(c_1, c_2, \ldots, c_n)}{s}.$$

Note that the increment s appears only in the *first* coordinate slot of the function f (in contrast with a general directional derivative). Hence f is being differentiated in its first variable only. Thus, in order to compute $\partial f/\partial x_1 (X_0)$, we may proceed as follows (note that $X_0 = (c_1, c_2, \ldots, c_n)$):

1. Form the function $f(x_1, c_2, \ldots, c_n)$ of the single variable x_1.
2. Differentiate this function with respect to x_1 by the ordinary rules of calculus.
3. Let $x_1 = c_1$ in this derivative to obtain a number $\partial f/\partial x_1 (X_0)$.

REMARK: It is clear that the other partial derivatives $\partial f/\partial x_2 (X_0), \ldots, \partial f/\partial x_n (X_0)$ are obtained by an entirely similar procedure. Observe, however, that we cannot compute arbitrary directional derivatives using this three-step procedure.

EXAMPLES: 1. Let $f: \mathbb{R}^2 \to \mathbb{R}$ be given by $f(X) = x_1^2 + 3x_1 x_2 + x_2^3$, and let $X_0 = (c_1, c_2) \in \mathbb{R}^2$. We compute $\partial f/\partial x_1 (X_0)$ as follows:

a. Form $f(x_1, c_2) = x_1^2 + 3x_1 c_2 + c_2^3$, a function of x_1 alone.
b. Differentiate with respect to x_1, obtaining $2x_1 + 3c_2$.
c. Let $x_1 = c_1$, obtaining $\partial f/\partial x_1 (X_0) = 2c_1 + 3c_2$.

For instance, if $X_0 = (1, -2)$, then $\partial f/\partial x_1 (X_0) = -4$.

In practice, one does not explicitly introduce $X_0 = (c_1, c_2)$, but just keep in mind the fact that in computing $\partial f(X)/\partial x_1$ one thinks of $x_2, \ldots, x_n$ as constants. Thus, in Example 1 we would think of x_2 as a constant and compute

$$\frac{\partial f}{\partial x_1}(X) = 2x_1 + 3x_2,$$

which, of course, is essentially the same procedure.

2. Now we compute $\partial f/\partial x_2 (X)$ for the same function. This time we must think of x_1 as a constant. We find

$$\frac{\partial f}{\partial x_2}(X) = 3x_1 + 3x_2^2.$$

In particular, at $X_0 = (1, -2)$, we have $(\partial f/\partial x_2)(X_0) = 15$.

NOTATION: The following alternate notations occur frequently:

$$\frac{\partial f}{\partial x_j}(X) = f_{x_j}(X) = D_{x_j}f(X) = D_jf(X) = f_j(X).$$

Each of these has its virtues and drawbacks. We use the first two notations mainly. Incidentally, when we use $f(x, y, z)$ instead of $f(x_1, x_2, x_3)$, then we write

$$\frac{\partial f}{\partial x}(x, y, z) = f_x(x, y, z),$$

with similar meanings for f_y and f_z.

3. Let $g: \mathbb{R}^3 \longrightarrow \mathbb{R}$ be given by $g(X) = g(x, y, z) = x \sin yz$. We compute g_x by thinking of y and z as constants. Then

$$g_x(x, y, z) = \sin yz.$$

Thus, if $X_0 = (1, \pi, \frac{1}{2})$, then $g_x(X_0) = 1$. Similarly we compute g_y by thinking of x and z as constants. Then

$$g_y(x, y, z) = xz \cos yz,$$

and so $g_y(X_0) = 0$ (recall that the derivative of $\sin ax$ is $a \cos ax$). We leave g_z to you.

5.2c   Rate of change

At this point we know the definition of the partial derivatives of a function $f(X)$ and also how to compute these using ordinary calculus. We now ask, "How do we interpret $f_{x_j}(X_0)$?"

To answer this, we go back to the definition. To be concrete, let $f(X) = f(x_1, x_2)$, $f: \mathbb{R}^2 \rightarrow \mathbb{R}$. The partial derivative $f_{x_1}(X_0)$ is, by definition, $\nabla_{e_1} f(X_0)$, where e_1 is the unit vector $(1, 0)$. From Sec. 5.1, we know that the directional derivative $\nabla_{e_1} f(X_0)$ measures the rate of change of $f(X)$ at X_0 as the variable X moves along the line of points $X = X_0 + se_1$ in the direction pointed out by e_1, that is, in the direction of increasing s.

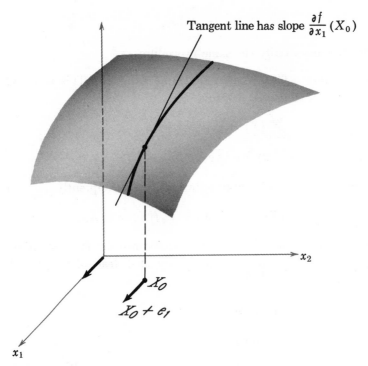

Tangent line has slope $\dfrac{\partial f}{\partial x_1}(X_0)$

x_2

X_0

$X_0 + e_1$

x_1

FIGURE 5.11

Letting $X = (x_1, x_2)$, $X_0 = (c_1, c_2)$, we have, as before, from $X = X_0 + se_1$ that $x_1 = c_1 + s$. It follows that x_1 increases as s increases (see Fig. 5.11), and so we conclude that $f_{x_1}(X_0)$ measures the instantaneous rate of change of $f(X)$ with respect to increasing x_1 at the fixed point X_0, that is, when $x_1 = c_1$, $x_2 = c_2$.

We recall also that there is a more geometric interpretation of $f_{x_1}(X_0)$. If we intersect the graph $z = f(X)$ with the vertical plane $x_2 = c_2$ in $x_1 x_2 z$-space, we obtain a curve. This is the graph of $z = f(x_1, c_2)$, a function of the single variable x_1. We see that the number $f_{x_1}(X_0)$ is actually the *slope* of this curve above the point X_0.

AN APPLICATION: Recall from physics or chemistry the formula

$$PV = nRT,$$

relating pressure P, volume V, and temperature T of an ideal gas enclosed in a cylinder. The numbers n and R are physical constants, fixed throughout our discussion. From the formula, we conclude that changes in any of the variables may cause changes in those remaining. For example, if the temperature T is held constant while the volume V is increased, then it appears that the pressure P must decrease, so that PV continues to equal nRT.

Let us examine this situation, using the calculus. First, we may write the pressure P as a function $P(V, T)$ of volume V and temperature T (= temperature above absolute zero, hence positive):

$$P = P(V, T) = \frac{nRT}{V}.$$

The notation $P(V, T)$ means that pressure depends on volume and temperature. Now we may ask, "What is the rate of change of pressure as the volume is increased, assuming that the temperature is held fixed, say $T = T_0 > 0$?" This is a question about partial derivatives. We have

$$\frac{\partial P}{\partial V}(V, T_0) = \frac{\partial}{\partial V}\left(\frac{nRT_0}{V}\right) = \frac{-nRT_0}{V^2},$$

where we have differentiated nRT_0/V with respect to V. We interpret this computation as follows. Since n, R, T_0, and V^2 are positive, $\partial P/\partial V (V, T_0)$ is negative for each V. Hence the rate of change of pressure with respect to volume (at a fixed temperature) is negative; that is, *pressure decreases as volume increases*. Just as we had expected.

In like manner, you may show that, for a fixed volume V_0, pressure increases as temperature increases by computing the exact rate of change $\partial P/\partial T (V_0, T)$ and verifying that it is positive.

5.2d Higher-order partial derivatives

If $f(x)$ is a differentiable function of one variable, then its first derivative $f'(x)$ may also be a differentiable function. Hence we are led to consider the second derivative $f''(x)$ and derivatives of higher order $f'''(x)$, $f^{(4)}(x)$, $\ldots$, $f^{(k)}(x)$, $\ldots$. A similar situation obtains with partial derivatives. Let $f: \mathbb{R}^2 \to \mathbb{R}, f(X) = f(x_1, x_2)$, for instance. If f has the partial derivatives $\partial f/\partial x_1 (x_1, x_2)$ and $\partial f/\partial x_2 (x_1, x_2)$, then each of them is a function of $X = (x_1, x_2)$. It is therefore natural to ask for the partial derivatives of $\partial f/\partial x_1$, say; that is, we seek $\partial/\partial x_1 (\partial f/\partial x_1)$ and $\partial/\partial x_2 (\partial f/\partial x_1)$.

EXAMPLES: 1. Let $f(x_1, x_2) = 3x_1^2 + x_1 x_2^2 + x_2^5$. Then

$$\frac{\partial f}{\partial x_1}(x_1, x_2) = 6x_1 + x_2^2.$$

Thus

$$\frac{\partial}{\partial x_1}\left(\frac{\partial f}{\partial x_1}\right)(x_1, x_2) = \frac{\partial}{\partial x_1}(6x_1 + x_2^2) = 6.$$

On the other hand,

$$\frac{\partial}{\partial x_2}\left(\frac{\partial f}{\partial x_1}\right)(x_1, x_2) = \frac{\partial}{\partial x_2}(6x_1 + x_2^2) = 2x_2,$$

where, as usual, x_1 is regarded as a constant when the differentiation with respect to x_2 is carried out.

NOTATION: The notation above is clear but cumbersome. Here are some alternates we illustrate with $f(x_1, x_2) = x_1^2 + x_2^5 e^{-x_1} + 9$:

$$f_{11} = f_{x_1 x_1} = \frac{\partial^2 f}{\partial x_1^2} = \frac{\partial}{\partial x_1}\left(\frac{\partial f}{\partial x_1}\right) = 2 + x_2^5 e^{-x_1},$$

$$f_{12} = f_{x_1 x_2} = \frac{\partial^2 f}{\partial x_2 \, \partial x_1} = \frac{\partial}{\partial x_2}\left(\frac{\partial f}{\partial x_1}\right) = -5x_2^4 e^{-x_1},$$

$$f_{22} = f_{x_2 x_2} = \frac{\partial^2 f}{\partial x_2^2} = \frac{\partial}{\partial x_2}\left(\frac{\partial f}{\partial x_2}\right) = 20x_2^3 e^{-x_1},$$

$$f_{21} = f_{x_2 x_1} = \frac{\partial^2 f}{\partial x_1 \, \partial x_2} = \frac{\partial}{\partial x_1}\left(\frac{\partial f}{\partial x_2}\right) = -5x_2^4 e^{-x_1}.$$

2. Let $g(x, y) = x^2 y^3 + x \cos y$. Then $g_x = 2xy^3 + \cos y$ and $g_y = 3x^2 y^2 - x \sin y$. Therefore

$$g_{xx} = \frac{\partial}{\partial x}\left(\frac{\partial g}{\partial x}\right) = 2y^3,$$

$$g_{xy} = \frac{\partial}{\partial y}\left(\frac{\partial g}{\partial x}\right) = 6xy^2 - \sin y,$$

$$g_{yx} = \frac{\partial}{\partial x}\left(\frac{\partial g}{\partial y}\right) = 6xy^2 - \sin y,$$

and g_{yy} is left to you.

There is no reason to stop at second partial derivatives. Sometimes third, fourth, and occasionally even higher-order derivatives appear. The next example illustrates the idea and exhibits various equivalent notation.

3. Let $f(x_1, x_2) = \sin x_1 x_2$. Then $f_{x_1} = x_2 \cos x_1 x_2$ and $f_{x_1 x_1} = -x_2^2 \sin x_1 x_2$, so that

$$f_{112} = f_{x_1 x_1 x_2} = \frac{\partial^3 f}{\partial x_2 \, \partial x_1^2} = \frac{\partial}{\partial x_2}\left(\frac{\partial^2 f}{\partial x_1^2}\right) = \frac{\partial}{\partial x_2}\left(-x_2^2 \sin x_1 x_2\right)$$

$$= -2x_2 \sin x_1 x_2 - x_1 x_2^2 \cos x_1 x_2,$$

$$f_{111} = f_{x_1 x_1 x_1} = \frac{\partial^3 f}{\partial x_1^3} = \frac{\partial}{\partial x_1}\left(\frac{\partial^2 f}{\partial x_1^2}\right) = \frac{\partial}{\partial x_1}\left(-x_2^2 \sin x_1 x_2\right) = -x_2^3 \cos x_1 x_2,$$

$$f_{1112} = f_{x_1 x_1 x_1 x_2} = \frac{\partial^4 f}{\partial x_2 \, \partial x_1^3} = \frac{\partial}{\partial x_2}\left(\frac{\partial^3 f}{\partial x_1^3}\right) = \frac{\partial}{\partial x_2}\left(-x_2^3 \cos x_1 x_2\right)$$

$$= -3x_2^2 \cos x_1 x_2 + x_1 x_2^3 \sin x_1 x_2.$$

In this collection of examples, one striking phenomenon may have attracted your attention. Whenever we computed $f_{x_1 x_2}$ and $f_{x_2 x_1}$ (or g_{xy} and g_{yx}) these "mixed partial derivatives" turned out to be equal. Is this a coincidence, or is there some theorem lurking in the background? There is a theorem.

Now $f_{x_i x_j}$ and $f_{x_j x_i}$ mean, respectively,

$$\frac{\partial}{\partial x_j}\left(\frac{\partial f}{\partial x_i}\right) \quad \text{and} \quad \frac{\partial}{\partial x_i}\left(\frac{\partial f}{\partial x_j}\right),$$

and so we are asking, is

$$\frac{\partial}{\partial x_j}\frac{\partial}{\partial x_i} = \frac{\partial}{\partial x_i}\frac{\partial}{\partial x_j} ?$$

The problem is if one can interchange, or *commute*, the partial derivative operators. In our examples, we could do so. To answer the problem, we must note that the definition of partial derivatives involves the taking of a limit. Thus, at the heart of the issue is the commuting of two limiting processes. It turns out that for essentially all functions that arise in practice, we have $f_{x_1 x_2} = f_{x_2 x_1}$. Here is the official theorem.

THEOREM: (Equality of mixed partials.) Let $f(x_1, \ldots, x_n)$ be a scalar-valued function, and suppose that the second partial derivatives $\partial^2/(\partial x_i \, \partial x_j) f(x_1, \ldots, x_n)$ and $\partial^2/(\partial x_j \, \partial x_i) f(x_1, \ldots, x_n)$ exist as functions of $x_1, \ldots, x_n$. If both of these are continuous functions, then they are equal.

REMARKS: 1. There are examples of functions whose mixed second partial derivatives fail to be continuous at a point and are, in fact, unequal at that point. See Exercise 12.

2. We do not offer a proof of this theorem. It is somewhat technical, although not beyond our means. The basic tool used in the proof is the Mean Value Theorem from ordinary calculus.

3. As mentioned before, most of the functions and their derivatives

we encounter in these chapters are continuous, guaranteeing equality of mixed partials.

4. Interchanging the order of differentiation enables us to compactify some notation. Thus, if f is suitably differentiable,

$$f_{yxy} = \frac{\partial^3 f}{\partial y\, \partial x\, \partial y}$$

may be rewritten as

$$\frac{\partial^3 f}{\partial x\, \partial y^2}, \quad \text{or} \quad \frac{\partial^3 f}{\partial y^2\, \partial x}, \quad \text{or} \quad f_{xyy}, \quad \text{or} \quad f_{yyx}.$$

Here we differentiate with respect to y twice and then with respect to x (or x first and then y twice, as suits us.)

5. In general, interchanging the order of partial differentiation allows us to do all differentiations with respect to one variable before going on to another variable. A typical kth-order derivative would be written

$$\frac{\partial^k}{\partial x_1^{k_1} \cdots \partial x_n^{k_n}} f(x_1, \ldots, x_n),$$

where $k_1 + k_2 + \cdots + k_n = k$. Here we would differentiate k_1 times with respect to x_1, then k_2 times with respect to x_2, and so on.

5.2e   Where we are

In discussing the differential calculus of a scalar-valued function f: $\mathbb{R}^n \to \mathbb{R}$, we have achieved the following thus far:

1. We have defined the directional derivative $\nabla_e f(X_0)$ and the somewhat more general derivative $\nabla_Z f(X_0)$ along the vector Z, although we do not yet have a rapid method of computing them.

2. We have defined the partial derivatives $\partial f/\partial x_1\,(X_0), \ldots, \partial f/\partial x_n\,(X_0)$, and we can compute them readily.

Still open are the following problems:

1. An easy way to compute the derivatives $\nabla_Z f(X_0)$.

2. Construction of a single object analogous to the first derivative of a function in ordinary calculus. This is the "total derivative" $f'(X_0)$, a very handy object; it carries *all* the information about the various partial derivatives and derivatives $\nabla_Z f(X_0)$.

3. Construction of the best affine approximation $T(X)$ to $f(X)$ at the point X_0.

We will see in the following section that all these problems yield to the same attack.

Exercises

1. Compute all the first partial derivatives for the following functions:

(a) $f(x_1, x_2) = x_1 x_2^2 + 2x_1 - 7$;   (f) $f(x, y) = x^y$, $(x > 0)$;

(b) $g(x_1, x_2) = x_1 + x_2 \sin 2x_1$;   (g) $g(x_1, x_2, x_3) = x_1 x_2 x_3$;

(c) $h(x, y) = xe^{x+2y} + y^2$;   (h) $u(x, y, z) = xye^{y+z} + 5$;

(d) $f(s, t) = (1 + 2s^2 + t^2)^{-1}$;   (i) $v(x, y, z) = \log(x^2 + y^2 + z^2 + 3)$;

(e) $g(\theta, \varphi) = \cos(\theta - 3\varphi)$;   (j) $p(r, \theta, \varphi) = r \sin \theta \cos \varphi$.

2. Compute all the second partial derivatives of the functions in Exercise 1.

3. Compute all the third partial derivatives for the functions in Exercise 1 (a), (b), (g), and (h).

4. Find an affine function $f(x, y)$ such that $f_x(x, y) = -1$, $f_y(x, y) = 3$ for all x, y and $f(-1, 2) = 4$.

5. (a) Let $f(x, y) = \sqrt{x^2 + y^2}$. Show that $xf_x + yf_y = f$ and that $yf_x - xf_y = 0$ for $(x, y) \neq O$.

   (b) Let $\varphi(s, t) = ste^{st}$. Show that $s\varphi_s - t\varphi_t = 0$.

6. Show that the following functions satisfy the two-dimensional **Laplace equation** $u_{xx} + u_{yy} = 0$:

   (a) $u(x, y) = x^2 - y^2 - 3xy + 5x - 6$;

   (b) $u(x, y) = \log(x^2 + y^2)$, $(x, y) \neq O$;

   (c) $u(x, y) = e^x \sin y$.

7. Show that the following functions satisfy the two- (space) dimensional **wave equation** $u_{tt} = u_{xx} + u_{yy}$:

   (a) $u(x, y, t) = e^{x-t} - 3e^{y+t} + 7xy - t + 13$;

   (b) $u(x, y, t) = (3x - 4y + 5t)^\alpha$, α a constant.

8. Pick constants α, β so that $u(x, y) = e^{\alpha x + \beta y}$ satisfies

$$u_{xx} - 5u_{xy} + 6u_{yy} = 0.$$

9. Let $f: \mathbb{R}^2 \to \mathbb{R}$ be defined by $f(X) = \langle X, AX \rangle$, where

$$A = \begin{bmatrix} a & b \\ b & c \end{bmatrix}$$

is a constant matrix. Show that

$$\frac{\partial f}{\partial x_1} = 2\langle AX, e_1 \rangle, \qquad \frac{\partial f}{\partial x_2} = 2\langle AX, e_2 \rangle,$$

where $e_1 = (1, 0)$, $e_2 = (0, 1)$. Note that this generalizes $f(x) = ax^2$.

10. (a) Let $f: \mathbb{R}^2 \to \mathbb{R}$ have a maximum at X_0, so that $f(X_0) \geq f(X)$ for all X. Show that if all relevant derivatives exist,

$$\frac{\partial f}{\partial x_1}(X_0) = 0, \quad \text{and} \quad \frac{\partial f}{\partial x_2}(X_0) = 0.$$

(Suggestion: If $X_0 = (a, b)$, consider how f varies on the line $x_2 = b$ and on $x_1 = a$.)

(b) Generalize this to $f: \mathbb{R}^n \to \mathbb{R}$.

11.   This is an example of a function $f: \mathbb{R}^2 \to \mathbb{R}$ such that f_x and f_y exist everywhere, but f is not continuous everywhere. Let

$$f(x, y) = \begin{cases} \dfrac{2xy^2}{x^2 + y^4}, & (x, y) \neq 0; \\ 0, & (x, y) = 0. \end{cases}$$

(a) Use the definition of f_x and f_y as limits to show that f_x and f_y both exist at the origin. These derivatives clearly exist for all $(x, y) \neq 0$ also.

(b) Prove that f is discontinuous at the origin by showing $\lim\limits_{X \to 0} f(X)$ has two different values as $X \to 0$ along the x-axis and along the curve $x = y^2$.

12.   This is an example of a function $f: \mathbb{R}^2 \to \mathbb{R}$ such that f_x, f_y, f_{xy}, and f_{yx} all exist but $f_{xy}(0) \neq f_{yx}(0)$. Define f by

$$f(x, y) = \begin{cases} \dfrac{xy(x^2 - y^2)}{x^2 + y^2}, & (x, y) \neq 0; \\ 0, & (x, y) = 0. \end{cases}$$

(a) Prove that $f_x(0, y) = -y$ for all y and $f_y(x, 0) = x$ for all x. (Suggestion: Use the definition of the derivative as a limit.)

(b) Prove that $(f_x)_y(0, y) = -1$ and $(f_y)_x(x, 0) = 1$ and conclude that $f_{xy}(0) \neq f_{yx}(0)$.

13.   (a) If $u(x, t)$ denotes the displacement, say in centimeters, of a vibrating string at a point x on the string at time t, how would you physically interpret the functions $u_t(x, t)$, $u_{tt}(x, t)$, and $u_x(x, t)$?

(b) Let $u(x, t) = 3 \sin 2x \cos 2\pi t$ be the displacement of a vibrating string of length π stretched between $x = 0$ and $x = \pi$. What can you deduce about the endpoints of the string? What is the initial (at $t = 0$) position of the string? What is the initial velocity of the string? What is the velocity at $x = \pi/4$ for $t = 3$? What is the slope of the string at $x = \pi/4$ for $t = 3$? Draw sketches showing the position of the string at $t = 0, t = 1/4, t = 1/2$, and $t = 3$.

5.3   TANGENCY AND AFFINE APPROXIMATION

5.3a   Introduction

We have pictured, in Fig. 5.12, the graph of the function $f: \mathbb{R}^2 \to \mathbb{R}$, $z = f(X) = 1 - x_1^2 - x_2^2$, where $X = (x_1, x_2)$. Above a point X_0 in $\mathbb{R}^2$ we have drawn what looks like the tangent plane to the surface $z = f(X)$. This tangent plane is the graph of an affine function $z = T(X)$. We do not yet know *which* affine function this is. In this section we make a reasonable definition of $T(X)$, the best affine approximation to $f(X)$ at the point X_0. We accomplish this by examining the one-dimensional tangent line closely and deciding how this should be generalized. The definition we make will apply to functions of $X = (x_1, \ldots, x_n)$, where n may be larger

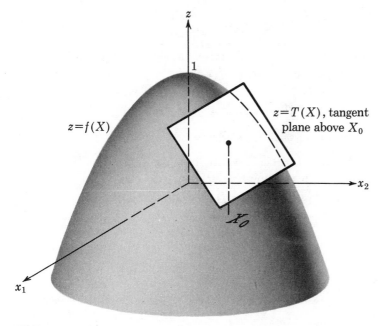

FIGURE 5.12

than two; however, most of our examples and discussion take place in two dimensions, $X = (x_1, x_2)$.

Even after defining $z = T(X)$ by the appropriate limit requirement, we will *not* yet be able to compute it in a particular case. The explicit computation of the best affine approximation is carried out later in this chapter, after we have reexamined the partial derivatives of $f(X)$.

Why do we want the best affine approximation $T(X)$? One reason is this: Since $T(X)$ is affine, its graph $z = T(X)$ is a plane, in the case of $X = (x_1, x_2)$. If we lived on the curved surface $z = f(X)$ near the point $(X_0, f(X_0))$, we would not be able to distinguish this surface from the "flat" plane $z = T(X)$ (the earth is flat, as every child knows). But the function $T(X)$, being affine, is much easier to handle than the nonlinear $f(X)$; we have all linear algebra working for us in the affine case.

5.3b Tangency in $\mathbb{R}^1$

Let us review some one-variable calculus. Let $y = f(x)$ be differentiable at x_0, an interior point of $\mathscr{D}(f)$. This means that

$$f'(x_0) = \lim_{x \to x_0} \frac{f(x) - f(x_0)}{x - x_0}$$

exists as a finite number. Then the tangent line to the graph of $y = f(x)$ at x_0, that is, through the point $(x_0, f(x_0))$, is the graph of the affine function

$$y = T(x) = f(x_0) + f'(x_0)(x - x_0).$$

To convince yourself of this, note that:

1. The graph $y = T(x)$ is a straight line. See Fig. 5.13.
2. It passes through the point $(x_0, f(x_0))$.
3. It has the same slope as $y = f(x)$ at x_0, namely, $f'(x_0)$.

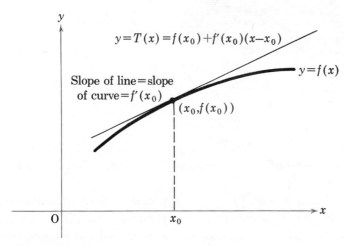

FIGURE 5.13

We have used this formulation of the tangent line before and shall do so again.

Let us be slightly more general in our discussion of differentiability. We seek criteria relating the differentiability of f on the one hand and its tangent line $y = T(x)$ on the other. Consider the following assertion.

THEOREM 1: (Tangency in one variable). Let $f: \mathbb{R}^1 \to \mathbb{R}^1$, with x_0 an interior point of $\mathcal{D}(f)$. Let $T: \mathbb{R}^1 \to \mathbb{R}^1$ be an affine map. Then the following statements are equivalent:

1. f is differentiable at x_0, and its tangent line at x_0 is the graph $y = T(x)$.
2. f is differentiable at x_0 and $T(x) = f(x_0) + f'(x_0)(x - x_0)$.
3. $f(x_0) = T(x_0)$, and $\lim\limits_{x \to x_0} \dfrac{f(x) - T(x)}{x - x_0} = 0$.

REMARK: The crucial point here is that statement 3 makes no mention of $f'(x_0)$, the differentiability of f, or the existence of a tangent line at x_0. But it implies all these things. Hence we may, and shall, use it in higher dimensions to *define* differentiability, the tangent plane, and so on.

Proof:

$1 \Leftrightarrow 2$. We saw this above.

$2 \Rightarrow 3$. Given $T(x) = f(x_0) + f'(x_0)(x - x_0)$, it is clear that $T(x_0) = f(x_0)$. Also, we have

$$\lim_{x \to x_0} \frac{f(x) - T(x)}{x - x_0} = \lim_{x \to x_0} \frac{f(x) - f(x_0) - f'(x_0)(x - x_0)}{x - x_0}$$

$$= \lim_{x \to x_0} \frac{f(x) - f(x_0)}{x - x_0} - f'(x_0)$$

$$= f'(x_0) - f'(x_0) = 0,$$

as claimed. This shows that $2 \Rightarrow 3$.

$3 \Rightarrow 2$. Since $T(x)$ is affine, it has the form

$$T(x) = y_0 + m(x - x_0)$$

for scalars y_0, m (think about this!). But $y_0 = T(x_0) = f(x_0)$. It remains to show that f is differentiable and $m = f'(x_0)$.

Now we are given that

$$0 = \lim_{x \to x_0} \frac{f(x) - T(x)}{x - x_0} = \lim_{x \to x_0} \left(\frac{f(x) - f(x_0)}{x - x_0} \right) - m,$$

using $y_0 = f(x_0)$. Since m is finite, the limit on the right must exist and equal m. But this limit is $f'(x_0)$, whence f is differentiable at x_0 and also $T(x) = f(x_0) + f'(x_0)(x - x_0)$. This shows that $3 \Rightarrow 2$ and completes the proof of the theorem. $\ll$

Statement 3 gives the analytic criteria for tangency. We encourage you to try to get a feeling for the meaning of $\lim\limits_{x \to x_0} \dfrac{f(x) - T(x)}{x - x_0} = 0$. This implies that $\lim\limits_{x \to x_0} (f(x) - T(x)) = 0$ but is much stronger; it shows that as the denominator $x - x_0$ gets small, the numerator $f(x) - T(x)$ gets smaller faster, so that even the ratio $\dfrac{f(x) - T(x)}{x - x_0}$ gets small.

Geometrically, this forces the graph $y = f(x)$ to be close to $y = T(x)$ when x is close to x_0. In other words, $y = f(x)$ has the tangent line $y = T(x)$.

REMARK: We could define two functions $f(x)$ and $g(x)$, neither of them affine, to be **tangent at** x_0 provided that $f(x_0) = g(x_0)$ and $\displaystyle\lim_{x \to x_0} \frac{f(x) - g(x)}{x - x_0}$ $= 0$. An example of the tangency of functions at $x_0 = 0$ is given by $f(x)$ $= x^2$, $g(x) = x^3$. See Fig. 5.14. Two differentiable functions that are tangent at x_0 have the same tangent line $y = T(x)$ there. In the case $f(x) = x^2$, $g(x) = x^3$, $x_0 = 0$, the common tangent line is the x-axis, the graph of $y = T(x) \equiv 0$.

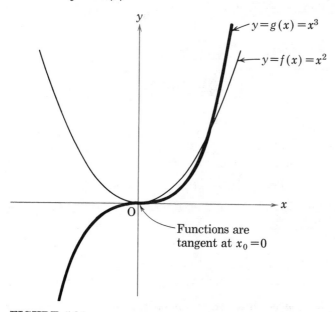

FIGURE 5.14

Our chief concern, however, is not this more general notion of tangency of arbitrary functions but rather the tangency of $f(x)$ with an affine function $T(x)$.

5.3c The basic definitions

Guided by our close analysis of the ordinary calculus, we are almost ready to define the best affine approximation and total derivative of a scalar-valued function of a vector.

First, however, let $T : \mathbb{R}^n \to \mathbb{R}$ be any affine map. Then we may put $T(X)$ in the form

$$T(X) = T(X_0) + L(X - X_0).$$

Here X_0 is a fixed point in $\mathbb{R}^n$, and $L\colon \mathbb{R}^n \to \mathbb{R}$ is a linear functional. Writing T this way facilitates our understanding of its properties for X near X_0 (note that if we write $y_1 = T(O)$, the usual representation $T(X) = y_1 + LX$ is useful near $X_0 = O$).

And now we proceed to the business at hand.

DEFINITION: Let $f\colon \mathbb{R}^n \to \mathbb{R}$, and let X_0 be an interior point of $\mathcal{D}(f) \subset \mathbb{R}^n$. We say that f is **differentiable at** X_0 if and only if there exists an affine map $T\colon \mathbb{R}^n \to \mathbb{R}$ that is tangent to f at X_0; that is,

1. $f(X_0) = T(X_0)$.

2. $\displaystyle \lim_{X \to X_0} \frac{f(X) - T(X)}{\|X - X_0\|} = 0.$

If this is the case, then $T(X)$ is the **best affine approximation**, or **tangent map**, to f at X_0. Moreover, $T(X) = T(X_0) + L(X - X_0)$, and the linear map $L\colon \mathbb{R}^n \to \mathbb{R}$ is called the **total derivative**, or **first derivative**, of f at X_0 and denoted $f'(X_0)$.

REMARKS: 1. This definition bears a strong resemblance to statement 3 in Theorem 1.

2. In statement 2 above, however, the denominator is $\|X - X_0\|$, a *number* and not a vector. We *never* divide by vectors.

3. If $T(X)$ is the best affine approximation to $f(X)$ at X_0, then $T(X_0) = f(X_0)$ and $L = f'(X_0)$, whence, in beautiful accordance with the one-variable case, recalling that

$$T(x) = f(x_0) + f'(x_0)(x - x_0),$$

we now have, in n dimensions,

$$T(X) = f(X_0) + f'(X_0)(X - X_0).$$

Now, of course, the right-hand term denotes operation of the linear map $f'(X_0)$ on the vector $X - X_0$.

4. The following questions now become urgent:

    a. How can we decide if a given f is differentiable at X_0?

    b. If f is differentiable there, is its best affine approximation $T(X)$ unique? Or could there be two "best" affine approximations?

    c. If $T(X)$ exists and is unique, how do we compute it?

The answer to question *b* is that $T(X)$ *is* unique. The proof is not particularly difficult, but we omit it. We will resume consideration of the other questions after examining some particular instances of the best affine approximation.

EXAMPLES: 1. Let $f(X) = f(x_1, x_2) = 1 - x_1^2 - x_2^2$, the function mentioned in the introduction to this section. Let $X_0 = (1, 2) \in \mathbb{R}^2 \ (= \mathcal{D}(f))$. Consider the map $T(X) = T(x_1, x_2) = -4 - 2(x_1 - 1) - 4(x_2 - 2)$. We claim that this map is the best affine approximation to f at X_0. (You may well ask where we obtained this $T(X)$. This is another story, to be told shortly.)

a. We must verify the criteria for best affine approximation. Note first that $T(X)$ is affine; it has the form

$$T(X) = T(X_0) + L(X - X_0),$$

with $T(X_0) = -4$, $X - X_0 = (x_1 - 1, x_2 - 2)$, and $L(X - X_0) = -2(x_1 - 1) - 4(x_2 - 2)$, so that $LX = -2x_1 - 4x_2$ is linear in x_1, x_2.

b. Now we observe that $T(X_0) = -4 = f(X_0)$, so that statement 1 of the definition is satisfied.

c. To check statement 2, namely, $\lim\limits_{X \to X_0} \dfrac{f(X) - T(X)}{\|X - X_0\|} = 0$, we first put $f(X)$ into a more useful form with respect to the point $X_0 = (1, 2)$; namely, we rewrite it in powers of $x_1 - 1$ and $x_2 - 2$. This is accomplished as follows: We have, surely,

$$x_1 = (x_1 - 1) + 1 \quad \text{and} \quad x_2 = (x_2 - 2) + 2,$$

so that

$$\begin{aligned}
f(X) &= 1 - \{(x_1 - 1) + 1\}^2 - \{(x_2 - 2) + 2\}^2 \\
&= 1 - \{(x_1 - 1)^2 + 2(x_1 - 1) + 1\} - \{(x_2 - 2)^2 + 4(x_2 - 2) + 4\} \\
&= -4 - 2(x_1 - 1) - 4(x_2 - 2) - (x_1 - 1)^2 - (x_2 - 2)^2.
\end{aligned}$$

Although this expression looks more cluttered, it will be simplified greatly in a moment.

Now note that $f(X) - T(X) = -(x_1 - 1)^2 - (x_2 - 2)^2$. In fact the constant (zero-th order) and first-order terms of $f(X)$ are identical to those of $T(X)$, whence T *is* an affine or first-order approximation to f at X_0. The limit has become

$$\begin{aligned}
\lim_{X \to X_0} \frac{-\{(x_1 - 1)^2 + (x_2 - 2)^2\}}{\|X - X_0\|} &= \lim_{X \to X_0} \frac{-\|X - X_0\|^2}{\|X - X_0\|} \\
&= \lim_{X \to X_0} -\|X - X_0\| = 0,
\end{aligned}$$

as expected.

Since statements 1 and 2 are verified, we may conclude that $T(X)$ $= -4 - 2(x_1 - 1) - 4(x_2 - 2)$ is the best affine approximation to $f(X)$ at the point $X_0 = (1, 2)$. Also, the map L equals the map $f'(X_0)$, the total derivative. Thus $f'(X_0)(X - X_0) = -2(x_1 - 1) - 4(x_2 - 2)$.

Since $L(X - X_0) = -2(x_1 - 1) - 4(x_2 - 2)$, we see that we have also found the total derivative $f'(X_0)$. It is, by definition, this linear map L. Now $L: \mathbb{R}^2 \to \mathbb{R}$, and so the matrix representing this map is $[L] =$ $[-2 \quad -4]$; that is, the total derivative matrix is

$$[f'(X_0)] = [-2 \quad -4].$$

2. Let us verify our geometric intuition regarding $f(X) = 1 - x_1^2 - x_2^2$. From the graph in Fig. 5.15, we expect that its best affine approximation at the point $X_0 = (0, 0)$, a different choice of X_0 from above, is the constant map $z = T(X) = 1$ whose graph is the horizontal plane through the point $(0, 0, 1)$ of $x_1 x_2 z$-space.

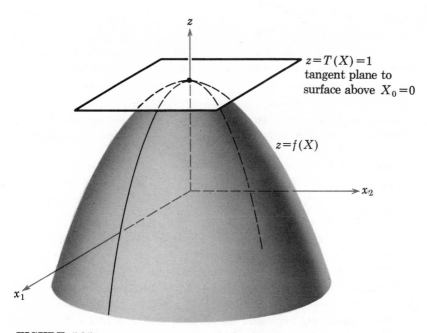

$z = T(X) = 1$
tangent plane to
surface above $X_0 = 0$

$z = f(X)$

FIGURE 5.15

a. We note first that $z = T(X) = 1$ is affine.

b. We have $T(0) = 1 = f(0)$, so that statement 1 is satisfied at $X_0 = 0$.

c. To verify statement 2, note that

$$\frac{f(X) - T(X)}{\|X - X_0\|} = \frac{-x_1^2 - x_2^2}{\|X\|} = -\frac{\|X\|^2}{\|X\|} = -\|X\|.$$

Clearly, then,

$$\lim_{X \to 0} \frac{f(X) - T(X)}{\|X\|} = 0.$$

Thus, computation and intuition agree, reassuring us that we have made the "correct" definition of best affine approximation (recall that definitions are made by men and are rejected if they do not prove worthy).

Finally, the linear part of $T(X) = 1$ is the linear map $L = 0$, since $T(X) = 1 + 0x_1 + 0x_2$. Therefore, the total derivative matrix of f is

$$[f'(0)] = [0 \quad 0].$$

It is pleasant that the total derivative is, as we shall see, always 0 whenever the tangent plane is horizontal, just as in elementary calculus. This is critical in max-min problems.

Exercises

1. For each of the functions $f \colon \mathbb{R} \longrightarrow \mathbb{R}$, find $f'(x_0)$ and the tangent line to the graph of the curve $y = f(x)$ at x_0. Then verify that assertion 3 in Theorem 1 is true. Draw a sketch illustrating the graphs of $f(x)$ and $T(x)$:

    (a) $f(x) = x^2$, $x_0 = 1$;   (c) $f(x) = 2/(1 + x^2)$, $x_0 = -1$;
    (b) $f(x) = 4/x$, $x_0 = 2$;   (d) $f(x) = 2 - x^{7/3}$, $x_0 = 0$.

2. For each of the following functions $f \colon \mathbb{R}^2 \longrightarrow \mathbb{R}$, verify that the given function $T(X)$ is the best affine approximation to f at X_0, and exhibit the matrix of the total derivative at X_0:

    (a) $f(x_1, x_2) = 3 + 2x_1^2 - 4x_1 + 2x_2^2$, $T(X) = 1$, $X_0 = (1, 0)$;
    (b) $f(x_1, x_2) = 1 + 2x_1^2 - 6x_1 + 2x_2^2$, $T(X) = 3 - 6x_1 + 4(x_2 - 1)$,
      $X_0 = (0, 1)$;
    (c) $f(x, y) = x^2 - 3x - 7$, $T(X) = -8 - x$, $X_0 = (1, -17)$;
    (d) $f(x, y) = x^4 + y^2$, $T(x, y) = -2y - 1$, $X_0 = (0, -1)$;
      (Suggestion: Observe that if $|x| \le 1$, then $x^4 \le x^2$.)
    (e) $f(x, y) = xy + 2x - y - 2$, $T(x, y) = -2 + 2x - y$, $X_0 = (0, 0)$.
      (Suggestion: Prove and use the easy inequality $2xy \le x^2 + y^2$.)

3. Let f and $g \colon \mathbb{R}^n \longrightarrow \mathbb{R}$ be differentiable at X_0, and let their total derivatives at X_0 be $f'(X_0)$ and $g'(X_0)$:

    (a) Guess an obvious expression for the total derivative of $h = f + g$ at X_0 in terms of $f'(X_0)$ and $g'(X_0)$. Then guess an expression for the best affine approximation $T(X)$ of h at X_0 in terms of f and g. (Use your intuition for maps $f\colon \mathbb{R} \longrightarrow \mathbb{R}$.)

(b) Use the guess above to prove that h is, in fact, differentiable at X_0 with best affine approximation $T(X)$.

4.   Let $f\colon \mathbb{R}^2 \longrightarrow \mathbb{R}$ have the form $f(x, y) = g(x)$; that is, f does not depend on y. If g, a function of one variable, is differentiable at x_0, prove that f is differentiable at $X_0 = (x_0, y_0)$ by verifying that the affine function $T(x, y) = g(x_0) + g'(x_0)(x - x_0)$ is the best affine approximation to f at X_0. What is $f'(X_0) = ?$

5.   (a) Let $f\colon \mathbb{R}^2 \longrightarrow \mathbb{R}$ have the form $f(x,y) = g(x) + h(y)$. Prove that f is differentiable at $X_0 = (x_0, y_0)$ if both g and h are differentiable at x_0 and y_0, respectively. You will need first to guess a formula for $f'(X_0)$ in terms of $g'(x_0)$ and $h'(y_0)$.

(b) Use the formula you just found for $f'(X_0)$ to obtain $f'(X_0)$ and the best affine approximations for the functions in Exercises 2(a) to (d).

6.   If a function $f\colon \mathbb{R}^n \longrightarrow \mathbb{R}$ is differentiable at X_0, prove that f is continuous there; that is, $\lim\limits_{X \to X_0} f(X) = f(X_0)$.

7.   Repeat Exercise 3, replacing the function h there by $h = fg$.

8.   Let $f\colon \mathbb{R}^n \longrightarrow \mathbb{R}$. Assume that there is an affine function $T(X)$ such that $T(X_0) = f(X_0)$ and

$$\lim_{X \to X_0} \frac{f(X) - T(X)}{\|X - X_0\|^2} = 0.$$

Must f be differentiable at X_0? Proof or counterexample.

5.4 THE MAIN THEOREM ON DIFFERENTIABLE FUNCTIONS

5.4a Introduction

Let $f\colon \mathbb{R}^n \longrightarrow \mathbb{R}$ be a scalar-valued function of the vector $X = (x_1, \ldots, x_n)$. In this chapter we have defined the following kinds of derivatives for f:

1. The directional derivative $\nabla_e f(X_0)$ with $\|e\| = 1$ and the slightly more general $\nabla_Z f(X_0)$
2. The partial derivatives $f_{x_1}(X_0), \ldots, f_{x_n}(X_0)$
3. The total derivative $f'(X_0)$—a linear map from $\mathbb{R}^n$ into $\mathbb{R}$

At the moment we may feel reasonably confident about our ability to compute partial derivatives, since they may be handled by the methods of ordinary calculus. The other derivatives still elude us, however. It is time to bring order out of this chaos.

5.4b The theorem

This will settle the problems above, namely, computation of the total derivative $f'(X_0)$—and, therefore, the best affine approximation—and the directional derivative as well.

If f is differentiable at X_0, then its best affine approximation is

$$T(X) = f(X_0) + f'(X_0)(X - X_0),$$

where $f'(X_0) : \mathbb{R}^n \to \mathbb{R}$ is a linear map. Given f, we can compute the number $f(X_0)$ readily enough. What about $f'(X_0)$?

Since $f'(X_0)$ is a linear map, it has an associated matrix of the form

$$[f'(X_0)] = [\alpha_1 \quad \alpha_2 \quad \cdots \quad \alpha_n].$$

This is a 1 by n matrix, since $f'(X_0)$ maps $\mathbb{R}^n$ into $\mathbb{R}^1$. We would like to compute $\alpha_1, \alpha_2, \ldots, \alpha_n$. They determine the map $f'(X_0)$ entirely. We saw all this in Chap. 3.

The following theorem, central to the entire development of the theory, gives us these α's (see statement 3 below) and much more.

THEOREM 2: (Main theorem on differentiable functions.) Let $f: \mathbb{R}^n \to \mathbb{R}$ be differentiable at X_0, an interior point of $\mathscr{D}(f)$. Then:

1. For each $Z \in \mathbb{R}^n$, the derivative $\nabla_Z f(X_0)$ exists; moreover,

$$\nabla_Z f(X_0) = f'(X_0) Z = \frac{\partial f}{\partial x_1}(X_0) z_1 + \cdots + \frac{\partial f}{\partial x_n}(X_0) z_n.$$

2. In particular, $f'(X_0) e_j$ is given by

$$f'(X_0) e_j = \nabla_{e_j} f(X_0) = f_{x_j}(X_0)$$

for the standard basis $e_1, \ldots, e_n$ in $\mathbb{R}^n$.

3. The 1 by n matrix of $f'(X_0)$ is given by

$$[f'(X_0)] = [f_{x_1}(X_0) \ldots f_{x_n}(X_0)].$$

4. *The best affine approximation $T(X)$ to $f(X)$ at $X_0 = (c_1, \ldots, c_n)$ is*

$$T(X) = f(X_0) + f'(X_0)(X - X_0)$$

$$= f(X_0) + \frac{\partial f}{\partial x_1}(X_0)(x_1 - c_1) + \cdots + \frac{\partial f}{\partial x_n}(X_0)(x_n - c_n).$$

REMARK: Statements 1 and 4 are especially useful, since they enable us to compute directional derivatives and the best affine approximation by way of the partial derivatives of f, reducing all these computations to ordinary calculus.

Proof of the main theorem: The crux here is statement 1. Statements 2 to 4 are straightforward consequences.

Let f be differentiable at X_0, and let Z be any vector in $\mathbb{R}^n$. Since the theorem claims that $\nabla_Z f(X_0)$ exists, let us form its related difference quotient and work from there. We have

$$\frac{f(X_0 + tZ) - f(X_0)}{t} = \frac{f(X_0 + tZ) - f(X_0) - f'(X_0)tZ + f'(X_0)tZ}{t},$$

by the standard trick of adding and subtracting $f'(X_0)tZ$, which exists because f is differentiable at X_0. Clearly this equals

$$\frac{f(X_0 + tZ) - [f(X_0) + f'(X_0)tZ]}{t} + f'(X_0)Z. \tag{A}$$

Now f differentiable implies that

$$\lim_{tZ \to 0} \frac{f(X_0 + tZ) - [f(X_0) + f'(X_0)tZ]}{\|tZ\|} = 0. \tag{B}$$

You might see this better by putting $X = X_0 + tZ$, so that $\|X - X_0\| = \|tZ\| = |t| \, \|Z\|$.

Since Z is fixed, so is $\|Z\|$; therefore we also have

$$\lim_{t \to 0} \frac{f(X_0 + tZ) - [f(X_0) + f'(X_0)tZ]}{t} = 0, \tag{C}$$

since the expressions in (B) and (C) differ only by a constant factor $\|Z\|$. But the quotient in (C) is the left-hand term in (A). Hence,

$$\begin{aligned}
\nabla_Z f(X_0) &= \lim_{t \to 0} \frac{f(X_0 + tZ) - f(X_0)}{t} \\
&= \lim_{t \to 0} \quad \text{expression (A)} \\
&= f'(X_0)Z, \quad \text{by (C).}
\end{aligned}$$

This proves that $\nabla_Z f(X_0) = f'(X_0)Z$ in statement 1 of the theorem.

Statement 2 is true because $\nabla_{e_j} f(X_0) = f_{x_j}(X_0)$.

Statement 3 is true because the jth column of $[f'(X_0)]$ is the "column vector" associated with $f'(X_0)e_j$.

To complete the proof of statement 1, we compute the number $f'(X_0)Z$ as

$$[f'(X_0)][Z] = \left[\frac{\partial f}{\partial x_1}(X_0) \cdots \frac{\partial f}{\partial x_n}(X_0) \right] \begin{bmatrix} z_1 \\ \vdots \\ \vdots \\ z_n \end{bmatrix}$$

$$= \frac{\partial f}{\partial x_1}(X_0)z_1 + \cdots + \frac{\partial f}{\partial x_n}(X_0)z_n,$$

using statement 3.

Statement 4 follows from the facts that the matrix $[f'(X_0)]$ is $\left[\dfrac{\partial f}{\partial x_1}(X_0) \cdots \dfrac{\partial f}{\partial x_n}(X_0) \right]$ and the vector $X - X_0$ is $(x_1 - c_1, \ldots, x_n - c_n)$. You should be able to fill in the details as a mental exercise. $\ll$

EXAMPLES: 1. Let us use the theorem to compute the total derivative and best affine approximation of the differentiable function $f(X) = 1 - x_1^2 - x_2^2$ at $X_0 = (1, 2)$.

The first thing to do is compute partial derivatives. We have $\partial f/\partial x_1 = -2x_1$ and $\partial f/\partial x_2 = -2x_2$, whence the total derivative $f'(X_0)$ has the matrix

$$[f'(X_0)] = \left[\frac{\partial f}{\partial x_1}(X_0) \quad \frac{\partial f}{\partial x_2}(X_0) \right] = [-2 \quad -4].$$

It follows that the best affine approximation at X_0 is

$$T(X) = f(X_0) + f'(X_0)(X - X_0)$$
$$= -4 - 2(x_1 - 1) - 4(x_2 - 2).$$

Note that we directly verified, in the preceding section, that this map is indeed the best affine approximation to the given f at the given X_0. The Main Theorem checks with our earlier limit computations.

2. Here is a problem: Compute $\nabla_e f(X_0)$, where $f(X) = x_1^2 + 2x_1x_2 + 3x_2$, $X_0 = (1, 0)$ and $e = (-1/\sqrt{2}, 1/\sqrt{2})$. Note that $\|e\| = 1$.

We apply the theorem, of course. The partial derivatives of f are $f_{x_1}(X) = 2x_1 + 2x_2$ and $f_{x_2}(X) = 2x_1 + 3$. Thus $f_{x_1}(X_0) = 2$, $f_{x_2}(X_0) = 5$. By statement 1 of the theorem,

$$\nabla_e f(X_0) = 2\left(\frac{-1}{\sqrt{2}} \right) + 5\left(\frac{1}{\sqrt{2}} \right) = \frac{3}{\sqrt{2}}.$$

Actually, we computed similar directional derivatives in Sec. 5.1 using only the definition. The present method obviates the taking of limits and is much faster.

3. An economic interpretation. Let $q(x_1, \ldots, x_n)$ be the steel output for production of an iron mill as a function of various input variables. For example, $x_1 =$ number of workers, $x_2 =$ amount of iron ore available, $x_3 =$ amount of capital investment (in, say, dollars), and so on. Then the derivative matrix is

$$[q'(X)] = \left[\frac{\partial q}{\partial x_1} \cdots \frac{\partial q}{\partial x_n} \right].$$

The partial derivative $\partial q/\partial x_1$ measures how the productivity changes as the number of workers is changed while all other variables are held constant. Economists refer to q_{x_1} as the **marginal product of labor**. Similarly q_{x_2} and q_{x_3} are, respectively, the **marginal product of raw materials** and **marginal product of capital**. The matrix $[q'(X)]$ might be called the marginal product matrix. Any time you see the word "marginal" in an economic setting, you can be certain that there is a derivative lurking in the background.

Now $q(X)/x_1$ represents the average production of a worker, which we might call the **productivity** of each worker. How does the productivity change as the number of workers is changed? To find out, we compute

$$\frac{\partial}{\partial x_1}\left(\frac{q(X)}{x_1}\right) = \frac{x_1 q_{x_1} - q}{x_1^2}.$$

From this we conclude that the productivity of each worker increases with the size of the labor force as long as

$$\frac{\partial q}{\partial x_1} > \frac{q}{x_1},$$

while the productivity of each worker decreases if the opposite inequality holds. The factory is operating most efficiently if equality holds, that is, if the marginal productivity of labor equals the average productivity.

We shall see more applications of the Main Theorem after discussing the gradient vector.

5.4c The gradient

We have just seen that $f'(X_0)$ is specified by the 1 by n matrix $[f_{x_1}(X_0) \cdots f_{x_n}(X_0)]$. For instance, if $f(X) = x_1^2 - x_1 x_2$ and $X_0 = (-1, 3)$, then $[f'(X_0)] = [-5 \quad 1]$, as you may readily verify.

It is customary to think of the 1 by n total derivative matrix as a *vector* in $\mathbb{R}^n$. This is easily done, since a 1 by n matrix consists essentially of n numbers. We call this vector the **gradient of f at X_0** and denote it $\nabla f(X_0)$, read "del f at X_0." Thus

$$\nabla f(X_0) = \left(\frac{\partial f}{\partial x_1}(X_0), \ldots, \frac{\partial f}{\partial x_n}(X_0)\right).$$

Another notation is grad $f(X_0)$. "Grad" rhymes with "bad."

In the case of $f(X) = x_1^2 - x_1 x_2$ and $X_0 = (-1, 3)$, the gradient vector $\nabla f(X_0) = (-5, 1)$.

Thus we have three closely related objects:

1. $f'(X_0)$, the total derivative, a linear map from $\mathbb{R}^n$ into $\mathbb{R}$,
2. $[f'(X_0)] = [\partial f/\partial x_1 (X_0) \cdots \partial f/\partial x_n (X_0)]$, the matrix of $f'(X_0)$,
3. $\nabla f(X_0) = (\partial f/\partial x_1 (X_0), \ldots, \partial f/\partial x_n (X_0))$, the gradient vector in $\mathbb{R}^n$.

Actually, the idea of associating a vector with a linear functional is not new to us. We did so in Chap. 3, when we introduced linear functionals. In the present situation, we know that $f'(X_0)X = f_{x_1}(X_0)x_1 + \cdots + f_{x_n}(X_0)x_n$ (statement 1 of the Main Theorem), which clearly equals the inner product $\langle \nabla f(X_0), X \rangle$.

Here is a consequence: The notation $\nabla f(X_0)$ for the gradient vector, coupled with statement 1 of the Main Theorem, yields the following aesthetically pleasing formula for the derivative along the vector $Z \in \mathbb{R}^n$. It is worth stating as a theorem.

THEOREM 3: The derivative of f at X_0 along the vector Z is given by

$$\nabla_Z f(X_0) = \langle \nabla f(X_0), Z \rangle.$$

In particular, if $\|e\| = 1$, the directional derivative is given by

$$\nabla_e f(X_0) = \langle \nabla f(X_0), e \rangle.$$

EXAMPLE: Let $f(X) = x_1^2 - x_1 x_2$, as mentioned above, and let $X_0 = (-1, 3)$. What is $\nabla_Z f(X_0)$ if $Z = (-4, 13)$?

Solution: We know that $\nabla f(X_0) = (-5, 1)$. It follows that $\nabla_Z f(X_0) = \langle \nabla f(X_0), Z \rangle = (-4)(-5) + (13)(1) = 33$. Done.

REMARK: You may object that introduction of the gradient vector (when we already have $f'(X_0)$, the total derivative) is more word building. But here is something to think about. Now that $f'(X_0)$ may be interpreted as a *vector* $\nabla f(X_0)$, we ask, "What is special about the direction pointed out by $\nabla f(X_0)$? What does this direction tell us about the function f?" We will return to this after some further basic applications of the Main Theorem (see Section 5.4g).

5.4d The Main Theorem and tangent planes

Having related the best affine approximation to partial derivatives, we are in the happy position of being able to calculate the tangent plane at any point to the surface given by $z = f(X)$, where $X = (x_1, x_2)$.

Here is what we mean. If the scalar-valued function $f(X)$ is differentiable at X_0, then by the **tangent plane** to the surface $z = f(X)$ above the point X_0 we mean the graph $z = T(X)$, where T is best affine approximation to

f at X_0. Statement 4 of the Main Theorem enables us to compute the equation of the tangent plane directly. We give an example.

Let $f(X) = x_1^2 + x_2^2$, $X_0 = (1, 2)$.

PROBLEM: Calculate the tangent plane to $z = f(X)$ above X_0.

Solution: We have $f_{x_1}(X) = 2x_1$, $f_{x_2}(X) = 2x_2$, whence $f_{x_1}(X_0) = 2$, $f_{x_2}(X_0) = 4$. Also $f(X_0) = 5$. Thus the equation of the tangent plane is

$$z = T(X) = f(X_0) + \frac{\partial f}{\partial x_1}(X_0)(x_1 - c_1) + \frac{\partial f}{\partial x_2}(X_0)(x_2 - c_2)$$

$$= 5 + 2(x_1 - 1) + 4(x_2 - 2).$$

This *does* give a plane in $x_1 x_2 z$-space, for it is the same as

$$2(x_1 - 1) + 4(x_2 - 2) - z = -5,$$

which is the same equation as

$$2x_1 + 4x_2 - z = 5,$$

and we know from Chap. 1 that the set of solutions (x_1, x_2, z) to this inhomogeneous linear equation is a plane in $\mathbb{R}^3$. See Fig. 5.16.

REMARK: Reducing the equation $2(x_1 - 1) + 4(x_2 - 2) - z = -5$ to the form $2x_1 + 4x_2 - z = 5$ by simple arithmetic is usually *not* advisable, because this reduction effaces all connection with the point $X_0 = (1, 2)$. As we have seen on several occasions, when investigating a function in the

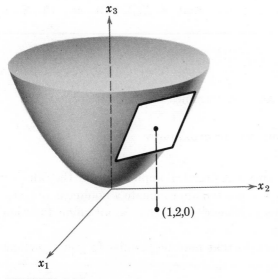

FIGURE 5.16

vicinity of a point $X_0 = (c_1, c_2)$, it is usually helpful to work with $x_1 - c_1$, $x_2 - c_2$, and $X - X_0$ rather than x_1, x_2, and X.

Higher dimensions: the tangent space.   Let $f: \mathbb{R}^n \to \mathbb{R}$, with $n \geq 2$. The graph $z = f(X)$ is the set of all points $(X, f(X)) = (x_1, \ldots, x_n, f(X))$ in $\mathbb{R}^{n+1}$ with $X \in \mathscr{D}(f)$. If $n > 2$, this is not a surface in the ordinary two-dimensional sense. In fact, there is a one-to-one correspondence between points X of $\mathscr{D}(f)$ and points of the graph, given by $X \leftrightarrow (X, f(X))$. This can be used to give meaning to the statement that if $\mathscr{D}(f) = \mathbb{R}^n$, or an open subset of $\mathbb{R}^n$, then the graph $z = f(X)$ is an n-dimensional subset of $\mathbb{R}^{n+1}$, commonly called a **hypersurface**. Since we cannot visualize $\mathbb{R}^n$ for $n > 3$, we cannot hope to visualize these hypersurfaces as subsets sitting in $\mathbb{R}^{n+1}$.

Nonetheless, given $f: \mathbb{R}^n \to \mathbb{R}$, we may think of a "flat" affine subspace tangent to the graph $z = f(X)$, with point of contact $(X_0, f(X_0))$. This is the **tangent space at** X_0, defined to be the graph $z = T(X)$ of the best affine approximation to f at X_0. If $n = 1$, this is the usual tangent line and, if $n = 2$, the tangent plane. If $n \geq 3$, the tangent space is usually termed a "hyperplane" inside $\mathbb{R}^{n+1}$.

EXAMPLE: Let $f: \mathbb{R}^3 \to \mathbb{R}$ be given by $f(X) = 3x_1^3 + 4x_2^3 + 5x_3^3$, and let $X_0 = (1, 2, -1)$. We compute the equation of the tangent space to $z = f(X)$ in $x_1 x_2 x_3 z$-space "above" the point X_0.

To get $z = T(X)$, we take partial derivatives. Thus $f_{x_1}(X) = 9x_1^2$, $f_{x_2}(X) = 12x_2^2$, $f_{x_3}(X) = 15x_3^2$. It follows that $[f'(X_0)] = [9 \quad 48 \quad 15]$. Since $f(X_0) = 30$, we get

$$z = T(X) = f(X_0) + f'(X_0)(X - X_0)$$
$$= 30 + 9(x_1 - 1) + 48(x_2 - 2) + 15(x_3 + 1).$$

This equation in x_1, x_2, x_3, and z determines the tangent space in $\mathbb{R}^4$.

5.4e   The Main Theorem and approximate numerical values

This is one of the most important practical applications of the calculus. It illustrates very well why we so often approximate a nonlinear object—function, curve, surface—by a "linear" one—that is, an affine function, a straight line, a flat plane.

Let us begin with an example that may be familiar to you from one-variable calculus.

PROBLEM: Calculate a close approximation to the number $\sqrt{4.031}$.

Solution: We seek a number closer to the true value of $\sqrt{4.031}$ than the obvious rough approximation 2.000 ($= \sqrt{4.000}$):

1. Let $y = f(x) = \sqrt{x}$, and observe that if $x_0 = 4$, then $f(x_0) = 2$ and, since $f'(x) = 1/2\sqrt{x}$, $f'(x_0) = 1/4$. We choose $x_0 = 4$ because it is close to 4.031, and also its square root is known.

2. The tangent line above $x_0 = 4$ to $y = f(x) = \sqrt{x}$ is the graph of $y = T(x) = f(x_0) + f'(x_0)(x - x_0) = 2 + \frac{1}{4}(x - 4)$.

3. Now we make the crucial observation that *if x is close to x_0, then $T(x)$ is close to $f(x)$*. See Fig. 5.17. If $x = 4.031$, then $\sqrt{4.031} = f(4.031) \approx T(4.031)$. (Note: $\approx$ means "approximately equal").

But in contrast to $f(4.031)$, the value of $T(4.031)$ is easy to calculate, since $T(x)$ is affine. We have

$$T(4.031) = 2 + \tfrac{1}{4}(4.031 - 4) = 2 + \tfrac{1}{4}(0.031) \approx 2.008.$$

Hence we conclude that

$$\sqrt{4.031} \approx 2.008.$$

This is our approximate value of the square root. (Since the datum only has three significant decimal figures, we have only kept three decimal figures in our approximation).

4. As a check we square our approximate square root, obtaining

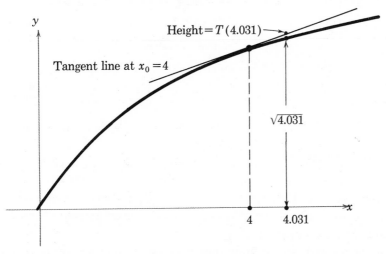

Height$= T(4.031)$

Tangent line at $x_0 = 4$

$\sqrt{4.031}$

4   4.031

$\sqrt{4.031} \approx T(4.031)$

FIGURE 5.17

$$(2.008)^2 \approx 4.032,$$

which is very close to 4.031.

REMARKS: 1. Because we use $T(4.031)$ to approximate $f(4.031)$, this method is usually called *linear*, or *first-order*, *approximation*.

2. It is by this method and refinements of it that tables of square roots, sines, cosines, logrithms are constructed.

3. Modern scientific work is full of approximations of various kinds. For one thing, measurements of nature—temperatures, distances, half-lives of radioactive elements—can be made only with approximate accuracy. Again, electronic computers do not handle irrational numbers, infinitely long nonrepeating decimals, such as $e = 2.718\ldots$ or $\pi = 3.14159\ldots$; rather, they approximate these important numbers by rational numbers, whence we often hear that π "is" 22/7, or 3.14, or 3.1416, these being closer and closer rational approximations to the true value. Thus, the numbers of applied science and technology are approximate.

Approximation in several variables. Again we illustrate the method by an example.

PROBLEM: Calculate $\sqrt{(3.01)^2 + (3.98)^2}$.

Solution: 1. First we need a function. Let $f(X) = \sqrt{x_1^2 + x_2^2}$. Choose $X_1 = (3.01, 3.98)$. We want a number close to $f(X_1)$. Choose $X_0 = (3, 4)$. This is close to X_1, and also we can compute $f(X_0) = 5$ exactly.

2. The tangent plane to $z = f(X)$ above $X_0 = (3, 4)$ has the equation

$$z = T(X) = f(X_0) + f'(X_0)(X - X_0).$$

To get $f'(X_0)$, compute $f_{x_1}(X) = x_1(x_1^2 + x_2^2)^{-1/2}$ and $f_{x_2}(X) = x_2(x_1^2 + x_2^2)^{-1/2}$. Thus $f_{x_1}(X_0) = \frac{3}{5}$ and $f_{x_2}(X_0) = \frac{4}{5}$. These numbers were readily computed also. Hence

$$z = T(X) = 5 + \tfrac{3}{5}(x_1 - 3) + \tfrac{4}{5}(x_2 - 4),$$

the best affine approximation to $f(X)$ at $X_0 = (3, 4)$.

3. Now we make the crucial observation that, since X_1 is close to X_0, the exact value $f(X_1)$ is close to $T(X_1)$. See Fig. 5.18. This is because $f(X_1)$ is the height of the curved surface $z = f(X)$ above X_1, and $T(X_1)$ is the height above X_1 of the tangent plane to a nearby point; that is,

$$\sqrt{(3.01)^2 + (3.98)^2} = f(X_1) \approx T(X_1).$$

We may readily compute $T(X_1)$, because T is affine. We have

$$T(X_1) = 5 + \tfrac{3}{5}(3.01 - 3) + \tfrac{4}{5}(3.98 - 4) = 4.99;$$

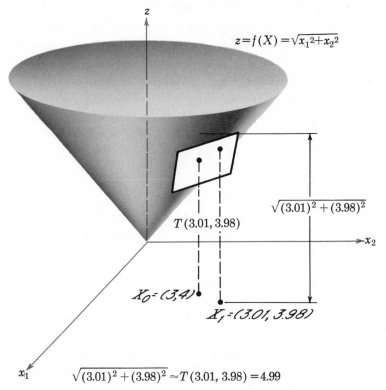

$$z = f(X) = \sqrt{x_1{}^2 + x_2{}^2}$$

$$\sqrt{(3.01)^2 + (3.98)^2}$$

$T(3.01, 3.98)$

$X_0 = (3,4)$

$X_1 = (3.01, 3.98)$

x_1 $\sqrt{(3.01)^2 + (3.98)^2} \approx T(3.01, 3.98) = 4.99$

FIGURE 5.18

hence we conclude that

$$\sqrt{(3.01)^2 + (3.98)^2} \approx 4.99.$$

This number is our approximation.

SUMMARY: Success with this method hinges on our ability to find a point X_0 with the properties:

  1. X_0 is close to X_1, where $f(X_1)$ is the value we wish to approximate.
  2. The number $f(X_0)$ can be computed exactly.
  3. The total derivative $f'(X_0)$, that is, the partials $f_{x_1}(X_0)$ and $f_{x_2}(X_0)$, may be computed exactly (compare $X_0 = (3, 4)$ in the above).

Having found such an X_0, we construct the best affine approximation (a function) $T(X)$ to $f(X)$ at X_0. We then take the number $T(X_1)$ as a close approximation to the desired value $f(X_1)$.

5.4f   When does the total derivative exist?

The point of the Main Theorem is that the existence of the best affine approximation and total derivative to f at X_0 guarantees the existence of all the partial derivatives $f_{x_j}(X_0)$, and, moreover, the partial derivatives can be used to compute the total derivative explicitly, $[f'(X_0)] = [f_{x_1}(X_0) \cdots f_{x_n}(X_0)]$. Hence it is natural to wonder whether the existence of all the partial derivatives $f_{x_j}(X_0)$ assures that the total derivative $f'(X_0)$ and best affine approximation $T(X)$ to f at X_0 exist. The answer is "No, not quite." However, if all the partial derivatives exist *and are continuous*, then the best affine approximation and total derivative exist. Consider

THEOREM 4:   Let all partial derivatives $f_{x_j}(X)$ exist and be continuous functions of X near the point X_0. Then the function f has a best affine approximation $T(X) = f(X_0) + f'(X_0)(X - X_0)$ at X_0,   with $[f'(X_0)] = [f_{x_1}(X_0) \cdots f_{x_n}(X_0)]$.

We omit the proof. As mentioned before, most of the functions, including partial derivatives, we encounter in this book are continuous. Hence the function $f(X)$ is generally differentiable (has a total derivative) at a point X_0 of its domain. Combine Exercise 11 in Sec. 5.2 and Exercise 6 in Sec. 5.3 for an interesting example of a non-differentiable function.

5.4g   The heat-seeking bug

We close this chapter with an amusing problem that also points the way to our study of maxima and minima in the following chapter. Suppose that you are a heat-seeking bug located at the point X_0 on the plane $\mathbb{R}^2$. At each point X on the plane, the temperature is given by some scalar-valued function $f(X)$. You want to travel toward warmth as efficiently as possible. In fact, you *refuse* to visit any point that is cooler than any previous point you have visited. Thus, if the temperature at your present position, namely $f(X_0)$, is not less than the temperature $f(X)$ at any nearby point X, if $f(X_0) \geq f(X)$, then you'll stay where you are. But if not, you'll head for warmth the fastest way possible. Which way?

Knowing a bit of vector calculus, you realize that this is a directional derivative problem. You want to find a direction, pointed out by some unit vector e, that has these properties:

1. The rate of change of temperature with respect to distance in the e direction is positive, $\nabla_e f(X_0) > 0$; in brief, e points toward relative warmth.

2. The rate of increase in the e direction is greater than in any other direction; in brief, e points the quickest way to warmth. In mathematical language, given X_0 and $f(X)$, you want to find e such that the value $\nabla_e f(X_0)$ is maximized.

Let us examine the directional derivative $\nabla_e f(X_0)$. We know from Theorem 3 that

$$\nabla_e f(X_0) = \langle \nabla f(X_0), e \rangle,$$

where $\nabla f(X_0)$ is the gradient vector $(f_x(X_0), f_y(X_0))$.

Now we recall from Chap. 2 that

$$\langle \nabla f(X_0), e \rangle = \| \nabla f(X_0) \| \| e \| \cos \theta$$
$$= \| \nabla f(X_0) \| \cos \theta,$$

where θ is the angle between $\nabla f(X_0)$, considered a vector in $\mathbb{R}^2$, and the unit vector e. Note that we have used $\| e \| = 1$. See Fig. 5.19.

Finally, observe that $\| \nabla f(X_0) \| \cos \theta$ is greatest precisely when $\cos \theta = 1$, that is, when the angle θ is zero (recall that X_0 and hence $\nabla f(X_0)$ are fixed and only θ varies.) This means that the directional derivative $\nabla_e f(X_0)$ is greatest when the vector e is chosen to point in the same direction as the gradient vector $\nabla f(X_0)$. See Fig. 5.20.

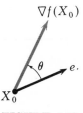

$\nabla f(X_0)$

FIGURE 5.19

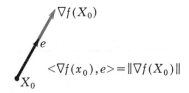

$\langle \nabla f(x_0), e \rangle = \| \nabla f(X_0) \|$

Choose e in same direction as $\nabla f(X_0)$

FIGURE 5.20

Hence, the vector e should be defined by $e = \nabla f(X_0)/\| \nabla f(X_0) \|$. This gives a unit vector pointing in the same direction as $\nabla f(X_0)$. Of course, all this requires that $\nabla f(X_0)$ be different from zero. We have proved

THEOREM 5: Fix $X_0 \in \mathscr{D}(f)$. Then the directional derivative of f at X_0 is largest in that direction pointed out by the gradient vector $\nabla f(X_0)$, provided that $\nabla f(X_0) \neq 0$.

Thus the heat-seeking bug, now situated at X_0, should move off in the gradient direction pointed out by $\nabla f(X_0)$. As he travels, at each point X he should recompute $\nabla f(X)$ and move in that direction toward increasing warmth.

But what if $\nabla f(X_0) = 0$ and therefore has no direction? What does this mean, and what might our heat-seeking bug do about it? Has he any hope of nearby warmth?

The answer for now is "Perhaps, perhaps not." It all depends on f, the temperature function, and whether or not the point X_0 is a so-called local maximum for f (the hottest spot around). To determine this, it would be helpful to know the second derivative of the temperature $f(X)$. But this is another story.

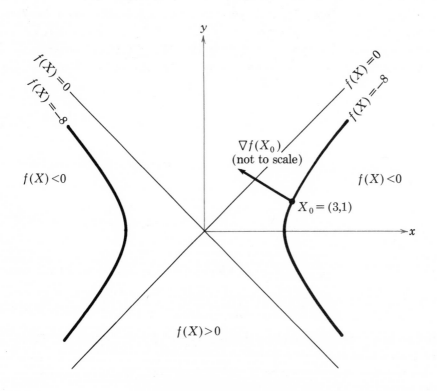

$\nabla f(X_0)$ points towards warmth

FIGURE 5.21

EXAMPLE: Let the temperature at $X = (x, y)$ in $\mathbb{R}^2$ be given by $f(x, y) = y^2 - x^2$. At $X_0 = (3, 1)$, the temperature is $f(X_0) = -8$. In fact, the temperature is precisely the same at all points (x, y) on the hyperbola $x^2 - y^2 = 8$. See Fig. 5.21.

To find the direction of fastest increase, we compute $\nabla f(X) = (f_x(X), f_y(X)) = (-2x, 2y)$, so that $\nabla f(X_0) = (-6, 2)$.

A unit vector e pointing in the same direction as $\nabla f(X_0)$ is obtained, of course, by dividing the vector $\nabla f(X_0)$ by a scalar equal to its length $\|\nabla f(X_0)\|$; that is, $e = \|\nabla f(X_0)\|^{-1}\nabla f(X_0)$. Since $\|\nabla f(X_0)\| = \sqrt{40} = 2\sqrt{10}$, $e = (-3/\sqrt{10}, 1/\sqrt{10})$. We have drawn this vector with tail end at the point X_0, indicating the initial direction of travel from X_0.

Of course, the bug should continually revise his direction as he moves, so that at each instant he is traveling in the direction of *fastest* temperature increase; that is, at each point X, the bug should aim himself in the direction of $\nabla f(X)$. See Fig. 5.22. If the bug does this, he will travel along the curve $xy = 3$, crossing the line $y = x$ (zero temperature) at the point $(\sqrt{3}, \sqrt{3})$ and continuing on to ever warmer climes.

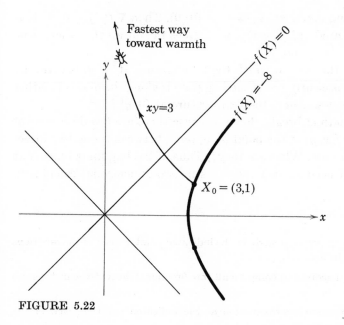

FIGURE 5.22

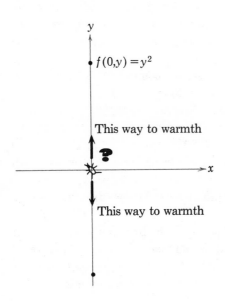

Two directions of fastest increase

FIGURE 5.23

Suppose the bug starts at $X_0 = O = (0, 0)$. Then $\nabla f(O) = 0$, so that no direction is pointed out. See Fig. 5.23. Does this mean that there is no nearby warmth to be found?

The answer, in the case in which the temperature at (x, y) is given by $f(x, y) = y^2 - x^2$, is clearly no. If the bug travels along the y-axis in either direction, things get warmer; the temperature at $(0, y)$ is y^2.

The gradient method breaks down, because there are *two* directions of fastest increase of $f(x, y)$ at the point $(0, 0)$ (and, between them, two directions of fastest *decrease*. What are they?) Thus, if the bug finds himself at the origin, he will need further resources in two-dimensional calculus in order to prevail.

Exercises

1.   Find the total derivative matrix at the indicated point of each of the functions in Exercise 1 in Sec. 5.1.

2.   Use the Main Theorem to compute all the directional derivatives in Exercise 1 in Sec. 5.1.

3.   Find the best affine approximation at the indicated point of each of the functions in Exercise 1 in Sec. 5.1.

4. (a) Sketch the graph of the surface in $\mathbb{R}^3$ defined by $z = f(x, y) = 4 - x^2 - y^2$.
   (b) Find the best affine approximation to f at $X_0 = (1, 1)$.
   (c) Draw the graph of this best affine approximation, in the sketch in Exercise 4(a).

5. Repeat Exercise 4 for $z = f(x, y) = -1 + (x - 1)^2 + (y - 1)^2$.

6. Repeat Exercise 4 for $z = f(x, y) = y^2 - x^2$.

7. Let $z = f(x, y) = x^2 + y^2$. Draw a sketch of $\mathbb{R}^2$ indicating the points where $z = 1$ and $z = 4$. On this same sketch, draw the vector $\nabla f(X)$ at each point $X \in \mathbb{R}^2$ where $z = 1$.

8. Repeat Exercise 7 for $z = f(x, y) = xy$.

9. Use the best affine approximation to evaluate the following numbers approximately:

   $a = \sqrt{(1.1)^3 + (1.98)^3}$, $d = (1.01)^7(3.97)^{1.5}$,

   $b = 2.98e^{.01}$, $e = \sqrt{(2.01)^2 + (5.98)^2 + (2.99)^2}$,

   $c = \sqrt{3.9/4.1}$, $f = \dfrac{1}{2.01} + \dfrac{1}{2.98} + \dfrac{1}{6.03}$.

10. For each of the following functions, find the direction(s) one must go at the given point in order that the function (i) increase most rapidly, (ii) decrease most rapidly:

    (a) $f(x_1, x_2) = 3 - 2x_1 + 5x_2$ at $(2, 1)$;
    (b) $g(x, y) = e^{2x+y}$ at $(1, -2)$;
    (c) $\varphi(x, y, z) = 2x^2 + 3xy + 5z^2 + 4y - y^2 + 7$ at $(1, 0, -1)$;
    (d) $h(u, v, w) = uvw$ at $(1, -2, 3)$;
    (e) $p(x, y) = xy - x + y - 2$ at $(-1, 1)$.

11. Use the Main Theorem to prove the assertions in Exercise 6 in Sec. 5.1 again.

12. Let $f, g: \mathbb{R}^n \to \mathbb{R}$ be differentiable functions. Prove:

    (a) $\nabla(fg) = g\nabla f + f\nabla g$,
    (b) $\nabla_{(aZ+bW)}f = a\nabla_Z f + b\nabla_W f$.

THE WORLD OF FIRST DERIVATIVES

Function	First (total) derivative at a point	First derivative in matrix form	Best affine approximation or tangent map at a point
$f: \mathbb{R} \longrightarrow \mathbb{R}$ $y = f(x)$ (ordinary calculus)	$f'(x_0): \mathbb{R} \longrightarrow \mathbb{R}$ (usually interpreted as slope)	$f'(x_0)$, a number	$y = T(x) = f(x_0) + f'(x_0)(x - x_0)$ (tangent line at x_0)
$f: \mathbb{R}^n \longrightarrow \mathbb{R}$ $z = f(X) = f(x_1,\ldots,x_n)$ scalar-valued (Chap. 5)	Linear map (functional) $f'(X_0): \mathbb{R}^n \longrightarrow \mathbb{R}$	$[f'(X_0)] = \left[\dfrac{\partial f}{\partial x_1}(X_0) \cdots \dfrac{\partial f}{\partial x_n}(X_0)\right]$; also, gradient vector $\nabla f(X_0) = \left(\dfrac{\partial f}{\partial x_1}(X_0),\ldots,\dfrac{\partial f}{\partial x_n}(X_0)\right)$	$z = T(X) = f(X_0) + f'(X_0)(X - X_0)$ $= f(X_0) + \dfrac{\partial f}{\partial x_1}(X_0)(x_1 - c_1)$ $\quad + \cdots + \dfrac{\partial f}{\partial x_n}(X_0)(x_n - c_n)$ (here $X_0 = (c_1,\ldots,c_n)$)
$F: \mathbb{R} \longrightarrow \mathbb{R}^q$ $F(t) = (f_1(t),\ldots,f_q(t))$ parametrized curve, in q-space (Chap. 7)	Linear map $F'(t_0): \mathbb{R} \longrightarrow \mathbb{R}^q$ NOTE: Usually drawn as velocity vector $(f'_1(t_0),\ldots,f'_q(t_0))$ at point $F(t_0)$ on curve	$[F'(t_0)] = \begin{bmatrix} f'_1(t_0) \\ \vdots \\ f'_q(t_0) \end{bmatrix}$	$T(t) = F(t_0) + F'(t_0)(t - t_0)$ $= (f_1(t_0) + f'_1(t_0)(t - t_0),\ldots,$ $\quad f_q(t_0) + f'_q(t_0)(t - t_0))$
$F: \mathbb{R}^n \longrightarrow \mathbb{R}^q$ $Y = F(X)$ $= (f_1(x_1,\ldots,x_n),\ldots,$ $\quad f_q(x_1,\ldots,x_n))$ vector-valued function of a vector (Chap. 8)	Linear map $F'(X_0): \mathbb{R}^n \longrightarrow \mathbb{R}^q$	$[F'(X_0)] = \begin{bmatrix} \dfrac{\partial f_1}{\partial x_1}(X_0) \cdots \dfrac{\partial f_1}{\partial x_n}(X_0) \\ \vdots \qquad \vdots \\ \dfrac{\partial f_q}{\partial x_1}(X_0) \cdots \dfrac{\partial f_q}{\partial x_n}(X_0) \end{bmatrix}$	$Y = T(X) = F(X_0)$ $\quad + F'(X_0)(X - X_0)$ $Y = (y_1,\ldots,y_q)$ $y_j = f_j(X_0) + \dfrac{\partial f_j}{\partial x_1}(X_0)(x_1 - c_1)$ $\quad + \cdots + \dfrac{\partial f_j}{\partial x_n}(X_0)(x_n - c_n)$ $(j = 1,\ldots,q)$

6. Scalar-Valued Functions and Extrema

6.0 INTRODUCTION

This is a familiar story in one-variable calculus. Let $f: \mathbb{R} \longrightarrow \mathbb{R}$. Then the point $x_0 \in \mathscr{D}(f)$ is called a **local maximum** for f provided that (1) x_0 is an interior point of the domain $\mathscr{D}(f)$ and (2) for all x in some open interval of points on the x-axis containing x_0 we have $f(x) \leq f(x_0)$. The concept of **local minimum** for f is defined in a similar way. For brevity, we call a local maximum or local minimum a **local extremum**.

In the graph pictured in Fig. 6.1, the local maxima are at x_0, x_3, x_5 (the "mountain peaks") and the local minima at x_2 and x_4 (the "pits" or "valleys"). Note that $f(x)$ may have several local maxima and minima.

Why are we interested in finding the local extrema of $f(x)$? For one thing, knowing the maxima and minima of a function gives us a fair idea of what its graph looks like and its general behavior. Also, if the function represents a physical quantity like temperature (or an economic quantity such as utility), then it may be useful to find for which values of x the temperature (or utility) $f(x)$ is greatest and for which it is least.

Recall the following basic result from one-variable calculus.

THEOREM: (Local extrema in one variable.) If $f: \mathbb{R} \longrightarrow \mathbb{R}$ is a differentiable function with a local extremum at x_0, then:

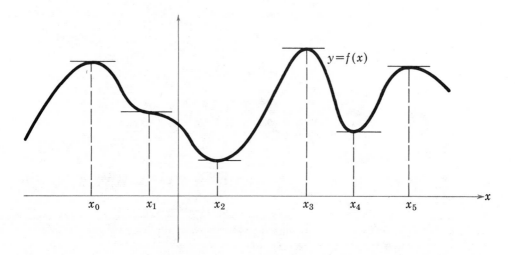

Local maxima at x_0, x_3, x_5, minima at x_2, x_4, non-extremal critical point at x_1

FIGURE 6.1

230

1. The tangent line to $y = f(X)$ above x_0 is horizontal.
2. $f'(x_0) = 0$.

We note that statements 1 and 2 are equivalent. Also, it is *not* true that $f'(x) = 0$ implies x is a local maximum or minimum, as the point $x = x_1$ in the graph in Fig. 6-1 illustrates: The slope $f'(x_1) = 0$, but the graph there has neither a peak nor a pit.

Thus, we find the local extrema of $y = f(x)$ by the following process:

1. Locate all points x_0 for which $f'(x_0) = 0$ (such points are called the **critical points** or **stationary points** of f).

2. Decide separately whether each critical point you have found is a maximum, a minimum, or neither. This may involve the *second* derivative $f''(x_0)$.

Just as in the one-variable case, the problem of locating the extrema of a scalar-valued function $f(X)$, $X = (x_1, \ldots, x_n)$, is of major importance. In the following sections, we examine this problem in some detail. We see that the statements about the first derivative $f'(x_0)$ quoted above in the one-variable case have their exact analogs here with regard to the total derivative $f'(X_0)$, or, if you prefer, the gradient vector $\nabla f(X_0)$. Moreover, the problem of deciding whether a point X_0 is a maximum or minimum leads us to the notion of the *second* derivative $f''(X_0)$. This, as we shall see, is neither a number, like $f(X_0)$, nor a vector, like $\nabla f(X_0)$, but a square (n by n) matrix.

Exercises

True or false, and why? Assume that all unspecified functions have as many derivatives as you want. We suggest lots of pictures.

1. $f(x) = (x - 1)^2$ has a local minimum at $x = 1$.

2. $f(x) = x^3$ has a local minimum at $x = 0$.

3. $f(x) = -x^4$ has a local maximum at $x = 0$.

4. $f(x) = (1 + x^2)^{-1}$ has no local minimum.

5. $f(x) = x^5$ has a critical point at $x = 0$.

6. $f(x) = 1 + x + x^3$ has no local maxima or minima.

7. If $f(x)$ has a local minimum at $x = 0$, then $f''(0) > 0$.

8. If $f(x)$ has a local maximum at $x = 0$, then $f''(0) \leq 0$.

9. If $f'(0) = 0$ and $f''(0) > 0$, then f has a local minimum at $x = 0$.

10. If $f'(0) = 0$ and $f''(0) = 0$, then f cannot have a local minimum at $x = 0$.

11. If f has local maxima at $x = -2$ and $x = 3$, then it must have a local minimum at some point $x = c$, $-2 < c < 3$.

12. If $f(-1) = 3$, $f(0) = 5$, $f(2) = -4$, then f has a local maximum at some point c in $-1 < c < 2$.

13. If both $f(x)$ and $g(x)$ have local maxima at $x = x_0$, then so does $h(x) = f(x) + g(x)$.

14. If both $f(x)$ and $g(x)$ have local minima at $x = x_0$, then so does $\varphi(x) = f(x)g(x)$.

15. $f(x, y) = x^2 + (y - 1)^2$ has a local minimum at $(0, 1)$.

16. $f(x, y) = 1 - (x - y)^2$ has a local minimum at $(1, 1)$.

17. $f(x, y) = y^2 - x^2$ has a local maximum at $(0, 0)$.

18. $f(x, y) = x^2 y^2$ has a local minimum at $(-1, 0)$.

19. $f(x, y) = 7 + xy$ has a local maximum at $(0, 0)$.

20. $f(x, y) = 3 - x + y$ has no local maxima or minima.

21. $f(x, y) = x + y^2$ has no local maxima or minima.

22. $f(x, y) = x^2 - \cos y$ has a local minimum at $(0, 0)$.

23. $f(x, y) = [x^2 + (y - 1)^2][x^2 + (y + 1)^2]$ has local minima at both $(0, 1)$ and $(0, -1)$.

24. If $f(X) \geq 0$ and f has a local minimum at X_0, then $g(X) = f^2(X)$ also has a local minimum at X_0.

25. If f satisfies $f'' - a(x)f = 0$, with $a(x) > 0$, then f cannot have a local minimum where f is negative.

6.1 LOCAL EXTREMA ARE CRITICAL POINTS

In this section we define the notions of local maximum and minimum for a function $f(X)$ and then obtain the analog of the one-dimensional theorem relating extrema with "first derivative equal to zero."

We are given a function $f: \mathbb{R}^n \to \mathbb{R}$.

DEFINITION: The point $X_0 \in \mathbb{R}^n$ is a **local maximum** for f if and only if:

1. X_0 is an interior point of the domain $\mathscr{D}(f)$,
2. For all X near X_0, we have $f(X) \leq f(X_0)$.

This second requirement means that there exists an open ball $B(X_0, r)$ around X_0 (of radius $r > 0$) such that if X is any point of this ball, the value $f(X)$ is not greater than $f(X_0)$. A **local minimum** for f is defined similarly, using $f(X) \geq f(X_0)$, of course. Again, brevity prompts us to call a local maximum or minimum a **local extremum**.

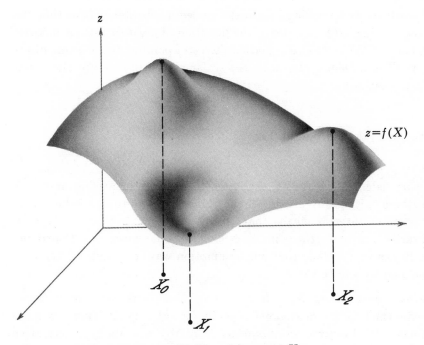

Local maxima at X_0, X_2, minimum at X_1.

FIGURE 6.2

Let's look at a picture. See Fig. 6.2. Suppose that $n = 2$; that is, f: $\mathbb{R}^2 \to \mathbb{R}$. Graphing $z = f(X)$ as usual, we see that the local maxima and minima are again the mountain peaks and the pits of the surface we obtain.

The following basic theorem is most useful for locating the extrema of $f(X)$. Note the similarity with one-variable calculus.

THEOREM 1: (Local extrema.) Let $f: \mathbb{R}^n \to \mathbb{R}$ be differentiable. If the point X_0 is a local maximum or minimum for f, then:

1. The total derivative $f'(X_0)$ is the zero map.
2. The gradient $\nabla f(X_0)$ is the zero vector.
3. The partial derivatives $f_{x_1}(X_0) = \cdots = f_{x_n}(X_0) = 0$.

REMARK: Statements 1, 2, and 3 are clearly equivalent.

Proof: We give two proofs, the first of them being "coordinate-free," not involving $x_1, \ldots, x_n$ or the partial derivatives.

First proof: Given X_0 is a maximum (the proof for X_0 a minimum is similar.) We show that $\nabla f(X_0) = O$. Assume that $\nabla f(X_0) \neq O$. Now our work with the heat-seeking bug in the preceding chapter tells us that the gradient vector $\nabla f(X_0)$ points in the direction of greatest rate of increase for f. Let $e = \nabla f(X_0)/\|\nabla f(X_0)\|$ be a unit vector pointing in the same direction as $\nabla f(X_0)$. Define $\varphi(s)$ to be the scalar-valued function of the scalar variable s given by

$$\varphi(s) = f(X_0 + se).$$

We used this notation in Sec. 5.1. There we saw that

$$\varphi'(0) = \nabla_e f(X_0) = \langle \nabla f(X_0), e \rangle = \|\nabla f(X_0)\| > 0.$$

Ordinary calculus assures us, therefore, that $\varphi(s)$ is increasing at $s = 0$. Thus there exist positive values of s arbitrarily close to $s = 0$ for which $\varphi(s) > \varphi(0)$. But this means that $f(X_0 + se) > f(X_0)$. Since $X_0 + se$ is arbitrarily close to X_0, the point X_0 is not a local maximum for f. Contradiction. Hence we conclude that our assumption was false and $\nabla f(X_0) = O$, as claimed by the theorem.

Second proof: Let $X_0 = (c_1, \ldots, c_n)$ be a local extremum for f. Consider the function of x_1 alone given by $f(x_1, c_2, \ldots, c_n)$. This new function has a local extremum at $x_1 = c_1$ (why?). Hence its first derivative with respect to x_1 is zero at $x_1 = c_1$, by ordinary calculus; that is, $f_{x_1}(X_0) = 0$. Likewise, we can show that each of the other partial derivatives vanishes at X_0, obtaining statement 3. Done. ≪

Just as in one-variable calculus, our search for extrema has led us to points at which the first derivative vanishes. We define X_0 to be a **critical point** for f if and only if:

1. X_0 is an interior point of $\mathscr{D}(f)$,
2. The total derivative $f'(X_0)$ is the zero map from $\mathbb{R}^n$ to $\mathbb{R}$. This second statement is, of course, equivalent to $\nabla f(X_0) = 0$ or the vanishing of all the partial derivatives of f at X_0. We remark that some people refer to critical points as **stationary** points.

Hence Theorem 1 establishes the title sentence of this section: *local extrema are critical points.* The second example below shows that the converse is false. Not all critical points are local extrema.

EXAMPLES: 1. Let $f(X) = f(x_1, x_2) = 1 - x_1^2 - x_2^2$. We locate the critical points by taking the partial derivatives

$$\frac{\partial f}{\partial x_1}(X) = -2x_1, \qquad \frac{\partial f}{\partial x_2}(X) = -2x_2$$

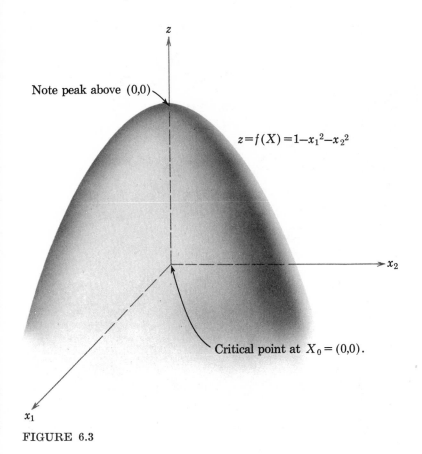

Note peak above (0,0)

$z = f(X) = 1 - x_1{}^2 - x_2{}^2$

z

x_2

x_1

Critical point at $X_0 = (0,0)$.

FIGURE 6.3

and setting them each equal to zero. See Fig. 6.3. This yields $X_0 = (0, 0)$; the origin is the only critical point for $f(X)$. It is clear from the graph, or from the fact that $f(X) \leq 1 = f(O)$, that the origin is a local maximum for f.

2. Let $g(x, y) = y^2 - x^2$. In this case $g_x = -2x$ and $g_y = 2y$, so that $X_0 = (0, 0)$ is again the only critical point. See Fig. 6.4.

In this case of $g(x, y)$, however, the origin $(0, 0)$ is not a local extremum. To see this, note that $g(0, 0) = 0$. Now $g(0, y) = y^2 > 0$ if $y \neq 0$. Hence there are points $(0, y)$ arbitrarily close to $(0, 0)$ (just take y small) at which the function g assumes larger values than $g(0, 0)$. Thus $(0, 0)$ is not a local *maximum* for g. Consideration of points $(x, 0)$ shows that the origin is not a local *minimum* either. This shows that not all critical points are extrema.

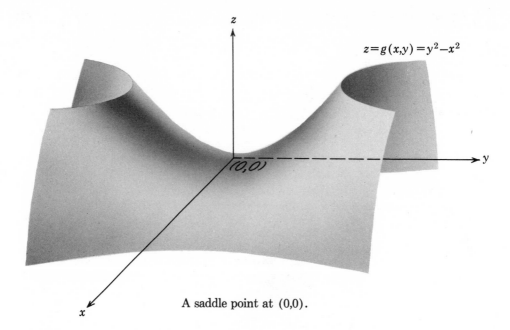

$$z = g(x,y) = y^2 - x^2$$

$(0,0)$

A saddle point at $(0,0)$.

FIGURE 6.4

The last phenomenon leads us to make the following definition. The point X_0 is a **saddle point** for $f: \mathbb{R}^n \rightarrow \mathbb{R}$ if and only if:

1. X_0 is a critical point for f; that is, $f'(X_0) = 0$.
2. X_0 is not a local extremum; that is, there exist points X_1, X_2 arbitrarily close to X_0 such that $f(X_1) < f(X_0) < f(X_2)$.

The second example above shows that the origin $(0, 0)$ is a saddle point for the function $g(x, y) = y^2 - x^2$.

SUMMARY: To locate the extrema of a given function $f(X)$:

1. Locate the critical points.
2. Classify each critical point as maximum, minimum, or saddle point.

To locate the critical points, we know that we must find all points X_0 satisfying the n equations $f_{x_1}(X_0) = 0, \ldots, f_{x_n}(X_0) = 0$. The *classification* of critical points may be more delicate, although in some cases (the two examples above) it is quite elementary. We turn to this in the following section.

Exercises

1. Find all the critical points of the following functions:

(a) $f(x_1, x_2) = x_1^2 - 6x_1 + 2x_2^2 + 10$,

(b) $g(x, y) = xy - x + y + 2$,

(c) $h(x, y) = x \sin y$,

(d) $\varphi(x, y) = (x^2 - 1)e^y$,

(e) $p(s, t) = (s^2 - 1)(t^2 - 1)$,

(f) $\psi(x, y) = (x - 2y)^2$,

(g) $u(\xi, \eta) = \xi^2 \eta^2$,

(h) $f(x, y) = \dfrac{xye^x}{1 + y^2}$,

(i) $f(x, y, z) = x^2 + 2x + 3y^2 + z^2 - 4z + 5$,

(j) $g(x, y, z) = xz + 5z^2$.

2. In order to examine the nature of a critical point $X_0 = (a, b)$ of $f(x, y)$, let $X = (a + h, b + k)$, and consider $\varphi(h, k) = f(a + h, b + k) - f(a, b)$ as a function of (h, k) for h, k small, so that X is near X_0. For example, $f(x, y) = x^2 + 2x + 5y^2$ has a critical point at $(-1, 0)$, and so we let $x = -1 + h$, $y = k$ and examine

$$\varphi(h, k) = (-1 + h)^2 + 2(-1 + h) + 5k^2 + 1 = h^2 + 5k^2 \geq 0;$$

that is, if $Z = (h, k)$, then

$$f(X_0 + Z) - f(X_0) = f(a + h, b + k) - f(a, b) = \varphi(h, k) \geq 0,$$

and so

$$f(X_0 + Z) \geq f(X_0);$$

that is, f has a local minimum at X_0. If φ can be either positive or negative for h, k small, we would have concluded that f had a saddle point. Use this method to determine if critical points in Exercises 1(a), (b), (e), (f), (g), (i), and (j) are local maxima, minima, or saddle points.

3. (a) Find the minimum distance from the origin to the plane $2x + y - 2z = 6$ in $\mathbb{R}^3$. (Remark: Minimizing the distance is equivalent to minimizing the square of the distance.)

(b) Find the minimum distance between the parallel planes

$$2x + y - 2z = 6 \quad \text{and} \quad 2x + y - 2z = -9.$$

(c) Find the points on the surface $z = 2xy + \frac{1}{4}$ nearest the origin.

(d) Find the points on the surface $z = 1/xy$ nearest the origin.

4. A game. Frank picks any real number x that he wants. Jerry then picks a real number y of his choice and pays Frank $xy - 3x + 2y$ dollars. What number should Frank pick?

6.2 MEAN VALUE THEOREM, SECOND DERIVATIVE, TAYLOR'S THEOREM

6.2a Introduction

In order to complete our study of local extrema, we must first return to more general matters. We shall discuss aspects of the following somewhat vague question. Suppose that we know the value $f(X_0)$ and we also have a good deal of information about the total derivative of f; this total derivative should tell us something about the way $f(X)$ varies as X varies. Consequently, we ask, "What is $f(X)$ for X near X_0?"

Consider the following answer to this question in the case of one-variable calculus.

MEAN VALUE THEOREM: Let $g : \mathbb{R} \longrightarrow \mathbb{R}$ be differentiable in an interval containing the points t_0 and t. Then there exists a point t^* between t_0 and t and depending on them such that

$$g(t) = g(t_0) + g'(t^*)(t - t_0).$$

This shows that the value $g(t)$ varies from $g(t_0)$ by an amount $g'(t^*)$ $(t - t_0)$, depending on the variation of t from t_0, that is, $t - t_0$, and the rate of change $g'(t^*)$ of the function g with respect to t. Such a statement is entirely reasonable and worthy of meditation. Note that the Mean Value Theorem is an "existence" theorem. In general, we cannot find t^* explicitly. Usually it is enough to know that it exists.

Instead of the formula given in the statement of the theorem, you may have seen

$$g'(t^*) = \frac{g(t) - g(t_0)}{t - t_0}.$$

Note that the two formulas are the same. This version may be interpreted

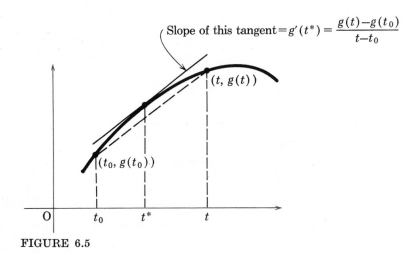

FIGURE 6.5

by noting that the slope of the tangent line to the graph of g directly above t^* equals the slope of the straight line through the two points $(t_0, g(t_0))$ and $(t, g(t))$, because the two lines are parallel. See Fig. 6.5. But the first slope is $g'(t^*)$, and the second is $(g(t) - g(t_0))/(t - t_0)$.

The Mean Value Theorem has many uses. For example, you should be able now to prove that if $g' = 0$ at all points, then g is a constant function; that is, $g(t) = g(t_0)$ for all t. This is a good example of knowledge of the derivative's yielding us knowledge of the original function. Do it.

We continue our informal discussion of one-variable calculus. In the next theorem, which may be familiar, we see that if we know a bit more about derivatives of $g(t)$, we may be able to say even more about the values of the function.

TAYLOR'S THEOREM WITH TWO TERMS: Let $g: \mathbb{R} \longrightarrow \mathbb{R}$ be twice differentiable in an interval containing the points t_0 and t. Then there exists a point t^* between t_0 and t and depending on them such that

$$g(t) = g(t_0) + g'(t_0)(t - t_0) + \tfrac{1}{2}g''(t^*)(t - t_0)^2.$$

Note how this improves the Mean Value Theorem. It tells us that the difference between the actual value $g(t)$ and the value $T(t) = g(t_0) + g'(t_0)(t - t_0)$ of the best affine approximation to the function g is measured by the second derivative of g and the quantity $(t - t_0)^2$. Note that $(t - t_0)^2$ is very small if $|t - t_0|$ is small, that is, if t is close to t_0.

Observe also that the expression $g(t_0) + g'(t_0)(t - t_0) + \tfrac{1}{2}g''(t^*)(t - t_0)^2$ is "almost a polynomial" in $t - t_0$. The numbers t_0, $g(t_0)$ and $g'(t_0)$ are fixed and act as coefficients. Only t and t^* may vary. The expression is useful because polynomials are more easily dealt with than more general functions.

If, as often happens, g'' is a continuous function, we may use Taylor's Theorem to obtain the following well-known result about extrema in the one-variable case.

SECOND DERIVATIVE TEST: Let t_0 be a critical point, $g'(t_0) = 0$:

1. If $g''(t_0) < 0$, then t_0 is a local maximum for g.
2. If $g''(t_0) > 0$, then t_0 is a local minimum for g.
3. If $g''(t_0) = 0$, then the test fails, and other methods must be used to decide the nature of t_0.

Now you see why we have discussed the one-variable case at such length. Our goal now is to prove a version of the Second Derivative Test for functions $f(X)$ where X is a vector. This will enable us to decide the nature of a given critical point X_0 and hence complete our study of local extrema. To do this, we will generalize the Mean Value Theorem, define a second derivative $f''(X_0)$, prove Taylor's Theorem and, at last, obtain a general Second Derivative Test for $f(X)$.

6.2b The Mean Value Theorem

Let us suppose for simplicity that the scalar-valued function $f(X)$ is defined for all $X \in \mathbb{R}^n$; that is, $\mathscr{D}(f) = \mathbb{R}^n$. The properties we will establish are true for more general functions, but we leave the details to you.

We need a simple version of the Chain Rule for $f(X)$ that generalizes the statement that $d/dt \, g(a + bt) = g'(a + bt)b$ from one-variable calculus.

LEMMA 1 :   (Simple Chain Rule.) Let f be defined and differentiable on all of $\mathbb{R}^n$, and let $X_0, Z \in \mathbb{R}^n$. Then

$$\frac{d}{dt} f(X_0 + tZ) = f'(X_0 + tZ)Z.$$

NOTE: The right-hand side is a number (depending on t) obtained by operating with a linear functional $f'(X_0 + tZ)$ on the vector Z. Hence it equals $\langle \nabla f(X_0 + tZ), Z \rangle$, which equals the derivative $\nabla_Z f(X_0 + tZ)$.

Proof: Observe that once X_0 and Z are fixed, $f(X_0 + tZ)$ is a scalar-valued function of the variable t. Thus we have

$$\frac{d}{dt} f(X_0 + tZ) = \lim_{\Delta t \to 0} \frac{f(X_0 + (t + \Delta t)Z) - f(X_0 + tZ)}{\Delta t}$$

$$= \lim_{\Delta t \to 0} \frac{f(X_0 + tZ + \Delta tZ) - f(X_0 + tZ)}{\Delta t}$$

$$= \nabla_Z f(X_0 + tZ) \qquad \text{by definition}$$

$$= f'(X_0 + tZ)Z.$$

by the Main Theorem on Differentiable Functions in Chap. 5. ≪

EXAMPLE: Let $f(X) = x_1^2 + 3x_2^2$, $X_0 = (1, 1)$, $Z = (2, 3)$.   What is $d/dt \, f(X_0 + tZ)$? Note that it should depend on t.

In order to apply the Chain Rule just proved, we should first compute $f'(X_0 + tZ)$. We know that $f'(X)$ is the linear map with matrix $[f'(X)] = [2x_1 \quad 6x_2]$. Since $X_0 + tZ = (1 + 2t, 1 + 3t)$, it follows that $f'(X_0 + tZ)$ has the matrix $[f'(X_0 + tZ)] = [2(1 + 2t) \quad 6(1 + 3t)]$.

Now the chain rule gives

$$\frac{d}{dt}f(X_0 + tZ) = f'(X_0 + tZ)Z$$

$$= [f'(X_0 + tZ)][Z] \quad \text{(matrices!)}$$

$$= [2 + 4t \quad 6 + 18t]\begin{bmatrix} 2 \\ 3 \end{bmatrix}$$

$$= (2 + 4t)2 + (6 + 18t)3 = 22 + 62t.$$

This is verified by noting that $f(X_0 + tZ) = (1 + 2t)^2 + 3(1 + 3t)^2$ as a function of t. By ordinary differentiation, therefore,

$$\frac{d}{dt}f(X_0 + tZ) = 4(1 + 2t) + 18(1 + 3t) = 22 + 62t,$$

agreeing with our result above.

We use the above in our proof of the following basic result.

THEOREM 2: (Mean Value Theorem.) Let f be defined and differentiable in all of $\mathbb{R}^n$, and let $X_0, X \in \mathbb{R}^n$. Then there exists a point X^* on the line segment between X_0 and X and depending on them such that

$$f(X) = f(X_0) + f'(X^*)(X - X_0).$$

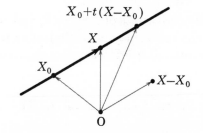

$X_0 + t(X - X_0)$

X

X_0

$X - X_0$

O

As t increases, the point $X_0 + t(X - X_0)$ traces out a line through X_0 and X in the direction indicated by $X - X_0$

FIGURE 6.6

NOTE: The second (linear) term on the right-hand side denotes the operation of the linear map $f'(X^*)$ on the vector $X - X_0$. It is a word-for-word generalization of the one-variable theorem discussed in the introduction.

Proof: The line segment from X_0 to X consists of all points $X_0 + t(X - X_0)$ with $0 \le t \le 1$. See Fig. 6.6. Putting $t = 0$ yields X_0, and $t = 1$ yields X.

Now define $g(t) = f(X_0 + t(X - X_0))$, a function of t alone. The one-variable Mean Value Theorem guarantees the existence of a number t^*, $0 < t^* < 1$, such that

$$g(1) = g(0) + g'(t^*)(1 - 0) = g(0) + g'(t^*). \tag{A}$$

But $g(1) = f(X)$, $g(0) = f(X_0)$, and, by the Chain Rule proved above,

$$g'(t) = \frac{d}{dt}f(X_0 + t(X - X_0)) = f'(X_0 + t(X - X_0))(X - X_0).$$

Putting $X^* = X_0 + t^*(X - X_0)$, we find that $g'(t^*) = f'(X^*)(X - X_0)$. On substituting these values in (A), we are done. $\ll$

We leave it to you to prove the following familiar-looking statement.

COROLLARY 3: Let $f : \mathbb{R}^n \to \mathbb{R}$ be a differentiable function. If the total derivative $f'(X) = $ zero for all X, then f is a constant function.

6.2c The second derivative of $f(X)$

We will see that the "proper" generalization of the second derivative to scalar-valued functions $f : \mathbb{R}^n \to \mathbb{R}$ is a linear map from $\mathbb{R}^n$ to $\mathbb{R}^n$; that is, the second derivative will be written as an n by n matrix.

DEFINITION: The **second derivative matrix**, or **Hessian**, of the function $f : \mathbb{R}^n \to \mathbb{R}$ at the point X_0 is the n by n matrix whose i, jth entry is the number $f_{ij}(X_0) \equiv f_{x_i x_j}(X_0)$.

We denote it $[f''(X_0)]$.

Thus, if $n = 3$, we have

$$[f''(X_0)] = \begin{bmatrix} f_{11}(X_0) & f_{12}(X_0) & f_{13}(X_0) \\ f_{21}(X_0) & f_{22}(X_0) & f_{23}(X_0) \\ f_{31}(X_0) & f_{32}(X_0) & f_{33}(X_0) \end{bmatrix}.$$

Note that there are n^2 second partial derivatives of $f : \mathbb{R}^n \to \mathbb{R}$, and we might expect the "total" second derivative to display them all, just as the "total" first derivative is constructed from all n first partial derivatives, $[f'(X_0)] = [f_{x_1}(X_0) \cdots f_{x_n}(X_0)]$.

EXAMPLE: Let $f(X) = x_1^2 x_2 + \sin x_2$. Then $f_{x_1}(X) = 2x_1 x_2$, $f_{x_2}(X) = x_1^2 + \cos x_2$. Now we take *second* partials:

$$f_{x_1 x_1}(X) = 2x_2, \qquad f_{x_1 x_2}(X) = 2x_1,$$

$$f_{x_2 x_1}(X) = 2x_1, \qquad f_{x_2 x_2}(X) = -\sin x_2.$$

Hence

$$[f''(X)] = \begin{bmatrix} 2x_2 & 2x_1 \\ 2x_1 & -\sin x_2 \end{bmatrix}.$$

For instance, if $X_0 = (-1, \pi/2)$, $X_1 = (0, \pi)$, we obtain

$$[f''(X_0)] = \begin{bmatrix} \pi & -2 \\ -2 & -1 \end{bmatrix}, \qquad [f''(X_1)] = \begin{bmatrix} 2\pi & 0 \\ 0 & 0 \end{bmatrix}.$$

Of course the second derivative matrix is different at different points X_0, X_1, and so on.

REMARK: For the functions we are considering, the mixed second partials are equal, $f_{x_i x_j}(X) = f_{x_j x_i}(X)$ for all i, j. This assures us that the second-derivative matrix is "symmetric" in the sense defined below.

DEFINITION: An n by n matrix $A = [\alpha_{ij}]$ is **symmetric** if and only if $\alpha_{ij} = \alpha_{ji}$ for all pairs of indices i, j.

The two second-derivative matrices above are symmetric. Some other examples:

$$\text{Symmetric:} \quad \begin{bmatrix} 3 & 1 & 0 \\ 1 & -1 & 2 \\ 0 & 2 & 0 \end{bmatrix}, \qquad \text{Nonsymmetric:} \quad \begin{bmatrix} 3 & 1 & 0 \\ 1 & -1 & 2 \\ 1 & 2 & 0 \end{bmatrix}.$$

6.2d Taylor's Theorem for $f(X)$

Just as in the one-variable case, when

$$g(t) = g(t_0) + g'(t_0)(t - t_0) + \tfrac{1}{2}g''(t^*)(t - t_0)^2,$$

we will obtain an exact expression for $f(X)$, not a quadratic *approximation*, involving the second derivative.

THEOREM 4: (Taylor's Theorem with two terms.) Let the twice-differentiable function $f(X)$ be defined for all $X \in \mathbb{R}^n$, and let $X_0, X \in \mathbb{R}^n$. Then there exists a point X^* on the line segment between X_0 and X such that

$$f(X) = f(X_0) + f'(X_0)(X - X_0) + \tfrac{1}{2}\langle f''(X^*)(X - X_0), X - X_0 \rangle.$$

NOTE: We obtain an exact expression for $f(X)$ at the expense of not knowing exactly where the point X^* is.

Proof: We use ideas we have seen already. Given X_0 and X, define

$$g(t) = f(X_0 + t(X - X_0)),$$

a scalar-valued function of the scalar t. We have $g(0) = f(X_0), g(1) = f(X)$, and by the Chain Rule (Lemma 1), $g'(0) = f'(X_0)(X - X_0)$. Now we know that $g(1) = g(0) + g'(0)(1 - 0) + \tfrac{1}{2}g''(t^*)(1 - 0)^2$ by the one-variable case of Taylor's Theorem; that is, $g(1) = g(0) + g'(0) + \tfrac{1}{2}g''(t^*)$. This means that

$$f(X) = f(X_0) + f'(X_0)(X - X_0) + \tfrac{1}{2}g''(t^*).$$

Hence we must show that $g''(t^*) = \langle f''(X^*)(X - X_0), X - X_0 \rangle$.

We know that $g'(t) = f'(X_0 + t(X - X_0))(X - X_0)$, by Lemma 1. Suppose for simplicity of notation that $n = 2$; the general case proceeds in exactly the same way.

Let $X = (x_1, x_2)$, $X_0 = (c_1, c_2)$, $[f'(X)] = [f_{x_1}(X) \; f_{x_2}(X)]$. We rewrite $g'(t)$ thus:

$$g'(t) = f_{x_1}(X_0 + t(X - X_0))(x_1 - c_1) + f_{x_2}(X_0 + t(X - X_0))(x_2 - c_2).$$

We differentiate again, applying the Chain Rule to both f_{x_1} and f_{x_2}, to get

$$\begin{aligned}
g''(t) = {} & f_{x_1 x_1}(X_0 + t(X - X_0))(x_1 - c_1)^2 \\
& + f_{x_1 x_2}(X_0 + t(X - X_0))(x_1 - c_1)(x_2 - c_2) \\
& + f_{x_2 x_1}(X_0 + t(X - X_0))(x_2 - c_2)(x_1 - c_1) \\
& + f_{x_2 x_2}(X_0 + t(X - X_0))(x_2 - c_2)^2.
\end{aligned}$$

Now for $t = t^*$, define $X^* = X_0 + t^*(X - X_0)$. Using the matrix notation for the second derivative, we obtain, as required,

$$g''(t^*) = \langle f''(X^*)(X - X_0), X - X_0 \rangle.$$

This completes the last step in the proof. $\ll$

REMARKS: 1. In the following section we use Taylor's Theorem to prove the Second Derivative Test, which is used in classifying critical points as maxima, minima, or saddle points.

2. The proofs of the Mean Value Theorem and Taylor's Theorem given above both depended on the truth of these theorems for functions $g(t)$ of one variable.

Exercises

1. Find the second derivative matrix for each of the following functions:

(a) $f(x_1, x_2) = x_1^2 - 4x_1 x_2 + x_2^2$,

(b) $g(x, y) = 2xy$,

(c) $h(x, y) = e^{3x - 2y}$,

(d) $\varphi(x_1, x_2) = \sin(x_1 x_2)$,

(e) $f(x_1, x_2, x_3) = x_1 x_2 x_3$,

(f) $u(x, y, z) = xe^{yz}$.

2. Let $f, g: \mathbb{R}^2 \to \mathbb{R}$ be given functions, and assume that they have arbitrarily many continuous partial derivatives:

(a) If $f'(X) = 0$ for all $X \in \mathbb{R}^2$, prove that $f(X) \equiv$ constant.

(b) If $f'(X) = g'(X)$ for all $X \in \mathbb{R}^2$ and $f(X_0) = g(X_0)$ for some fixed X_0, what can you conclude? Proof?

(c) If $f'(X) = (3, -2)$ for all $X \in \mathbb{R}^2$ and $f(1, -1) = 4$, find f. Is f uniquely determined? Proof?

3. Let $f: \mathbb{R}^2 \to \mathbb{R}$ have arbitrarily many continuous partial derivatives:

(a) If $f''(X) = 0$ for all $X \in \mathbb{R}^2$, prove that f is an affine map.

(b) If

$$f''(X) = \begin{bmatrix} 1 & 2 \\ 2 & -3 \end{bmatrix}$$

for all $X \in \mathbb{R}^2$, $f(1, 0) = 7$, and $f'(1, 0) = (-2, 5)$, find f. Is f uniquely determined? Proof?

4. Show that the surface $z = 2x^2 + 3y^2$ in $\mathbb{R}^3$ lies above its tangent plane at any point by showing that the second derivative term, in the Taylor expansion there, is never negative.

5. Repeat Exercise 4 for the surface $z = x^4 + 3x + e^{-y} - 2$.

6. Let $f: \mathbb{R} \longrightarrow \mathbb{R}$ be twice continuously differentiable:

    (a) If $f''(x) \geq 0$ for all x, prove that the graph of the curve $y = f(x)$ lies above its tangent line at any point.

    (b) f is called **convex** if $f\left(\dfrac{a+b}{2}\right) \leq \dfrac{1}{2}[f(a) + f(b)]$ for every a, b. If $f''(x) \geq 0$ for all x, prove that f is convex.

7. Let $f: \mathbb{R}^n \longrightarrow \mathbb{R}$ be a differentiable function.

    (a) Let S denote the straight line segment joining two points, X_1 and X_2, in $\mathbb{R}^n$. If the total derivative, $f'(X) = 0$ for every point X in S, prove that $f(X_1) = f(X_2)$.

    (b) Let A be a connected open set in $\mathbb{R}^n$ (see Sec. 4.2). If $f'(X) = 0$ for every point X in A, prove that f is constant in A; that is, $f(X) = f(Z)$ for any X, Z in A. Also, give an (easy) example showing this may be false if A is not connected.

6.3 THE SECOND DERIVATIVE TEST FOR LOCAL EXTREMA

6.3a Introduction

In this section we discuss a method that, in many cases, enables us to decide whether a critical point X_0 is a local maximum, minimum, or saddle point. You should not be surprised to hear that this method is the generalization to functions $f(X)$, with X a vector, of a method from the ordinary calculus of one real variable.

Let us recall the proof of one part of the Second-Derivative Test in the one-variable case, namely:

If $g''(t_0) < 0$, then t_0 is a local maximum for $g(t)$. Here, of course, t_0 is a critical point, $g'(t_0) = 0$. The proof is as follows. Since $g'(t_0) = 0$, Taylor's Theorem gives $g(t) = g(t_0) + \frac{1}{2}g''(t^*)(t - t_0)^2$. Given $g''(t_0) < 0$, then if t is close enough to t_0 but different from it, and g'' is continuous, we may conclude also that $g''(t^*) < 0$ (recall that t^* is between t and t_0 and hence close to t_0 if t is close.) But this implies that $\frac{1}{2}g''(t^*)(t - t_0)^2 < 0$, because $(t - t_0)^2 > 0$. Hence $g(t) = g(t_0) + a$ negative number $< g(t_0)$. This means that t_0 is a local maximum. Done.

In the case of $f(X)$, $X = (x_1, \ldots, x_n)$, the second derivative is given by an n by n matrix, not a number, and hence cannot be positive or negative in the usual sense of numbers. What we do in the following paragraph is define what we mean for a second derivative matrix to be "positive" or "negative." Actually, matters are somewhat more complicated in the case of matrices, which need not be either positive or negative or zero (see below).

Once we have these notions for the second derivative matrix, the proof of the generalized Second Derivative Test will be much the same as in the one-variable case of ordinary calculus.

6.3b Positivity in symmetric matrices

Let $A = [\alpha_{ij}]$ be an n by n symmetric matrix; that is, $\alpha_{ij} = \alpha_{ji}$ for all i, j (see Sec. 6.2c). We know that the matrix A determines a linear map $A: \mathbb{R}^n \to \mathbb{R}^n$ (for brevity we use the same symbol for both matrix and linear map.) Using the map A, we may define a "quadratic" function ("quadratic form") $h: \mathbb{R}^n \to \mathbb{R}$ by the formula

$$h(X) = \langle X, AX \rangle.$$

This equals $\langle AX, X \rangle$, of course, since $\langle X, Y \rangle = \langle Y, X \rangle$ for all X, Y. Note that if $A = I$, the identity, then $h(X) = \langle X, IX \rangle = \langle X, X \rangle$ is just the usual sum of squares $h(X) = x_1^2 + \cdots + x_n^2$. It is clear that $X = O$ implies that $h(X) = \langle X, AX \rangle = 0$.

Now let us restrict ourselves to vectors X different from O. In this case, precisely one of the following five phenomena occur for a given nonzero A:

1. $h(X) = \langle X, AX \rangle > 0$ for all $X \neq O$. In this case, A is said to be **positive definite.**

2. $h(X) = \langle X, AX \rangle \geq 0$ for all X, and, in particular, there exists some $X \neq O$ such that $\langle X, AX \rangle = 0$. In this case, A is **positive nondefinite.**

3. $h(X) = \langle X, AX \rangle$ assumes both positive and negative values. In this case, A is **indefinite.**

4. $h(X) = \langle X, AX \rangle \leq 0$ for all X, and, in particular, there exists some $X \neq O$ such that $\langle X, AX \rangle = 0$. In this case, A is **negative nondefinite.**

5. $h(X) = \langle X, AX \rangle < 0$ for all $X \neq O$. In this case, A is said to be **negative definite.**

You should convince yourself that these cases are mutually exclusive for $A \neq 0$.

EXAMPLES: 1. The n by n identity I is positive definite, because $\langle X, IX \rangle = \langle X, X \rangle = x_1^2 + \cdots + x_n^2 > 0$ if at least one of the x_i is nonzero.

2. The matrix $-I$ is negative definite.

3. The matrix

$$A = \begin{bmatrix} 2 & 0 \\ 0 & 3 \end{bmatrix}$$

is positive definite, because $\langle X, AX \rangle = 2x_1^2 + 3x_2^2 > 0$ for $X \neq (0, 0)$.

4. The matrix

$$B = \begin{bmatrix} 2 & 0 \\ 0 & -3 \end{bmatrix}$$

is indefinite, since $\langle X, BX \rangle = 2x_1^2 - 3x_2^2$, which clearly assumes both positive and negative values.

5. The matrix

$$A = \begin{bmatrix} 1 & 0 \\ 0 & 0 \end{bmatrix}$$

leads to $\langle X, AX \rangle = x_1^2$. This is zero if X has the form $(0, x_2)$ and positive otherwise. Hence A is positive nondefinite.

An important issue in linear algebra is that of deciding whether a given symmetric matrix A is positive definite, positive nondefinite, or whatever. For example, we are concerned with just these properties of the matrix $[f''(X_0)]$ in the following subsection. The question of positivity for a general n by n matrix A may be quite complicated. In the relatively uncomplicated case where A is a 2 by 2 symmetric matrix, there are some usable criteria (see Theorem 6 below.) Consequently, we emphasize the 2 by 2 case.

First, however, we establish one very useful result that applies to symmetric matrices of any size. This is a sufficient condition for indefiniteness.

THEOREM 5: Let $A = [\alpha_{ij}]$ be an n by n symmetric matrix. If A has two diagonal entries of opposite sign, say $\alpha_{ii} > 0$ and $\alpha_{jj} < 0$, then A is indefinite.

Proof: You may check that $\langle e_k, Ae_k \rangle = \alpha_{kk}$ for each standard basis vector e_k. Thus $\langle e_i, Ae_i \rangle = \alpha_{ii} > 0$, and $\langle e_j, Ae_j \rangle = \alpha_{jj} < 0$, so that A is indefinite as claimed. $\ll$

EXAMPLES: Theorem 5 enables us to conclude that the following symmetric matrices are indefinite:

$$\begin{bmatrix} 2 & 3 \\ 3 & -1 \end{bmatrix}, \quad \begin{bmatrix} 3 & 0 & 2 \\ 0 & -1 & 1 \\ 2 & 1 & -5 \end{bmatrix}, \quad \begin{bmatrix} 6 & 2 & 3 & 0 & 1 \\ 2 & 0 & -1 & 4 & -1 \\ 3 & -1 & 4 & -4 & 0 \\ 0 & 4 & -4 & 0 & 2 \\ 1 & -1 & 0 & 2 & -4 \end{bmatrix}.$$

CAUTION: The converse of the theorem above is false. If A is indefinite, it does not follow that two diagonal entries of A have opposite sign. For example,

$$A = \begin{bmatrix} 1 & -5 \\ -5 & 2 \end{bmatrix}$$

is indefinite, because $\langle e_1, Ae_1 \rangle = 1 > 0$, but if $X = (1, 1)$, then $\langle X, AX \rangle = -7 < 0$. Hence there are even more indefinite matrices than Theorem 5 indicates.

The 2 by 2 case. If

$$A = \begin{bmatrix} \alpha & \beta \\ \gamma & \delta \end{bmatrix}$$

is any 2 by 2 matrix, not necessarily symmetric, then we recall the determinant of A, written det A, is the number

$$\det A = \alpha\delta - \beta\gamma.$$

For example, if

$$A = \begin{bmatrix} 2 & -1 \\ -5 & 1 \end{bmatrix},$$

then det $A = -3$.

Note that A symmetric implies $\beta = \gamma$, and so $\det A = \alpha\delta - \beta^2$.

Now let A be symmetric,

$$A = \begin{bmatrix} \alpha & \beta \\ \beta & \delta \end{bmatrix}.$$

If $X = (x_1, x_2)$, then the quadratic form $h(X) = \langle X, AX \rangle = \alpha x_1^2 + 2\beta x_1 x_2 + \delta x_2^2$, as you may readily verify by matrix multiplication.

THEOREM 6: Let

$$A = \begin{bmatrix} \alpha & \beta \\ \beta & \delta \end{bmatrix}$$

be a 2 by 2 symmetric matrix.

Case I: $\det A > 0$. Then the diagonal entries α, δ are either both positive or both negative. Further, in this case:

    1. $\alpha > 0 \Rightarrow A$ is positive definite.

    2. $\alpha < 0 \Rightarrow A$ is negative definite.

Case II: $\det A < 0$. Then A is indefinite.

Case III: $\det A = 0$. Then A is either positive nondefinite or negative nondefinite or the zero matrix.

NOTE: To test a given 2 by 2 matrix A for positivity, we therefore begin by computing $\det A$.

Proof: We know that $h(X) = \langle X, AX \rangle = \alpha x_1^2 + 2\beta x_1 x_2 + \delta x_2^2$. Hence $h(X)$ has two possible shapes, namely, (1) at least one of α, δ different from zero or (2) $\alpha = \delta = 0$. Each shape prompts a different approach:

1. α or $\delta \neq 0$. We may as well assume $\alpha \neq 0$; the proof is the same if $\delta \neq 0$. Completing the square gives

$$h(X) = \frac{1}{\alpha}[(\alpha x_1 + \beta x_2)^2 + (\alpha\delta - \beta^2)x_2^2],$$

as you may verify by multiplying out to get $\alpha x_1^2 + 2\beta x_1 x_2 + \delta x_2^2$ again.

Case I: $\det A > 0$. Thus $\alpha\delta - \beta^2 > 0$, whence $\alpha\delta > 0$, so that α and δ are positive or negative together, as claimed. Since $\det A$ is the coefficient of x_2^2 in the expression above, it is easy to see that α positive or negative implies that $h(X)$ is positive or negative when $X \neq 0$. This yields case I.

Case II: $\det A < 0$. It is easy to see that we may choose $X = (x_1, x_2)$ so that $h(X) > 0$ and then choose a different $X = (x_1, x_2)$ so that $h(X) < 0$. Thus A is indefinite.

Case III: $\det A = 0$. Then $h(X) = (1/\alpha)(\alpha x_1 + \beta x_2)^2$. We can choose x_1, x_2 not both zero, so that $\alpha x_1 + \beta x_2 = 0$. Hence A is not definite. Also the value $h(X)$ is nonnegative or nonpositive according to whether α is positive or negative. It follows that A must be either positive nondefinite or negative nondefinite, as claimed.

2. $\alpha = \delta = 0$, thus $h(X) = 2\beta x_1 x_2$. Then $\det A = -\beta^2$, so that case I cannot occur.

Case II: $\det A < 0$. This means that $\beta \neq 0$. Clearly $h(X) = 2\beta x_1 x_2$ may be positive or negative, depending on our choice of x_1, x_2. Thus A is indefinite.

Case III: det A = 0. Thus $\beta = 0$, so that A is the zero matrix, as claimed. This completes the proof of the theorem. $\ll$

EXAMPLES: Let us test

$$A = \begin{bmatrix} 2 & 3 \\ 3 & 1 \end{bmatrix}, \qquad B = \begin{bmatrix} 4 & -3 \\ -3 & 5 \end{bmatrix}, \qquad C = \begin{bmatrix} 1 & 1 \\ 1 & 1 \end{bmatrix}$$

for positivity.

1. We compute first det $A = -7 < 0$. By Theorem 6, A is indefinite.

2. We compute first det $B = 11 > 0$. According to Theorem 6, the two diagonal entries of B should have the same sign. Indeed, both are positive, and B is therefore positive definite.

3. We compute first det $C = 0$. Theorem 6 implies that C is either positive nondefinite or negative nondefinite. But $\langle e_1, Ce_1 \rangle = 1$, so that C must be *positive* nondefinite. Note also that $\langle X, CX \rangle = x_1^2 + 2x_1x_2 + x_2^2 = (x_1 + x_2)^2 \geq 0$, so that C is clearly positive nondefinite.

6.3c The Second Derivative Test

At last we can state and prove this and thereby resolve, in theory at least, the problem of classifying a great number of critical points. In the case of $n = 2, f: \mathbb{R}^2 \rightarrow \mathbb{R}$, we are able to offer a complete solution (Theorem 8). But first the general statement.

THEOREM 7: (Second Derivative Test.) Let $f: \mathbb{R}^n \rightarrow \mathbb{R}$ be twice continuously differentiable, and let X_0 be a critical point for f.

1. If $[f''(X_0)]$ is negative definite, then X_0 is a local maximum for f.
2. If $[f''(X_0)]$ is positive definite, then X_0 is a local minimum for f.
3. If $[f''(X_0)]$ is indefinite, in particular, if two diagonal entries $f_{x_ix_i}(X_0)$ and $f_{x_jx_j}(X_0)$ are of opposite sign, then X_0 is a saddle point for f.
4. Otherwise, that is, if $[f''(X_0)]$ is positive nondefinite, negative nondefinite, or identically zero, then the test fails, and other methods must be used to determine the nature of X_0.

Proof: The idea is this. If the point X^* is sufficiently close to X_0, then $f''(X^*)$ is "close" to $f''(X_0)$; that is, the value $\langle f''(X^*)(X - X_0), X - X_0 \rangle$ is close to the value $\langle f''(X_0)(X - X_0), X - X_0 \rangle$ for all vectors $X - X_0$. This follows from continuity. It implies that if $f''(X_0)$ is positive definite, negative definite, or indefinite, then $f''(X^*)$ is also. (Caution: If $f''(X_0)$ is nondefinite or zero, it does not follow that $f''(X^*)$ is likewise.)

Now let us prove statement 1. Taylor's Theorem gives
$$f(X) = f(X_0) + \tfrac{1}{2}\langle f''(X^*)(X - X_0), X - X_0\rangle,$$
since $f'(X_0) = 0$. For all X sufficiently close to X_0, the point X^* is even closer; by the paragraph above, therefore, $f''(X^*)$ is negative definite; that is, $\tfrac{1}{2}\langle f''(X^*)(X - X_0), X - X_0\rangle < 0$ if $X \neq X_0$. Hence $f(X) < f(X_0)$ and X_0 is a local maximum, as claimed in statement 1.

The proofs of statements 2 and 3 are similar. Note that statement 3 includes the useful Theorem 5.

Statement 4 is discussed in the examples. This completes the proof of the theorem. $\ll$

Reading the statement of the theorem, you will understand why it may be important to decide whether a given matrix is positive definite, negative definite, or whatever. If n is large, then it may be difficult to decide the nature of the second derivative matrix $[f''(X_0)]$, although statement 3 gives an easily applied sufficient criterion for indefiniteness. If $n = 2$, then Theorem 6 applies, and we have the following analysis. We write $X = (x, y)$ for variety.

THEOREM 8: (Second Derivative Test in two variables.) Let $f: \mathbb{R}^2 \to \mathbb{R}$ be twice continuously differentiable, and let X_0 be a critical point for f.

   Case I: $\det [f''(X_0)] > 0$. Then the diagonal entries $f_{xx}(X_0)$ and $f_{yy}(X_0)$ are both positive or both negative. Further, in this case:

   1. $f_{xx}(X_0) > 0 \Rightarrow X_0$ is a local minimum for f.
   2. $f_{xx}(X_0) < 0 \Rightarrow X_0$ is a local maximum for f.

   Case II: $\det [f''(X_0)] < 0$. Then X_0 is a saddle point for f.

   Case III: $\det [f''(X_0)] = 0$. Then f may have a maximum, a minimum, or a saddle point at X_0. This test fails to decide and other tests must be used.

Proof: Combine Theorems 6 and 7, and the result is immediate. $\ll$

EXAMPLES OF LOCAL EXTREMA: In each of these we are to find *all* the local maxima, minima, and saddle points. The procedure we have developed is this:

First step: Locate all critical points by solving $f'(X) = 0$ for X.

Second step: Test each critical point X_0 by the Second Derivative Test (or other methods) to decide its nature.

1. Let $f: \mathbb{R}^2 \to \mathbb{R}$, $f(X) = 1 - x_1^2 - x_2^2$. We found in Sec. 6.1 that $X_0 = (0, 0)$ is the only critical point of f. We readily compute

$$[f''(X)] = \begin{bmatrix} -2 & 0 \\ 0 & -2 \end{bmatrix}$$

for all X, not just X_0. Since this is negative definite, Theorem 7 or 8 implies that $X_0 = (0, 0)$ is a local maximum. Of course, we could have decided this from the graph of f as well.

2. Let $g(x, y) = 3 - 2x + x^2 + 4y^2$.

First step: We set the partial derivatives equal to zero:

$$g_x(x, y) = -2 + 2x = 0, \qquad g_y(x, y) = 8y = 0,$$

obtaining one critical point $X_0 = (x, y) = (1, 0)$.

Second step: We compute

$$[g''(X)] = \begin{bmatrix} 2 & 0 \\ 0 & 8 \end{bmatrix}$$

for all X. This is positive definite (why?), and so the crtical point $(1, 0)$ is a local minimum.

3. Let $f(x, y) = \{x^2 + (y + 1)^2\}\{x^2 + (y - 1)^2\}$.   Before computing blindly, we note that $f(x, y) \geq 0$ and $f(x, y) = 0$ only if $(x, y) = (0, -1)$ or $(0, 1)$. Hence, we deduce that $X_0 = (0, -1), X_1 = (0, 1)$ are local minima (why?). But there may be others. It is not clear at the moment what the surface $z = f(x, y)$ looks like. So we must be systematic.

First step: We compute the partial derivative with respect to x:

$$f_x(x, y) = 2x\{x^2 + (y - 1)^2\} + \{x^2 + (y + 1)^2\}2x.$$

Likewise, we compute

$$f_y(x, y) = 2(y + 1)\{x^2 + (y - 1)^2\} + 2(y - 1)\{x^2 + (y + 1)^2\}.$$

Now $f_x(x, y) = 0$ implies that $x = 0$. With $x = 0$, $f_y(0, y) = 0$ implies that $y = -1$ or $y = 1$ or $y = 0$. Check this.

Hence the critical points of f are $X_0 = (0, -1)$, $X_1 = (0, 1)$, both found above, and also $X_2 = (0, 0)$. This third critical point is a revelation. Is it a maximum, minimum, or saddle point?

Second step: We must examine the second-derivative matrices. We compute directly

$$f_{xx}(x, y) = 12x^2 + 4y^2 + 4,$$

$$f_{xy}(x, y) = 8xy,$$

$$f_{yy}(x, y) = 4x^2 + 12y^2 - 4.$$

Thus for X_0, X_1, X_2 above, we obtain by substitution

$$[f''(X_0)] = [f''(X_1)] = \begin{bmatrix} 8 & 0 \\ 0 & 8 \end{bmatrix}, \qquad [f''(X_2)] = \begin{bmatrix} 4 & 0 \\ 0 & -4 \end{bmatrix}.$$

We see that the left-hand matrix is positive definite, and so X_0, X_1 are local minima, as expected. On the other hand, $X_2 = (0, 0)$ yields an indefinite matrix (see the theorems above), and so this point is a saddle point for f. We note that $f(X_2) = f(0, 0) = 1$. Now that we have the extrema of f, we may sketch a portion of its graph $z = f(x, y)$. See Fig. 6.7. Note the two minima and the saddle point.

4. Now we examine a function of three variables. Let $f: \mathbb{R}^3 \to \mathbb{R}$, $f(x, y, z) = 1 - 2x + 3x^2 - xy + xz - z^2 + 4z + y^2 + 2yz$.

First step: We set the partial derivatives $f_x = f_x(x, y, z)$, f_y, f_z equal to zero:

$$f_x = -2 + 6x - y + z = 0,$$

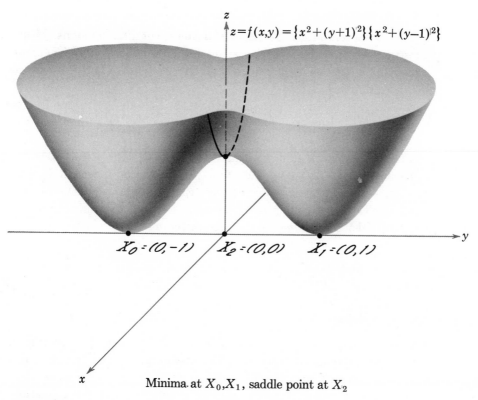

$$z = f(x,y) = \{x^2 + (y+1)^2\}\{x^2 + (y-1)^2\}$$

$X_0 = (0, -1)$ $X_2 = (0, 0)$ $X_1 = (0, 1)$

Minima at X_0, X_1, saddle point at X_2

FIGURE 6.7

$$f_y = \qquad -x + 2y + 2z = 0,$$
$$f_z = \quad 4 + \ x + 2y - 2z = 0.$$

This is a system of three linear equations in the unknowns x, y, z. By eliminating variables as usual, we find that the unique solution to this system is $(x, y, z) = (0, -1, 1)$. Hence $X_0 = (0, -1, 1)$ is the unique critical point of f.

Second step: To compute the second derivative matrix, we take all second partials, thus:

$$f_{xx} = 6, \qquad f_{xy} = -1, \qquad f_{xz} = \quad 1,$$
$$f_{yy} = \quad 2, \qquad f_{yz} = \quad 2,$$
$$f_{zz} = -2.$$

Note that $f_{xy} = f_{yx}$, and so on. Also, for all (x, y, z), the matrix is constant:

$$[f''(x, y, z)] = \begin{bmatrix} 6 & -1 & 1 \\ -1 & 2 & 2 \\ 1 & 2 & -2 \end{bmatrix}.$$

Now we observe that the diagonal terms are of different signs. Theorem 5 assures us that the matrix is indefinite, therefore, and Theorem 7, statement 3, says that the critical point $X_0 = (0, -1, 1)$ is a *saddle point* for the function f.

5. Now we consider a situation where the Second Derivative Test fails to apply (see Exercise 9 for others). Consider the two functions $f(x, y) = x^4 + y^4$, $g(x, y) = -f(x, y) = -x^4 - y^4$.

First step: Both functions have the same critical point $X_0 = (0, 0)$.

Second step: At $(0, 0)$, each function has the same second-derivative matrix, namely, the zero matrix—an easy computation. On the other hand, f clearly has a minimum, since $f \geq 0$, and g a maximum at $X_0 = (0, 0)$. Hence the *same* second derivative matrix

$$\begin{bmatrix} 0 & 0 \\ 0 & 0 \end{bmatrix}$$

could not be expected to inform us of the *different* natures of f and g. The Second Derivative Test *had* to fail in this situation.

REMARK: The examples above illustrate the method for locating and classifying extrema of $f(X)$. Roughly speaking, the search for extrema has been reduced to solving equations $(f'(X) = 0)$ and linear algebra (studying a certain symmetric matrix). In carrying out this reduction of the problem, however, we made a general study of functions involving the directional derivative, the Chain Rule, the Mean Value Theorem, the second derivative $f''(X_0)$, and Taylor's Theorem—all monuments of calculus.

Exercises

1. Determine if the following matrices are positive definite, indefinite, and so on:

(a) $\begin{bmatrix} 1 & 0 \\ 0 & 2 \end{bmatrix}$,

(b) $\begin{bmatrix} 1 & 1 \\ 1 & -2 \end{bmatrix}$,

(c) $\begin{bmatrix} 3 & 5 \\ 5 & 4 \end{bmatrix}$,

(d) $\begin{bmatrix} -3 & 2 \\ 2 & -4 \end{bmatrix}$,

(e) $\begin{bmatrix} 1 & 2 \\ 2 & 4 \end{bmatrix}$,

(f) $\begin{bmatrix} -1 & 2 \\ 2 & -4 \end{bmatrix}$,

(g) $\begin{bmatrix} 0 & 0 \\ 0 & 3 \end{bmatrix}$,

(h) $\begin{bmatrix} 1 & 1 \\ 1 & 1 \end{bmatrix}$,

(i) $\begin{bmatrix} 1 & 2 & 3 \\ 2 & -4 & 5 \\ 3 & 5 & 6 \end{bmatrix}$,

(j) $\begin{bmatrix} 1 & 0 & 0 \\ 0 & 2 & 0 \\ 0 & 0 & 4 \end{bmatrix}$,

(k) $\begin{bmatrix} 3 & 2 & 0 \\ 2 & 4 & 0 \\ 0 & 0 & 1 \end{bmatrix}$.

2. Find and classify the critical points of the following functions, even if the second derivative test fails:

(a) $f(x_1, x_2) = x_1^2 - 4x_1 + 2x_2^2 + 10$,
(b) $\varphi(x, y) = 3 - 2x + 2y + x^2 y^2$,
(c) $f(x, y) = xy - x + y - 2$,
(d) $u(x_1, x_2) = x_1^2 - x_2^2$,
(e) $g(x, y) = y^3 - 2x^2 - 2y^2 + y$,
(f) $v(x, y) = (x^2 + y^2)^2 - 8y^2$,
(g) $w(x, y) = x^2 - 2xy + \frac{1}{3}y^3 - 3y$,
(h) $\psi(x_1, x_2) = \dfrac{\cos x_2}{1 + x_1^2}$,
(i) $h(x, y) = \dfrac{e^{2x} + e^{-2x}}{1 + 2y^2}$,

(j) $f(x, y) = \dfrac{3x^4 + 4x^3 - 12x^2 + 6}{1 + y^2}$,
(k) $\psi(y, z) = \dfrac{z^3 - 3z}{1 + y^2}$,
(l) $k(x_1, x_2) = (x_1^2 - 1)e^{x_2}$,
(m) $g(x, y, z) = \dfrac{xye^y}{1 + x^2} + z^2 + 2z$,
(n) $h(x, y, z) = \dfrac{3x^4 - 6x^2 + 1}{1 + z^2} - y^2$,
(o) $\varphi(x, y, z) = x^3 - 3x + y^2 + z^2$,
(p) $r(x, y, z) = (1 + 2x + 3y - z)^2$.

3. If $f(x, y)$ has a local minimum at $X_0 = (a, b)$, then $f_{xx}(X_0) \geq 0$ and $f_{yy}(X_0) \geq 0$. (Suggestion: Consider the behavior of f on the lines $x = a$ and $y = b$.)

4. If $f(x, y)$ satisfies $3f_{xx} + 4f_{yy} = -1$ at every point, then f cannot have a local minimum anywhere.

5. Let $f: \mathbb{R}^2 \to \mathbb{R}$ have $f''(X_0)$ positive definite for every $X_0 \in \mathbb{R}^2$. Prove that the surface $z = f(X)$ lies above its tangent plane at any point (compare Exercises 4, 5, and 6 in Sec. 6.2).

6.   A function $f: \mathbb{R}^2 \longrightarrow \mathbb{R}$ is called **convex** if

$$f\left(\frac{X+Y}{2}\right) \leq \frac{1}{2}[f(X) + f(Y)]$$

for all X, Y. Prove that a (sufficiently differentiable) function f is convex if $f''(X)$ is positive nondefinite for all X.

7.   Let A be a square symmetric matrix, Z a fixed vector, and define f by

$$f(X) = \langle X, AX\rangle - 2\langle X, Z\rangle.$$

(a) If f has a critical point at X_0, prove that $AX_0 = Z$.
(b) If A is positive definite and X_0 satisfies $AX_0 = Z$, prove that f has a minimum at X_0.

8.   Let

$$A = \begin{bmatrix} a & b \\ b & c \end{bmatrix}$$

be positive definite. Prove that

$$B = \begin{bmatrix} a & b & 0 \\ b & c & 0 \\ 0 & 0 & \alpha \end{bmatrix}$$

is positive definite $\Leftrightarrow \alpha > 0$.

9.   This exercise shows that if $f''(X_0)$ is positive nondefinite, then f may possibly have a minimum or saddle point:

(a) Classify the critical points of $f(x, y) = x^2 + y^3$.
(b) Classify the critical points of $f(x, y) = x^2 + y^4$.
(c) If f has a critical point at X_0 and $f''(X_0) \neq 0$ is positive nondefinite, prove that f cannot have a local maximum at X_0; that is, if f has a local maximum at X_0, then $f''(X_0)$ is negative nondefinite or negative definite. (This is not particularly easy.)

10. The aim of this exercise is to show that the nature of a critical point cannot be determined by merely approaching it along straight lines. Let $f(x, y) = (y - 4x^2)(y - x^2)$:

(a) Make a sketch of $\mathbb{R}^2$, indicating the points where $f(x, y) = 0, f(x, y) > 0$, and $f(x, y) < 0$.
(b) Show that the origin is a critical point of f.
(c) Show that on any straight line through the origin, the function f has a local minimum at the origin.
(d) Use some other path to show that the origin is actually a saddle point of f.

11. Proof or counterexample:

(a) If a matrix is positive definite, then all its elements are positive.
(b) If all the elements of a symmetric matrix are positive, then the matrix is positive definite.
(c) A diagonal matrix is positive definite $\Leftrightarrow$ all of the diagonal elements are positive (a matrix $[\alpha_{ij}]$ is *diagonal* if $\alpha_{ij} = 0$ for all $i \neq j$).

6.4 CONSTRAINED EXTREMA

6.4a Introduction

A standard problem in applications is the following. Let f be a scalar-valued function, defined in all of $\mathbb{R}^n$, say, and let $\mathscr{K}$ be a given "small" subset of $\mathbb{R}^n$. Locate and compute the extreme values of $f(X)$, where X is constrained to be a point of $\mathscr{K}$. Here are some examples.

EXAMPLES: 1. Extremes of temperature. Let $\mathscr{K}$ be the unit disk in $\mathbb{R}^2$,

$$\mathscr{K} = \{(x, y) \in \mathbb{R}^2 \,|\, x^2 + y^2 \leq 1\},$$

representing a heated metal plate of radius 1. Suppose that the temperature at (x, y) in the plate is given by $f(x, y) = 60(y^2 - x^2)$. Which are the hottest and coldest points on the plate? We mention that basic theory, not discussed here, allows us to suppose that there are, somewhere, points of highest and lowest temperatures on the disk.

2. Maximum economic utility. The cost per unit of commodities A and B is a and b, respectively. The total amount you are to spend is c. You will purchase amounts x and y of commodities A and B, respectively. Your purchase is therefore constrained by the **budget equation**, which says that the total amount spent must be c:

$$ax + by = c.$$

Suppose that you are given a function $f(x, y)$ whose numerical value is the **utility index**. This is a pure number that economists use to indicate the satisfaction or utility one receives from having quantities x and y of commodities A and B, respectively. Constructing such a utility index function $f(x, y)$ requires economic judgments, of course. Given such a function, however, it is natural to ask for the values of x and y satisfying the budget equation that maximize this function. That is, how much of A and how much of B should be acquired to produce the greatest satisfaction $f(x, y)$, while adhering to the limited budget?

In this problem, of course, the set $\mathscr{K}$ is the straight line in $\mathbb{R}^2$ given by $ax + by = c$.

REMARKS: 1. As these examples indicate, it is natural to be concerned with only a subset of the variables (x, y). The physicist or engineer is concerned with temperature only in the plate he is studying; the economist must deal with a situation in which resources are limited by a budget of some kind.

2. These two examples are representative of two kinds of constrained

extremal problems. In the temperature example, the points (x, y) of interest were constrained to be inside the unit disk, $x^2 + y^2 \le 1$. Here the
constraint takes the form of an *inequality*, and the hottest and coldest
points may be either inside the disk or on its boundary. On the other
hand, the problem of maximum utility involved an *equality constraint*,
since the points (x, y) have to satisfy $ax + by = c$. Different tools are often
used, depending on the nature of the constraint. Moreover, the economics
example would have involved an inequality constraint if we said that you
are to spend at most c dollars; that is, $ax + by \le c$.

6.4b Solution of the temperature problem

We wish to locate the point (or points) (x_0, y_0) in the disk $\mathscr{K}$ such that
$f(x_0, y_0) = 60(y_0^2 - x_0^2)$ is the highest temperature attained in the entire
disk; likewise for the lowest temperature.

To do this, we note that the hottest point (x_0, y_0) occurs either on the
boundary (unit circle) of the disk or in the inside. If the second alternative
were the case, then the temperature $f(x_0, y_0) \ge f(x, y)$ for all nearby (x, y).
Why? This means that (x_0, y_0) is a *local maximum* of the function f. Hence
(x_0, y_0) must be a critical point of f. Now the only critical point of $f(x, y)$
$= 60(y^2 - x^2)$ occurs at $(x, y) = (0, 0)$. This is because all critical points
(x, y) are given by the simultaneous solutions of

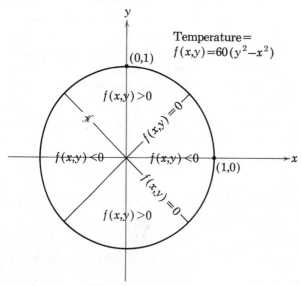

FIGURE 6.8

$$f_x(x, y) = -120x = 0, \qquad f_y(x, y) = 120y = 0.$$

Moreover, it is easy to check that $(0, 0)$ is a *saddle point* for $f(x, y)$ and not a local maximum. Hence we conclude that the hottest points on the disk $\mathscr{K}$ cannot occur inside and must occur *on the boundary*, the unit circle. See Fig. 6.8.

We now ask, "Precisely where on the unit circle does the hottest point or points occur?"

To answer this, note that the unit circle is given by $x^2 + y^2 = 1$, so that $y^2 = 1 - x^2$. If we restrict $f(x, y) = 60(y^2 - x^2)$ to the unit circle, we may eliminate the y^2 and obtain a new function, defined for $-1 \leq x \leq 1$,

$$\varphi(x) = 60(1 - 2x^2).$$

At each point (x, y) on the unit circle, the temperature is $\varphi(x)$.

It is easy to see, now by graphing $\varphi(x)$, say, that:

1. When $x = 0$, $\varphi(x) = 60$.
2. When $x = \pm 1$, $\varphi(x) = -60$.
3. When x satisfies $-1 < x < 1$ and also $x \neq 0$, then $-60 < \varphi(x) < 60$. See Fig. 6.9.

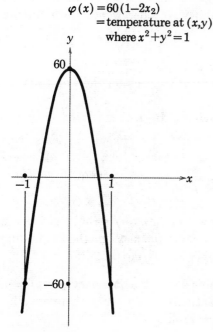

$\varphi(x) = 60(1 - 2x_2)$
$= $ temperature at (x, y)
where $x^2 + y^2 = 1$

FIGURE 6.9

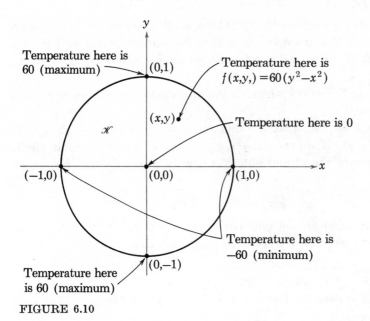

Temperature here is
60 (maximum)

Temperature here is
$f(x,y,) = 60(y^2 - x^2)$

$(0,1)$

(x,y)

Temperature here is 0

$\mathcal{K}$

$(-1,0)$

$(0,0)$

$(1,0)$

Temperature here is
-60 (minimum)

$(0,-1)$

Temperature here
is 60 (maximum)

FIGURE 6.10

Thus the maximum temperature on the disk, which is 60, occurs at the two points $(0, 1)$ and $(0, -1)$, and, by similar considerations, the minimum occurs at the points $(1, 0)$ and $(-1, 0)$. See Fig. 6.10.

NOTE: It was crucial to:

1. Realize that the extreme temperatures occur on the boundary of the disk.

2. Eliminate y from $f(x, y)$ by replacing y^2 by $1 - x^2$.

The ideas that appeared here are typical of extremal problems with inequality constraints. Say that we want to maximize $z = f(x, y)$, where $g(x, y) \geq 0$. The constraint $g(x, y) \geq 0$ appears as an inequality. There are two steps.

Step 1.   Check for local maxima and minima of f strictly inside the constraint region, $g(x, y) > 0$. This is done by the usual method of finding the critical points of f and seeing which satisfy $g(x, y) > 0$.

Step 2.   Check for maxima and minima of f on the boundary, $g(x, y) = 0$, of the constraint region. Since this is actually a problem with an equality constraint, it falls under the next paragraph too.

6.4c Equality constraints

We begin by solving the problem of maximum utility raised in the introduction. Let the budget equation be

$$2x + 3y = 10.$$

This is a *linear constraint*, since it determines a straight line in $\mathbb{R}^2$.

Suppose that the utility index function is

$$f(x, y) = 2xy + 3.$$

This gives a reasonable index of satisfaction, because (1) if the amount x is fixed and the amount y increases, then satisfaction $f(x, y)$ increases, and (2) if $f(x, y)$ is held constant and x increases, then y must decrease. Property 2 says that if satisfaction stays constant while we obtain more of one commodity, it must be because we are losing the other commodity.

To find the maximum utility subject to our budgetary constraint, use the fact that we may solve for y,

$$y = \tfrac{1}{3}(10 - 2x),$$

to eliminate y from $f(x, y)$, obtaining

$$f(x, y) = \tfrac{2}{3}x(10 - 2x) + 3$$
$$= -\tfrac{4}{3}x^2 + \tfrac{20}{3}x + 3.$$

Call this new function $h(x)$. It is crucial to note that if x_0 is an extremum of $h(x)$ in the sense of one-variable calculus, then (x_0, y_0), where $y_0 = \tfrac{1}{3}(10 - 2x_0)$, is an extremum of the same kind for f restricted to the line $2x + 3y = 10$.

Setting $h'(x) = -\tfrac{8}{3}x + \tfrac{20}{3} = 0$, we see that $x_0 = \tfrac{5}{2}$ is the only extremum of $h(x)$ and that it is a maximum.

By the remark above, $(x_0, y_0) = (\tfrac{5}{2}, \tfrac{5}{3})$ is the unique maximum point for $f(x, y)$ on the line $2x + 3y = 10$. Thus the greatest utility or satisfaction accrues when one devotes his resources to obtaining $\tfrac{5}{2}$ units of commodity A and $\tfrac{5}{3}$ of B. The utility index is then $f(x_0, y_0) = 11\tfrac{1}{3}$.

It is wise to recapitulate. We want to maximize

$$z = f(x, y),$$

where $g(x, y) = 0$. The constraint equation appears as an equality (for us, $g(x, y) = 2x + 3y - 10$).

Step 1. Solve the equation $g(x, y) = 0$ for $y = \varphi(x)$ or $x = \psi(y)$.

Step 2. Substitute into f, to get $z = h(x) \equiv f(x, \varphi(x))$, which is then solved as an ordinary maximum problem for a function of one variable.

The identical procedure extends to maximize (or minimize)

$$z = f(x_1, \ldots, x_n), \quad \text{where} \quad g(x_1, \ldots, x_n) = 0.$$

Here the constraint equation is $g = 0$.

EXAMPLE: Potatoes cost 5 cents per lb, wheat 4 cents per lb, and corn 9 cents per lb. A farmer buys x lb of potatoes, y lb of wheat, and z lb of corn each day. He is indifferent to what he buys except that $x^2yz = 225$ (this is plausible, since he will accept some wheat and almost no potatoes or corn). What should his weekly purchase be to minimize cost?

Now the cost C is

$$C = 5x + 4y + 9z,$$

which we must minimize subject to the constraint

$$x^2yz = 225.$$

It is convenient to solve this equation for y and substitute into C,

$$C = 5x + \frac{900}{x^2z} + 9z.$$

The next step is to minimize this function of two variables, so that

$$0 = C_x = 5 - \frac{1{,}800}{x^3z}, \qquad 0 = C_z = -\frac{900}{x^2z^2} + 9.$$

Consequently, $5x/2 = 900/x^2z = 9z$. We use this to eliminate z from the first equation, and we find that $x^4 = 2^4 \cdot 3^4$, whence $x = 6$. Therefore $z = 5x/18 = 5/3$ and $y = 225/x^2z = 15/4$, and so the farmer should buy 6 lb potatoes, $3\frac{3}{4}$ lb wheat, and $1\frac{2}{3}$ lb corn each day.

REMARK: The key to these problems is that the constraints are quite simple. For more complicated problems other tools are sometimes needed. These go under the names "Lagrange multipliers," "linear programming," "steepest descent." You can have a glimpse at one of these in the section at the end of this chapter entitled "Lagrange multipliers."

Exercises

1. Locate the points in the given set where each of the following functions have their absolute maximum and minimum values:

(a) $f(x, y) = 3x + 4y$, $x^2 + y^2 \le 1$;
(b) $g(x, y) = x^2 + y^2$, $x^2 + y^2 \le 4$;
(c) $h(x, y) = x^2 + y^2 - 1$, $|x| \le 1$, $|y| \le 1$;
(d) $k(x, y) = 5x^2 - 6x + 10y^2$, $x^2 + y^2 \le 1$.

(e) $\varphi(x, y) = 4 + x - y,\quad -1 \le x \le 2,\quad 3 \le y \le 5;$

(f) $\psi(x, y) = (3 + 2\cos y)\sin x,\quad 0 \le x \le 2\pi,\quad 0 \le x_2 \le 2\pi.$

2. Locate the point(s) where the following functions have their absolute minimum value:

(a) $f(x, y) = x^2 + y^2$ on $x - y = 3$;

(b) $g(x, y) = x + y$ on $xy = 1$;

(c) $h(x, y, z) = xyz$ on $x^2 + y^2 + z^2 = 12$;

(d) $\varphi(x, y, z) = (x - y)^2 + z^2$ on $x^2 + y^2 + z^2 = 18$.

3. (a) Let x, y, and z be three positive numbers such that $x + y + z = a$. How large can their product xyz be?

(b) If x, y, z are nonnegative, prove that

$$(xyz)^{1/3} \le \frac{x + y + z}{3}.$$

4. A post office regulation states that parcels can have a maximum size of 6 ft. in combined length and girth. What dimensions maximize the volume?

5. A rectangular box without a top is to hold 4 cu ft. How should the box be designed to use the least amount of material?

6. Repeat Exercise 5, but this time assume that the box has a top and that the volume is 8 cu ft.

7. Design a cylindrical can with a circular base but without a top so that its volume is 1 qt and the minimum amount of material is used.

8. If your utility function is $g(x, y) = 4xy - x^2 - 3y^2$ and the budget equation is $2x + 3y = 45$, find the values of x and y that maximize the utility.

9. Ten years from now you are on a local playground committee with the task of designing an oval-shaped playground consisting of a rectangle with two half-circles at the opposite ends of the playground. The rectangular part is to have an area of 5,000 sq yd, and the entire oval is to be enclosed by a fence. How should you design the playground to minimize the amount of fencing needed?

10. A wine dealer has space for 100 cases of California, French, and German wines in her cellar. Customers have already ordered 10 cases of California wine, but she knows that she cannot sell more than 20 cases of it. Although she wants no more than 60 cases of French wine, she wants at least as much French as California wine. If her profit is $10 per case for California, $15 per case for French, and $12 per case for German wines, how much of each should she order to maximize her profit, assuming that she ends up selling the entire stock?

LAGRANGE MULTIPLIERS

Imagine for a moment that you own a steel factory, and let x and y represent the quantity, in tons, of two different kinds of steel you produce.

Let $p = f(x, y)$ be the profit you make from selling them. The cost of producing the steel might be given by $c = h(x, y)$. Say that you have a certain amount of capital c_0, and so, operating at full capacity, the quantity of each kind of steel you can produce is determined by the *side condition*, or *constraint*, $h(x, y) - c_0 = 0$. The problem: How should x and y be determined in order to maximize your profit?

This is a *constrained extremal problem*:

$$\text{Maximize} \quad p = f(x, y) \quad \text{subject to} \quad g(x, y) = 0,$$

where, in our example, the constraint $g(x, y) = h(x, y) - c_0$. We saw that one method of solving this was to solve the constraint equation $g(x, y) = 0$ for, say, y as a function of x, giving $y = \varphi(x)$. Then substitute this into f. This reduces the problem to maximizing a function of the single variable x, $p = f(x, \varphi(x))$, which can be treated by the methods of elementary calculus.

But there is another method of treating constrained extremal problems that does not require solving $g(x, y) = 0$. Moreover, it has the virtue of treating both variables on an equal basis.

Let $f: \mathbb{R}^n \to \mathbb{R}$, $g: \mathbb{R}^n \to \mathbb{R}$ be given functions. We maximize $z = f(X)$ subject to $g(X) = 0$. Introduce a new function of the $n + 1$ variables $(X, \lambda) = (x_1, \ldots, x_n, \lambda)$,

$$F(X, \lambda) = f(X) - \lambda g(X). \tag{A}$$

Note that at those places where $g(X) = 0$, we have $F = f$. We claim, without proof, that the maximum (or minimum) of f subject to $g(X) = 0$ can be found by the following recipe:

1. Find the critical points of F as a function of X and λ.
2. Test them to see which actually gives the maximum (or minimum).

Now at a critical point of F, we have the following conditions:

$$F_{x_1} = 0, \qquad F_{x_2} = 0, \qquad \ldots, \qquad F_{x_n} = 0, \qquad F_\lambda = 0;$$

that is, we must solve the following $n + 1$ equations for X and λ:

$$f_{x_1} - \lambda g_{x_1} = 0, \qquad \ldots, \qquad f_{x_n} - \lambda g_{x_n} = 0, \qquad g = 0.$$

Note that the constraint equation $g(X) = 0$ actually has reappeared as a restatement of $F_\lambda = 0$. From Eq. (A), we see that λ can be interpreted as the penalty one has for violating the constraint $g(X) = 0$.

EXAMPLES: 1. A cylindrical metal can is to contain 2 qt of liquid. How should it be designed to minimize the quantity of metal used?

If the height of the can is y and its radius is x, we want to minimize the area = top + bottom + sides = $2\pi x^2 + 2\pi xy$ subject to $2 = \pi x^2 y$, which we write as $\pi x^2 y - 2 = 0$. Then let

$$F(x, y, \lambda) = 2\pi x^2 + 2\pi xy - \lambda(\pi x^2 y - 2).$$

The equations $F_x = 0$, $F_y = 0$, $F_\lambda = 0$ are, after simplifying,

$$2x + y - \lambda xy = 0, \qquad 2x - \lambda x^2 = 0, \qquad \pi x^2 y - 2 = 0.$$

From the second equation, we have $x = 0$ or $\lambda x = 2$. But $x = 0$ violates the third equation. Hence $\lambda x = 2$. Substituting this in the first equation, we find that $2x + y - 2y = 0$; that is, $y = 2x$. This means that the *height y equals the diameter $2x$.* The third equation thus yields $\pi 2x^3 = 2$; that is, $x = \pi^{-1/3}$, $y = 2(\pi)^{-1/3}$, and the minimum area $A = 6\pi^{1/3}$. We could solve for λ quite easily, but there is no need.

2. A steel factory makes a profit, in dollars per ton, of 1, 2, and 2 on three different kinds of steel. It costs $x^2 + 2y^2 + 4z^2$ to make x, y, and z tons, respectively, of these three kinds of steel. If the firm has \$1,600 capital (per hour) how much should they make of each kind of steel (per hour) to maximize profits?

The problem is to maximize

$$p = x + 2y + 2z \quad \text{subject to} \quad x^2 + 2y^2 + 4z^2 = 1,600.$$

Let

$$F(x, y, z, \lambda) = x + 2y + 2z - \lambda(x^2 + 2y^2 + 4z^2 - 1,600).$$

The conditions $F_x = F_y = F_z = F_\lambda = 0$ are

$$1 = 2\lambda x, \qquad 2 = 4\lambda y, \qquad 2 = 8\lambda z, \qquad x^2 + 2y^2 + 4z^2 = 1,600.$$

From the first two equations (eliminate λ) we find that $x = y$, and from the first and third equations we find that $2z = x$. Substituting this into the constraint, we conclude that

$$16z^2 = 1,600; \text{ that is, } z = 10, \text{ and so } x = y = 20.$$

The profit $p = \$80$, which is 5% on \$1,600 capital. This is small so a price increase should likely be made.

EXERCISES: Use Lagrange multipliers to work the examples in Sec. 6.4c as well as Exercises 2 to 9 in that section.

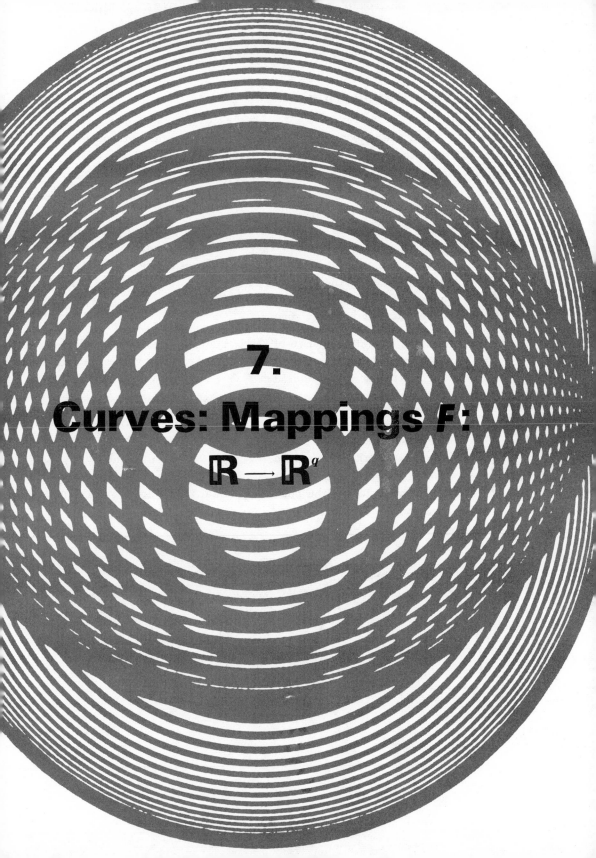

7.
Curves: Mappings F:
$\mathbb{R} \rightarrow \mathbb{R}^q$

7.0 INTRODUCTION

Now the object under study is the vector-valued mapping of a scalar variable,

$$F: \mathbb{R} \longrightarrow \mathbb{R}^q,$$

with $q \geq 1$. We use a capital letter for the mapping, because it is vector-valued.

Let $t \in \mathbb{R}$ be the scalar variable, and $F(t)$ the image point (or vector) in $\mathbb{R}^q$. We should think of t varying along a one-dimensional t-axis. As t moves along the axis, the point $F(t)$ moves in the target space $\mathbb{R}^q$ and traces out a curve there. Hence, we are really studying curves in Euclidean space. We are particularly interested in the case of $q = 2$ or 3, curves in the plane or in ordinary space.

EXAMPLES: The first two of these, the circle and the straight line, are by far the most important of all curves. Learn all you can about them.

1. *The circle in the plane.* Let $F: \mathbb{R} \longrightarrow \mathbb{R}^2$ be given by

$$F(t) = (a \cos t, a \sin t),$$

with $a > 0$. As t varies from 0 to 2π, the point $F(t)$ travels once counter-clockwise around the circle of radius a. See Fig. 7.1. Note $F(0) = (a, 0)$, $F(\pi/2) = (0, a)$, $F(\pi) = (-a, 0)$, $F(\frac{3}{2}\pi) = (0, -a)$, and $F(2\pi) = F(0) = (a, 0)$; we are back where we started. If we chose to let t continue varying from 2π to 4π, the point $F(t)$ would travel around the circle once again.

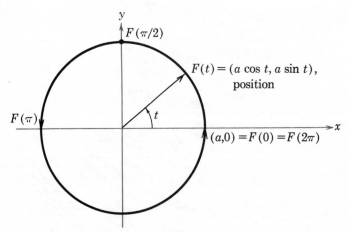

FIGURE 7.1

268

Note that if we interpret $F(t)$ as an arrow with the tail at the origin, then the angle, in radians, measured counterclockwise from the positive x-axis to the arrow is precisely equal to the variable t.

NOTE: Another word for variable is **parameter**; in the present situation we say that we have *parametrized* the circle by t, or by the angle t.

2. *Straight lines in space.* Let X_0 be any point in $\mathbb{R}^q$, and let $Z \neq O$ be in $\mathbb{R}^q$. Define a map $F: \mathbb{R} \to \mathbb{R}^q$ by

$$F(t) = X_0 + tZ.$$

As t varies from $-\infty$ to ∞, the point $F(t)$ traces out a straight line in $\mathbb{R}^q$. As in Fig. 7.2, it is helpful to imagine the vector Z with the tail end at X_0, or as an arrow from X_0 to $X_0 + Z$. At $t = 0$, $F(0) = X_0$. As t increases from 0, $F(t)$ moves on a straight line path in the direction pointed out by the vector Z. At $t = 1$, $F(1) = X_0 + Z$.

We note in passing that $F(t)$ is an affine map and that the mapping $L: \mathbb{R} \to \mathbb{R}^q$ given by

$$L(t) = tZ$$

is linear.

If, as in Fig. 7.2, $q = 3$, $X_0 = (c_1, c_2, c_3)$, and $Z = (a_1, a_2, a_3)$, then we also may write

$$F(t) = (c_1 + a_1 t, c_2 + a_2 t, c_3 + a_3 t),$$

or even

$$x_1 = c_1 + a_1 t, \qquad x_2 = c_2 + a_2 t, \qquad x_3 = c_3 + a_3 t$$

to describe the mapping F. All these forms occur in practice.

3. *Graphs as curves.* If $y = f(x)$ is a function graphed in the xy-plane, then we may regard its graph as the image of the mapping $F: \mathbb{R} \to \mathbb{R}^2$,

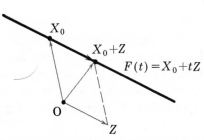

X_0

$X_0 + Z$

$F(t) = X_0 + tZ$

O

Z

y

$F(t) = (t, t^2)$
(note $t = x$)

$y = f(x)$

x

FIGURE 7.2   FIGURE 7.3

where

$$F(t) = (t, f(t)).$$

This is the same as saying that

$$x = t, \quad y = f(t) = f(x).$$

Thus the standard parabola $y = x^2$ is traced out by the mapping $F(t) = (t, t^2)$. See Fig. 7.3.

4. *A helix.* This may be given by $F: \mathbb{R} \to \mathbb{R}^3$,

$$F(t) = (\cos t, \sin t, t).$$

At $t = 0$, $F(0) = (1, 0, 0)$. See Fig. 7.4. As t increases from 0, the first two coordinates indicate that $F(t)$ moves in a circle with respect to a horizontal plane. Meanwhile the vertical z-coordinate is increasing, so that the point $F(t)$ is constantly rising. At $t = 2\pi$, the point $F(2\pi) = (1, 0, 2\pi)$ is 2π units of length directly above $F(0) = (1, 0, 0)$.

REMARKS: 1. *Motion of a particle.* It is helpful to think of the point $F(t)$ as a particle—atom, electron, as you see fit—moving through space. The variable t may be thought of as time: at time t, the particle is at point $F(t)$.

Hence it is natural to ask for the velocity of the particle at time t, that is, its direction and speed at a given instant. Also, we might ask for the distance traveled during a certain time interval. We attend to both of these questions later.

2. It is essential to note that we are dealing with more sensitive objects than curves, namely, **parametrized curves**. A single curve has many parametrization. For example, the equation $y = x$ determines a wellknown curve in the plane, the straight line through the origin whose slope is 1.

Here are three *different* functions that parametrize the line $y = x$:

$$F(t) = (t, t),$$

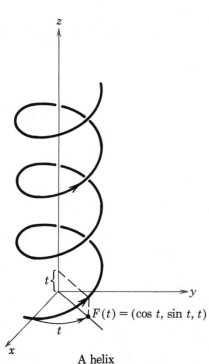

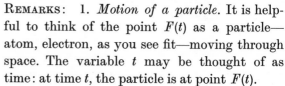

$F(t) = (\cos t, \sin t, t)$

A helix

FIGURE 7.4

$$G(t) = \left(\frac{t}{\sqrt{2}},\ \frac{t}{\sqrt{2}} \right),$$

$$H(t) = (-t, -t).$$

Let us start at $t = 0$, so that $F(0) = G(0) = H(0) = (0, 0)$. As t increases, $F(t)$ moves along the line $y = x$ in the direction of positive x and positive y. We have, in particular, $F(1) = (1, 1)$. See Fig. 7.5.

Meanwhile $G(t)$ has moved in the same direction but more slowly. In fact $G(1) = (1/\sqrt{2}, 1/\sqrt{2})$. If we think of t as time, then $G(t)$ moves one unit of length in one unit of time; its *speed* is 1. The speed of $F(t)$ is $\sqrt{2}$.

Note further that the point $H(t)$ moves in the opposite direction from $F(t)$ and $G(t)$. In fact $H(t) = -F(t)$.

In summary, the moving particles $F(t)$, $G(t)$, $H(t)$ all traverse the line $y = x$ as t varies from $-\infty$ to $+\infty$, but they do so at different speeds or in different directions. Another entirely different parametrization of the same line is

$$K(t) = (\log t, \log t), \qquad 0 < t < \infty.$$

3. *Coordinate form.* If $F: \mathbb{R} \longrightarrow \mathbb{R}^q$, then $F(t)$ is a vector, say $F(t) = X$, and we have, as usual, $X = (x_1, \ldots, x_q) \in \mathbb{R}^q$. As t varies, so does $X = F(t)$. We conclude that each coordinate of X is a scalar function of t; that is,

$$x_1 = f_1(t),\ x_2 = f_2(t),\ \ldots,\ x_q = f_q(t).$$

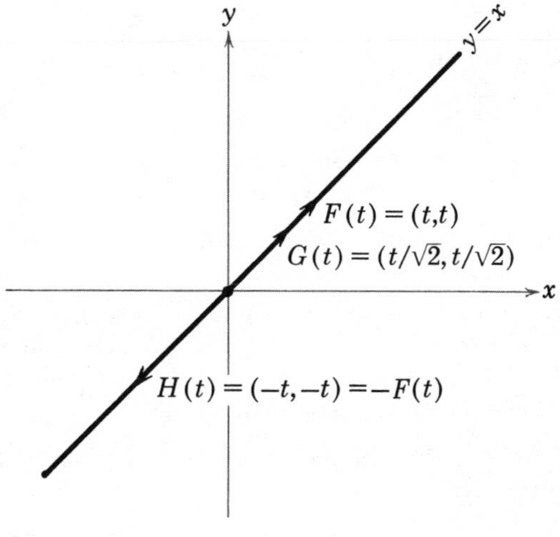

Different parametrizations of $y = x$

FIGURE 7.5

Hence the vector function F is composed of q **coordinate functions**, and so we arrive at the customary notation

$$F(t) = (f_1(t), \ldots, f_q(t)).$$

For example, if $F: \mathbb{R} \to \mathbb{R}^2$ is the function that traces out the circle of radius $a > 0$ (see Example 1), then $F(t) = (f_1(t), f_2(t))$, where

$$f_1(t) = a \cos t, \qquad f_2(t) = a \sin t.$$

4. *What's ahead.* In the sections to come we deal mainly with two problems:

1. The tangent problem (see Sec. 7.1)
2. The problem of arc length (see Sec. 7.3)

The first of these yields an answer to the question "How fast and in what direction is the particle $F(t)$ moving at each precise instant t_0?" The problem hinges on the computation of the best affine approximation to the map F at the point t_0.

The second deals with the question "How far along the curve does the moving particle $F(t)$ travel as t varies from t_0 to t_1?"

Exercises

1. Sketch the following parametrized curves for $0 \le t \le 2\pi$:
    (a) $F(t) = (a \cos 2t, a \sin 2t)$,
    (b) $G(t) = (a \sin t, a \cos t)$
        ($a > 0$ is a constant). Compare them with the circle in Example 1, and interpret the differences in terms of the motion of a particle.

2. Sketch the following curves, indicating with arrows the direction in which the parameter increases:
    (a) $F(t) = (2t, 3 - t)$,   (d) $H(\theta) = (\theta, \sin \theta)$,
    (b) $G(t) = (2t, |3 - t|)$,   (e) $V(t) = (t^2, 1 + t^4)$,
    (c) $\Phi(s) = (s^2, 1 + s^2)$,   (f) $F(t) = (t^2 \cos t, t^2 \sin t)$.

3. The helix in Example 4 goes up counterclockwise. Find the parametrized equation of a helix that goes up clockwise.

4. A particle moves along some parabola $y = cx^2$ in such a way that for $t \ge 0$ the y-coordinate is proportional to the cube of the time elapsed, and so $y = kt^3$ for some constant k. If the particle is at $(2, 1)$ when $t = 1$, find the function $F(t) = (x(t), y(t))$. Where is the particle when $t = 4$?

5. Repeat Exercise 4, but this time say that the particle moves along the cubic $y = x^3 + c$. Where is the particle when $t = 4$?

6. Write a parametric equation for the straight-line segments in $\mathbb{R}^2$ (or $\mathbb{R}^3$ for parts (c), (d)):

(a) From $(0, 0)$ to $(1, 1)$,
(b) From $(1, 2)$ to $(3, -2)$,
(c) From $(1, 1, 1)$ to $(2, 4, -6)$,
(d) From $(2, 4, -6)$ to $(1, 1, 1)$.

7. (a) Show that the curve $F(t) = (\cos t, \cos 2t)$, $0 \leq t \leq \pi$, lies on a parabola, and sketch the curve; that is, if $x = \cos t$, $y = \cos 2t$, find some relation between x and y showing the point is on a parabola.

(b) Show that the curve $F(t) = (a \cos t, b \sin t)$, $0 \leq t \leq 2\pi$, $a > 0$, $b > 0$, lies on the ellipse $x^2/a^2 + y^2/b^2 = 1$.

8. Sketch the plane curve $F(t) = (t(t^2 - 1), t^2 - 1)$. Note that the curve crosses itself. (Suggestion: Plot some points, and ask what, roughly, does the curve look like for $|t|$ large?)

9. Show that the curve $G(s) = (s \cos s, s \sin s, s^2 + 1)$ lies on the paraboloid $x^2 + y^2 = z - 1$. Draw a rough sketch.

10. (a) Find the parametric equation of a straight line that passes through $P = (-3, 1, -2)$ at $t = -1$ and $Q = (0, 1, 1)$ at $t = 2$.

(b) Let P and Q be distinct points in $\mathbb{R}^n$. Find the parametric equation of a straight line that passes through the point P at $t = t_1$ and Q at $t = t_2$.

7.1 THE TANGENT MAP

7.1a Introduction

Let $F: \mathbb{R} \to \mathbb{R}^q$ as above. Given $t_0 \in \mathbb{R}$, we wish to define and compute the best affine approximation or, more simply, tangent map T to F at t_0.

This should be a map $T: \mathbb{R} \to \mathbb{R}^q$ with the following reasonable properties:

1. $T(t)$ traces out the tangent line that contacts the curve at the point $F(t_0)$. See Fig. 7.6.

2. $T(t_0) = F(t_0)$.

3. At the instant $t = t_0$, the speed and direction of a particle moving along the tangent line, with position $T(t)$ at time t, is the same as the speed and direction of the particle moving along the curve $F(t)$. See Fig. 7.7.

This says that the straight-line motion $T(t)$ gives a good approximation to motion along the curve $F(t)$ near the point $F(t_0)$. Once again we are linearizing.

REMARK: Let us verify that a nonconstant affine map $T: \mathbb{R} \to \mathbb{R}^q$ does trace out a *straight line* in $\mathbb{R}^q$. Such a map is, by definition, of the form

$$T(t) = Y_1 + L(t),$$

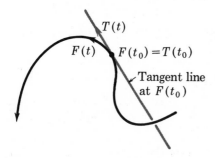

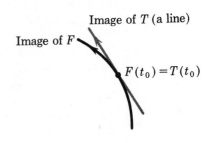

The image of the tangent map at t_0

FIGURE 7.6

$F(t) \approx T(t)$ for $t \approx t_0$

FIGURE 7.7

where Y_1 is a fixed vector in $\mathbb{R}^q$ and $L: \mathbb{R} \to \mathbb{R}^q$ is linear and not the zero map.

Look at L more closely. Letting $t = 1$, we have $L(1)$ a nonzero vector in $\mathbb{R}^q$, say $L(1) = Z$. Now $L(t) = L(t1) = tL(1)$, by homogeneity of the linear map. Therefore we may say that

$$L(t) = tL(1) = tZ.$$

Thus we have

$$T(t) = Y_1 + tZ,$$

and we saw in Example 2 that the image of such a map is a straight line through the point Y_1 in the same direction as the vector Z.

7.1b The basic definitions

Let $F: \mathbb{R} \to \mathbb{R}^q$ be defined in some open interval containing t_0 on the t-axis.

DEFINITION: F is **tangent at** t_0 to the affine map $T: \mathbb{R} \to \mathbb{R}^q$ if and only if

$$F(t_0) = T(t_0) \quad \text{and} \quad \lim_{t \to t_0} \frac{F(t) - T(t)}{t - t_0} = O.$$

If this is the case, we then say that F is **differentiable at** t_0, or a **differentiable curve at** t_0, and that the map T is the **best affine approximation** or **tangent map** to F at t_0.

This parallels closely the definitions for $f(X)$, the scalar-valued function of a vector. In that case also we approximated the nonlinear f by an affine map.

Having made our definition of differentiability, we are faced with two natural questions:

1. Given $F(t)$, how can we tell if it is differentiable at t_0?

2. If $F(t)$ is differentiable at t_0, how can we compute explicitly its best affine approximation $T(t)$ there?

Let us give a tentative answer to the second question. There are some things we can say about the best affine approximation (tangent map) $T(t)$.

Since $T(t)$ is affine, it is of the form

$$T(t) = Y_1 + tZ,$$

with $Y_1, Z \in \mathbb{R}^q$. We showed this in the introduction. Now we have $T(t_0) = Y_1 + t_0 Z$, whence $Y_1 = T(t_0) - t_0 Z$. Thus we may write

$$T(t) = T(t_0) - t_0 Z + tZ$$
$$= T(t_0) + Z(t - t_0).$$

Here $Z(t - t_0)$ denotes multiplication of the vector Z by the scalar $t - t_0$. This form is especially relevant to the behavior of $T(t)$ near $t = t_0$.

DEFINITION: If $T(t) = T(t_0) + Z(t - t_0)$ is the best affine approximation to $F: \mathbb{R} \longrightarrow \mathbb{R}^q$ at t_0, then the vector $Z \in \mathbb{R}^q$ is called the **total derivative, (first derivative, velocity vector)** of $F(t)$ at t_0. We write

$$Z = F'(t_0).$$

Some other notations for the total derivative are

$$\dot{F}(t), \qquad \frac{dF}{dt}(t), \qquad D_t F(t).$$

The length $\| F'(t_0) \|$ is called the **speed** of $F(t)$ at t_0 when F refers to particle motion.

REMARKS: 1. If $F(t)$ is differentiable at t_0 with tangent map $T(t)$ there, then $F(t_0) = T(t_0)$, so that we may write the tangent map thus:

$$T(t) = F(t_0) + F'(t_0)(t - t_0).$$

This looks very much like the formula for the best affine approximation of the scalar-valued functions we studied earlier.

2. Given the map $F(t)$ and the specific t_0, the image point $F(t_0)$ is readily computed (just plug in t_0). Hence computation of the tangent map $T(t)$ reduces to computation of the velocity vector $F'(t_0)$. We tend to this in Sec. 7.1c.

3. *Speed.* Our definition of the instantaneous speed of the particle at $F(t_0)$ as $\|F'(t_0)\|$ is reasonable. For suppose that at the instant t_0 the particle left the curve $F(t)$ and flew off on the tangent $T(t)$. At the end of one unit of time, the particle would be at

$$T(t_0 + 1) = F(t_0) + F'(t_0)((t_0 + 1) - t_0)$$
$$= F(t_0) + F'(t_0).$$

Hence the particle has moved from $F(t_0)$ to $F(t_0) + F'(t_0)$, a distance of $\|F'(t_0)\|$, in one unit of time. This coincides with our familiar notion of speed as distance per unit time.

7.1c The Main Theorem on differentiable curves

This enables us to compute the total derivative or velocity vector $F'(t_0)$ of a differentiable map $F(t)$. Happily, this computation reduces to ordinary differentiation.

First, let us recall that if $F: \mathbb{R} \to \mathbb{R}^q$, then F is of the form

$$F(t) = (f_1(t), \ldots, f_q(t)),$$

where $f_1, \ldots, f_q$ are the **coordinate functions**; these are scalar-valued functions of a scalar; that is, $f_j: \mathbb{R} \to \mathbb{R}$, as encountered in ordinary calculus.

If $F(t)$ is differentiable at t_0, then $F'(t_0)$ is a vector in $\mathbb{R}^q$ of the form

$$F'(t_0) = (\alpha_1, \ldots, \alpha_q).$$

We know $F'(t_0)$ if we know the coordinates $\alpha_1, \ldots, \alpha_q$. The following tells us how to obtain them.

THEOREM 1: (Main Theorem on differentiable curves.) Let $F: \mathbb{R} \to \mathbb{R}^q$, $F(t) = (f_1(t), \ldots, f_q(t))$ be a curve differentiable at t_0. Then:

1. Its best affine approximation at t_0 is the map

$$T(t) = F(t_0) + F'(t_0)(t - t_0).$$

2. The coordinate functions $f_1, \ldots, f_q$ are differentiable at t_0.
3. The velocity vector at t_0 is given by

$$F'(t_0) = (f_1'(t_0), \ldots, f_q'(t_0)).$$

Proof: 1. We proved this part in Sec. 7.1b.

2. and 3. Let $F'(t_0) = (\alpha_1, \ldots, \alpha_q)$. We show now that each $f_j'(t_0)$ exists and, further, $\alpha_j = f_j'(t_0)$.

Since F is differentiable at t_0, we have

$$\lim_{t \to t_0} \frac{F(t) - [F(t_0) + F'(t_0)(t - t_0)]}{t - t_0} = 0;$$

that is,

$$\lim_{t \to t_0} \frac{F(t) - F(t_0)}{t - t_0} = F'(t_0).$$

Now for any vector $G(t) = (g_1(t), \ldots, g_q(t))$ we have $|g_j(t)| \leq \|G(t)\|$, $j = 1, \ldots, q$. This implies that if, as $t \to t_0$, $\lim G(t) = G(t_0)$, then $\lim g_j(t) = g_j(t_0)$, because $|g_j(t) - g_j(t_0)| \leq \|G(t) - G(t_0)\| \to 0$. Thus, the limit of $g_j(t)$ exists and equals $g_j(t_0)$. Applied to the vector $(F(t) - F(t_0))/(t - t_0)$, this proves that $\lim (f_j(t) - f_j(t_0))/(t - t_0)$ exists—so that f_j is differentiable at t_0—and equals the jth coordinate of $F'(t_0)$. In other words, $F'(t_0) = (f'_1(t_0), \ldots, f'_q(t_0))$. $\ll$

REMARKS ON THE THEOREM: 1. Thus the derivative of $F(t) = (f_1(t), \ldots, f_q(t))$ is finally seen to be the most natural candidate, namely, $F'(t) = (f'_1(t), \ldots, f'_q(t))$. We could have *defined* $F'(t)$ to be this, rather than defining it as a certain vector in $\mathbb{R}^q$, determined by the affine map $T(t)$. Had we done this, however, the burden of proof would have been on us to show that $F'(t_0)$, defined computationally as $(f'_1(t_0), \ldots, f'_q(t_0))$, is *interesting*, that it helps us understand the curve $F(t)$. Our point of view is that the tangent map $T(t)$ is clearly interesting and relevant from a geometric point of view, and therefore we should prove a theorem that tells us how to compute it. This we have just done.

2. You can show that *if $F(t) = (f_1(t), \ldots, f_q(t))$ and if $f'_1(t_0), \ldots, f'_q(t_0)$ exist as finite numbers, then $F(t)$ is differentiable at t_0.* In fact, it is tangent there to the affine map $T(t) = X_0 + Z(t - t_0)$, where $X_0 = F(t_0)$ and $Z = F'(t_0) = (f'_1(t_0), \ldots, f'_q(t_0))$.

This reduces to ordinary one-variable calculus a question raised in Sec. 7.1b, "How can we tell if a given $F(t)$ is differentiable at t_0?"

Almost all the curves we encounter in this book are differentiable. In fact, they are what mathematicians call "smooth." This means, in particular, that the coordinate functions $f_j(t)$ have derivatives of all orders $f'_j(t_0), f''_j(t_0)$, and so on, for all t_0. No worry here about lack of derivatives.

3. If $F(t) = (f_1(t), \ldots, f_q(t))$, then $F'(t_0)$ is a linear map from $\mathbb{R}$ to $\mathbb{R}^q$. Thus its matrix is

$$[F'(t_0)] = \begin{bmatrix} f'_1(t_0) \\ \cdot \\ \cdot \\ \cdot \\ f'_q(t_0) \end{bmatrix}.$$

NOTE: As a vector we write $F'(t_0) = (f_1'(t_0), \ldots, f_q'(t_0))$, but as a matrix it is vertical. This will become important only when we discuss maps $F: \mathbb{R}^n \rightarrow \mathbb{R}^q$ and the Chain Rule.

EXAMPLES OF THE TANGENT MAP: Let us go back to the four examples of curves we saw in the introduction. We will compute the velocity vector and tangent map in each case.

1. *The circle.* This is traced out by

$$F(t) = (a \cos t, a \sin t).$$

Thus

$$F'(t_0) = (-a \sin t_0, a \cos t_0)$$

for each $t_0 \in \mathbb{R}$. Here we merely differentiated $a \cos t$ and $a \sin t$ at $t = t_0$ (note that $a \cos t$ and $a \sin t$ have derivatives of all orders; the circle map $F(t)$ is smooth (see Sec. 7.1e)).

For instance, if $t_0 = \pi/6$, then $F'(\pi/6) = (-a/2, \sqrt{3}\, a/2)$.

a. What is the best affine approximation or tangent map to $F(t)$ at $t_0 = \pi/6$?

We know that $T(t) = F(t_0) + F'(t_0)(t - t_0)$. Now

$$F(t_0) = F\left(\frac{\pi}{6}\right) = \left(\frac{\sqrt{3}a}{2}, \frac{a}{2}\right).$$

Using the velocity vector $F'(\pi/6)$ computed above, we have

$$T(t) = \left(\frac{\sqrt{3}a}{2}, \frac{a}{2}\right) + \left(\frac{-a}{2}, \frac{\sqrt{3}a}{2}\right)\left(t - \frac{\pi}{6}\right).$$

This may also be written

$$T(t) = \left(\frac{\sqrt{3}a}{2} - \frac{a}{2}\left(t - \frac{\pi}{6}\right), \frac{a}{2} + \frac{\sqrt{3}a}{2}\left(t - \frac{\pi}{6}\right)\right).$$

The image of the map $T(t)$ in $\mathbb{R}^2$ is the **tangent line** to the circle at the point $F(\pi/6)$. See Fig. 7.8.

b. It is easy to check that the velocity vector $F'(t_0)$ is orthogonal to the position vector $F(t_0)$ at each t_0; that is,

$$\langle F'(t_0), F(t_0) \rangle = 0.$$

c. What is the speed of a particle traveling around the circle of radius $a > 0$ whose position at time t is $F(t)$?

At a typical instant t_0 the velocity vector is, as we know, $F'(t_0) = (-a \sin t_0, a \cos t_0)$. The direction of the vector $F'(t_0)$ is the instantaneous direction of the moving particle; its length is the particle's speed. Now

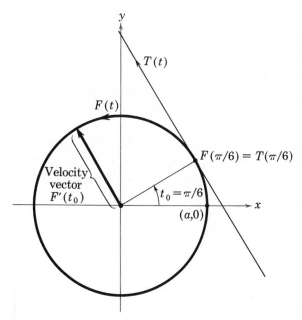

FIGURE 7.8

$$\| F'(t_0) \|^2 = \langle F'(t_0),\ F'(t_0) \rangle$$
$$= a^2(\sin^2 t_0 + \cos^2 t_0)$$
$$= a^2,$$

using the Pythagorean Theorem. It follows that the speed of the particle at time t_0 equals a (units of length per unit of time). This is a constant, independent of the time t_0.

Intuitively, it is not surprising that speed $= a$. The perimeter of the circle of radius a equals $2\pi a$ units of length, and the particle at $F(t)$ tranverses this circle in 2π units of time t (why?). Thus the constant speed is $2\pi a/2\pi = a$ units of length per unit of time.

2. *Straight lines in space.* Say that we are given a parametrized straight line $F:\mathbb{R} \to \mathbb{R}^q$, $F(t) = X_0 + tZ$ (we agreed earlier that such maps do trace out straight lines).

First question: What is the velocity vector at some given t_0?

There are two methods here.

Method 1. Observe simply that $T(t)$ defined by $T(t) = X_0 + tZ$ is an affine map; that is, $F(t)$ is itself affine, and so it is trivially tangent to itself at each t_0. Just note that $F(t) - T(t) = O$, and apply the definition of best affine approximation. To get the velocity vector at t_0, note that

$$T(t) = X_0 + (t - t_0 + t_0)Z = (X_0 + t_0 Z) + (t - t_0)Z$$
$$= F(t_0) + (t - t_0)Z,$$

and so, by definition, Z is the velocity vector for any time t_0, $Z = F'(t_0)$. In a sense, straight lines have constant "slope."

Method 2. Let us convince ourselves that the Main Theorem yields the same result: $F'(t_0) = Z$. To be concrete, suppose that $q = 3$. Thus $F(t)$ traces out a straight line in ordinary 3-space.

Suppose that $X_0 = (c_1, c_2, c_3)$ and $Z = (z_1, z_2, z_3)$. These are fixed vectors. Then we have $F(t) = (f_1(t), f_2(t), f_3(t))$, where $f_j(t) = c_j + t z_j$, with $j = 1, 2, 3$. Differentiating as in ordinary calculus, we see that

$$f_j'(t) = z_j$$

for all t. By the Main Theorem, therefore,

$$F'(t_0) = (z_1, z_2, z_3) = Z,$$

as we suspected. This is a constant vector, independent of t_0.

Second question: How fast is a particle traveling whose position at time t is $F(t)$?

The speed is $\| F'(t_0) \|$, that is, $\| Z \|$ units of distance per unit of time. Note that at $t = 0$, the particle is at the point X_0; at $t = 1$, it is at $X_0 + Z$; at $t = 2$, it is at $X_0 + 2Z$, and so on.

SUMMARY: The derivative $F'(t)$ of $F(t) = X_0 + tZ$ is equal to Z for all $t = t_0$, just as if we had differentiated $X_0 + tZ$ with respect to t, as in ordinary calculus, treating X_0 and Z as constant numbers rather than vectors.

3. *The graph of* $y = f(x)$. Say that $y = f(x) = x^2$. We realize its graph as a parametrized curve in $\mathbb{R}^2$ by defining $F: \mathbb{R} \to \mathbb{R}^2$, $F(t) = (x, y)$ $= (t, t^2)$. Let us compute the tangent map to this curve at $t_0 = 1$. See Fig. 7.9.

We have, first of all, $F(1) = (1, 1)$. Further, $F'(t) = (1, 2t)$, whence $F'(1) = (1, 2)$. It follows that the tangent map or best affine approximation is

$$T(t) = F(1) + F'(1)(t - 1)$$
$$= (1, 1) + (1, 2)(t - 1)$$
$$= (1 + (t - 1), 1 + 2(t - 1))$$
$$= (t, 1 + 2(t - 1)).$$

This is familiar. Let $g(x) = 1 + 2(x - 1)$. Then $T(t)$ traces out the

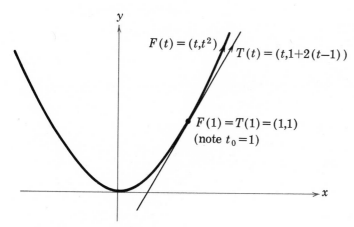

FIGURE 7.9

graph of $y = g(x)$, for

$$x = t, \qquad y = g(t) = 1 + 2(t - 1).$$

But also $y = g(x) = f(x_0) + f'(x_0)(x - x_0)$ in the case of $f(x) = x^2$, $x_0 = 1$, as you may check. *This is the ordinary tangent line in one-variable calculus.* Hence both senses of tangent—tangent *map* to a parametrized curve $F(t)$, tangent *line* to $y = f(x)$—coincide in the case of $F(t) = (t, f(t))$.

4. *How to fly off on a tangent.* Here is a problem we are now able to solve. A particle is traveling up a helical path in space, its position at time t given by $F(t) = (\cos t, \sin t, t)$. At time $t_0 = 13\pi/6$ it flies off on a tangent, leaving the helical path and traveling in a straight line at constant velocity. Discounting gravity, where is the particle at time $t_1 = 5\pi/2$? See Fig. 7.10.

Solution:

   a. At the start, $t = 0$, and the particle is at $F(0) = (1, 0, 0)$.

   b. At the instant the particle flies off on the straight-line tangent, it is at the point $F(t_0) = (\cos 13\pi/6, \sin 13\pi/6, 13\pi/6) = (\sqrt{3}/2, 1/2, 13\pi/6)$, because $\cos 13\pi/6 = \cos(2\pi + \pi/6) = \cos \pi/6$, and so on.

   c. Now we compute the tangent map (best affine approximation) to the helix $F(t)$ at the point t_0. This, of course, traces out the straight line path taken by the particle as it flies away "on the tangent."

   The tangent map is $T(t) = F(t_0) + F'(t_0)(t - t_0)$. We obtained $F(t_0)$ just above. Now we get $F'(t_0)$. We have

$$F'(t) = (-\sin t, \cos t, 1),$$

so that, when $t_0 = 13\pi/6$,

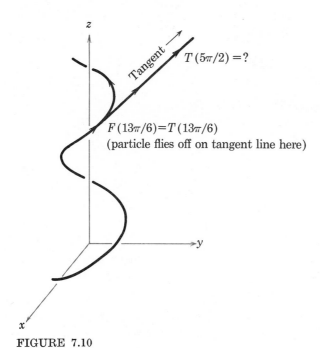

$T(5\pi/2) = ?$

$F(13\pi/6) = T(13\pi/6)$
(particle flies off on tangent line here)

FIGURE 7.10

$$F'(t_0) = \left(-\sin\frac{13\pi}{6}, \cos\frac{13\pi}{6}, 1\right)$$
$$= \left(-\frac{1}{2}, \frac{\sqrt{3}}{2}, 1\right).$$

Thus the tangent map at t_0 is given by

$$T(t) = F(t_0) + F'(t_0)(t - t_0)$$
$$= \left(\frac{\sqrt{3}}{2}, \frac{1}{2}, \frac{13\pi}{6}\right) + \left(-\frac{1}{2}, \frac{\sqrt{3}}{2}, 1\right)\left(t - \frac{13\pi}{6}\right).$$

We may also write this

$$T(t) = \left(\frac{\sqrt{3}}{2} - \frac{1}{2}\left(t - \frac{13\pi}{6}\right), \frac{1}{2} + \frac{\sqrt{3}}{2}\left(t - \frac{13\pi}{6}\right), \frac{13\pi}{6} + \left(t - \frac{13\pi}{6}\right)\right),$$

but there is no particular advantage in doing so.

  d. Now let us locate the particle at time $t_1 = 5\pi/2$. We note that $5\pi/2 = 15\pi/6$, and so

$$t_1 - t_0 = \frac{\pi}{3}.$$

Thus at time $t_1 = 5\pi/2$, the particle is located on the tangent line $T(t)$ at the point

$$T(t_1) = F(t_0) + F'(t_0)(t_1 - t_0)$$

$$= \left(\frac{\sqrt{3}}{2}, \frac{1}{2}, \frac{13\pi}{6}\right) + \left(-\frac{1}{2}, \frac{\sqrt{3}}{2}, 1\right)\frac{\pi}{3}$$

$$= \frac{1}{6}(3\sqrt{3} - \pi, 3 + \pi\sqrt{3}, 15\pi),$$

using the explicit $F(t_0)$ and $F'(t_0)$ obtained above. This point in $\mathbb{R}^3$ is our answer to the original question.

REMARKS: 1. You should verify that the *speed* of the particle, either on the helix or on the tangent line, is constantly equal to $\sqrt{2}$. On the tangent line, the *direction* of motion is, of course, constant as well.

2. The particle climbs at a constant rate, and this rate is the same either on the helix or on the tangent line. For on either curve the height (above the horizontal plane through the origin) at time t is t units of length. Check that the third (vertical) coordinate of $T(t)$ equals t.

7.1d Some properties of the velocity vector

Now we present some further results on the derivative $F'(t_0)$. These will be useful when we discuss acceleration.

If $F, G : \mathbb{R} \longrightarrow \mathbb{R}^q$, say $F(t) = (f_1(t), \ldots, f_q(t))$ and $G(t) = (g_1(t), \ldots, g_q(t))$, then we may form their sum:

$$(F + G)(t) = (f_1(t) + g_1(t), \ldots, f_q(t) + g_q(t)),$$

and their inner product, which is a scalar-valued function of t:

$$\langle F(t), G(t) \rangle = f_1(t)g_1(t) + \cdots + f_q(t)g_q(t),$$

as well as multiply by a scalar function $h(t)$:

$$h(t)F(t) = (h(t)f_1(t), \ldots, h(t)f_q(t)).$$

We leave it to you to prove

THEOREM 2:   Given $F, G : \mathbb{R} \longrightarrow \mathbb{R}^q$, differentiable maps, and $h(t)$ a differentiable function. Then :

1. $(F + G)'(t) = F'(t) + G'(t).$

2. $\dfrac{d}{dt} \langle F(t), G(t) \rangle = \langle F'(t), G(t) \rangle + \langle F(t), G'(t) \rangle.$

3. $(h(t)F(t))' = h'(t)F(t) + h(t)F'(t).$

These are clearly generalizations of the ordinary one-variable rules for differentiating sums and products:

$$(f+g)'(t) = f'(t) + g'(t),$$
$$(f(t)g(t))' = f'(t)g(t) + f(t)g'(t).$$

This theorem has a very useful corollary. Suppose that $G(t)$ is a vector-valued function with the property that the length $\|G(t)\|$ is a constant, independent of t. Say that $\|G(t)\| = \lambda \geq 0$ (we can think of $G(t)$ as the path of a particle moving on the surface of a sphere of radius λ).

Then we have $\langle G(t), G(t) \rangle = \|G(t)\|^2 = \lambda^2$. Let us differentiate both sides of this relation (a standard trick).

On the left, we have $(d/dt)\langle G(t), G(t) \rangle = 2\langle G'(t), G(t) \rangle$, as you may verify using Theorem 2. On the right, we have $(d/dt)\lambda^2 = 0$, since λ is constant. Now these derivatives are equal, whence

$$\langle G'(t), G(t) \rangle = 0.$$

But this says that $G(t)$ is orthogonal to its derivative $G'(t)$ for all t.

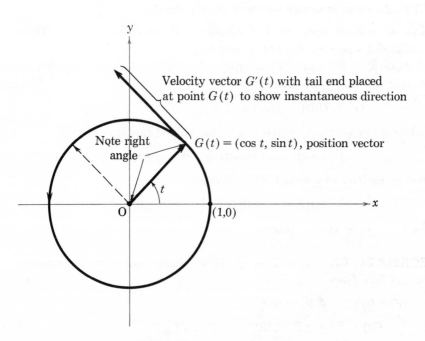

Showing $G(t) \perp G'(t)$ in the case $\|G(t)\| = 1$

FIGURE 7.11

COROLLARY 3: Let $G(t)$ be such that $\| G(t) \|$ is constant. Then for all t

$$G(t) \perp G'(t).$$

For example, this proves that if a particle moves on a circle or sphere, then its velocity vector is perpendicular to its position vector. See Fig. 7.11. However, the corollary is useful in other situations as well, as we shall see when we discuss acceleration.

7.1e Acceleration of a moving particle

This provides an introduction to further methods in the study of para-metrized curves $F(t)$. We have already discussed the velocity vector $F'(t)$ and the speed. We now turn to the acceleration vector $F''(t)$. The present discussion is *not* a prerequisite for the study of arc length in Sec. 7.3.

Let $F(t) = (f_1(t), \dots, f_q(t))$ determine a parametrized curve in $\mathbb{R}^q$ as usual.

DEFINITION: The curve $F(t)$ is **smooth** if and only if:

1. Each coordinate function $f_j(t)$ is infinitely differentiable; that is, all its derivatives $f'_j(t), f''_j(t), f'''_j(t), \dots$ exist for all t.
2. The velocity vector $F'(t)$ is not equal to zero for any t.

This second requirement says, of course, that the speed $\| F'(t) \| \neq 0$; that is, the moving particle never stops.

EXAMPLES: 1. The circle $F(t) = (a \cos t, a \sin t)$ is smooth, since:

a. We may differentiate $f_1(t) = a \cos t$ and $f_2(t) = a \sin t$ as often as we wish.

b. $\| F'(t) \| = a > 0$. The speed is never zero.

2. A curve that is *not* smooth is given by

$$G(t) = (t^3, t^2).$$

$G'(t) = (3t^2, 2t)$, and so $\| G'(0) \| = 0$ at $t = 0$. See Fig. 7.12. And, intuitively, this curve is not smooth as it passes through the origin. It has a corner there.

Now let $F(t) = (f_1(t), \dots, f_q(t))$ give a smooth curve. As usual, the velocity vector is

$$F'(t) = (f'_1(t), \dots, f'_q(t)).$$

To each t this assigns a vector in $\mathbb{R}^q$; hence F'

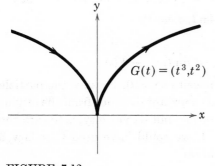

$G(t) = (t^3, t^2)$

FIGURE 7.12

is itself a mapping from $\mathbb{R}$ into $\mathbb{R}^q$, $F' : \mathbb{R} \to \mathbb{R}^q$. The mapping F' therefore has a first derivative, which is the same as the *second* derivative of the original F; thus

$$F''(t) = (f_1''(t), \ldots, f_q''(t)).$$

If F represents the position of a particle, then we call $F''(t)$ the **acceleration vector** of $F(t)$.

The acceleration vector $F''(t)$ is often pictured as extending from the tip of the velocity vector $F'(t)$. See Fig. 7.13. There it indicates the tendency of the velocity vector to vary, that is, to turn in some direction and to alter its length (speed of the moving particle). For since the acceleration vector $F''(t_0)$ is the first derivative at $t = t_0$ of $F'(t)$, it carries information about the change of $F'(t)$ at $t = t_0$.

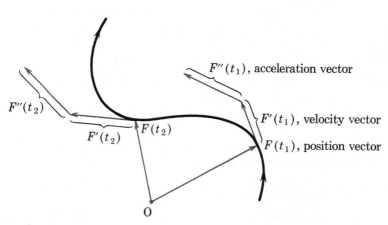

FIGURE 7.13

EXAMPLES OF ACCELERATION VECTORS: 1. Let $F(t) = (a \cos t, a \sin t)$ trace out the circle of radius $a > 0$ in $\mathbb{R}^2$. See Fig. 7.14. We then have

$$F'(t) = (-a \sin t, a \cos t),$$

whence

$$F''(t) = (-a \cos t, -a \sin t) = -F(t).$$

The vector $F''(t)$ points exactly opposite to $F(t)$, because the particle at $F(t)$ is always turning to maintain its constant distance from the origin; it is turning *inward*, so that its acceleration, not its velocity, is toward the center of the circle. Since $\| F'(t) \| = 1$, we could have used Corollary 3 to predict $F'' \perp F'$ without computation.

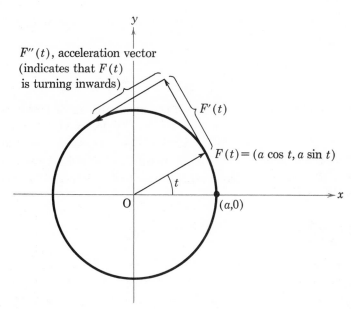

$F''(t)$, acceleration vector (indicates that $F(t)$ is turning inwards)

$F'(t)$

$F(t) = (a \cos t, a \sin t)$

t

O

$(a,0)$

FIGURE 7.14

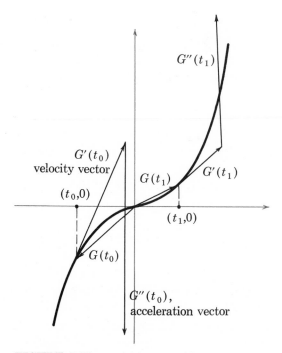

$G'(t_0)$ velocity vector

$(t_0,0)$

$G(t_0)$

$G(t_1)$

$G'(t_1)$

$G''(t_1)$

$(t_1,0)$

$G''(t_0)$, acceleration vector

FIGURE 7.15

287

2. Let $G(t) = (t, t^3)$ trace out the graph of $y = x^3$ in the xy-plane. See Fig. 7.15. Then

$$G'(t) = (1, 3t^2),$$

$$G''(t) = (0, 6t).$$

Note that $t_0 < 0$ implies that $G''(t_0)$ points downward, in accordance with the fact that, as t increases from $-\infty$ to 0, the velocity vector $G'(t)$ is tilting from almost vertical to horizontal—$G'(0) = (1, 0)$, horizontal. Likewise, for $t_1 > 0$ the acceleration vector points upward, because the velocity vector $G'(t)$ points more and more toward the vertical as $t > 0$ increases.

7.1f Curves with prescribed tangents or accelerations

So far we have been given $F(t)$ and asked to compute $F'(t_0)$ and $F''(t_0)$ for particular t_0. Now we reverse the question: Given the velocity vector $F'(t)$ for each t, or given $F''(t)$, plus some other information, compute $F(t)$. Here are some examples.

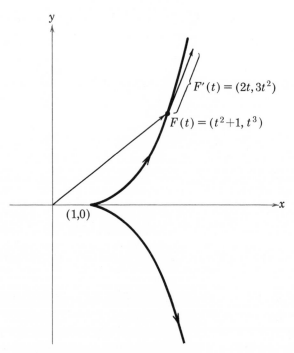

FIGURE 7.16

EXAMPLES: 1. *Given tangent and starting point.* Say that $F'(t) = (2t, 3t^2)$ and $F(0) = (1, 0)$. What is $F(t)$?

To get $F(t) = (x(t), y(t))$, note that $F'(t) = (x'(t), y'(t))$, and so

$$x'(t) = 2t, \qquad y'(t) = 3t^2.$$

Ordinary integration gives

$$x(t) = t^2 + c_1, \qquad y(t) = t^3 + c_2,$$

where the constants c_1, c_2 must be found. But we know that $x(0) = 1$ (why?), and so $c_1 = 1$. Likewise $c_2 = 0$. Thus we have

$$F(t) = (x(t), y(t)) = (t^2 + 1, t^3).$$

NOTE: You may sketch $F(t)$ by plotting points $F(0)$, $F(1)$, and so on. See Fig. 7.16. But also $x - 1 = t^2$, whence $(x - 1)^3 = y^2$; that is, $y = \pm(x - 1)^{3/2}$. This may help.

REMARK: If we know the velocity $F'(t)$ for all t and the position at one instant ($t = 0$ in the above), then we can find the position for *all* times t. Why is this reasonable?

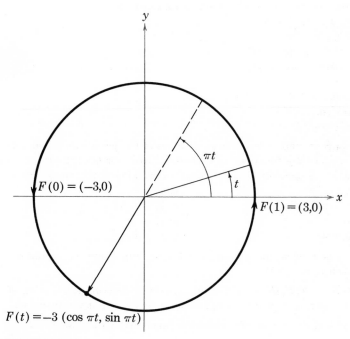

$$F(t) = -3\,(\cos \pi t, \sin \pi t)$$

FIGURE 7.17

2. *Given acceleration, starting point, and starting velocity.* Say that $F''(t)$ $= (3\pi^2 \cos \pi t, 3\pi^2 \sin \pi t)$, $F'(0) = (0, -3\pi)$, $F(0) = (-3, 0)$. We compute $F(t) = (x(t), y(t))$.

We have

$$x''(t) = 3\pi^2 \cos \pi t, \qquad y''(t) = 3\pi^2 \sin \pi t,$$

so that, by integrating once, we obtain

$$x'(t) = 3\pi \sin \pi t + c_1, \qquad y'(t) = -3\pi \cos \pi t + c_2.$$

Since $F'(0) = (x'(0), y'(0)) = (0, -3\pi)$, we conclude that $c_1 = c_2 = 0$. A second integration (the second derivative necessitates two integrations) yields

$$x(t) = -3 \cos \pi t + c_3, \qquad y(t) = -3 \sin \pi t + c_4.$$

Since $F(0) = (-3, 0)$, we conclude that $c_3 = c_4 = 0$. Finally

$$F(t) = (x(t), y(t)) = -3(\cos \pi t, \sin \pi t).$$

You might check that this $F(t)$ does have the prescribed starting point $F(0)$ and starting velocity $F'(0)$. See Fig. 7.17.

Exercises

1. Find the derivative of each of the following functions:

    (a) $F(t) = (1 - 7t, 5 + 9t)$;
    (b) $G(t) = (3t, t^2 - 7)$;
    (c) $H(\theta) = (2 + e^\theta, 3\theta - 7\theta^2)$;
    (d) $P(s) = (\sin 2s, \cos 2s, s)$;
    (e) $\Phi(t) = (3 - t, 1 + 4t, -9t)$;
    (f) $V(t) = (\sin \pi(1 + 3t), 4, \log \sqrt{1 + t^4})$;
    (g) $Q(r) = r^2(\sin r, e^{3r})$;
    (h) $K(t) = (3, t)/\|(3, t)\|$.

2. Find the tangent map at the indicated values of the parameter listed below for the corresponding functions in Exercise 1. Also, for (a), (b), (d), (e), and (h), draw a sketch showing the curve and its tangent line at the indicate points:

    (a) $t_0 = 0, 1, -2$;   (e) $t_0 = 0, -1, 3$;
    (b) $t_0 = 1, -3$;   (f) $t_0 = 0, 1$;
    (c) $\theta_0 = 0$;   (g) $r_0 = 0$;
    (d) $s_0 = 0, \pi/4$;   (h) $t_0 = 0, 4$.

3. Verify that the tangent maps found in Exercise 2(b) satisfy the conditions of the definition in Sec. 7.1b.

4. If the functions F, G, Φ, V, and K in Exercise 1 describe the motion of a particle at time t, find their position, velocity, and speed at $t = 0$ and $t = 1$.

5. Find the second derivative of each of the functions in Exercise 1.

6. Find functions $F(t) = (x(t), y(t))$, or $F(t) = (x_1(t), x_2(t), x_3(t))$, such that:

(a) $F'(t) = (1 + 2t, 3)$, $F(0) = (1, 3)$;

(b) $F'(t) = (2 \cos t, 3 \sin t)$, $F(0) = (0, 1)$;

(c) $F'(t) = (1 - 4t, 3 + 2t, 1 + 3t^2)$, $F(1) = (1, 4, 0)$;

(d) $F'(t) = (1 + e^{2t}, t - t^2, 5 + t^4)$, $F(0) = (0, 1, -2)$.

7. (a) A particle travels on a curve in $\mathbb{R}^3$ in such a way that its velocity is $V(t)$ $= (2, 5 + 6t, 0)$ and its position at $t = 0$ is $(1, 0, 3)$. Find its position at $t = 1$.

(b) Repeat Exercise 7(a) where $V(t) = (3t^2, -2t + 7, \cos \pi t)$ and the position at $t = 0$ is $(1, 5, 0)$.

8. Find functions $F(t)$ such that:

(a) $F''(t) = (1 + 2t, 3)$, $F(0) = (0, 0)$, $F'(0) = (1, 3)$;

(b) $F''(t) = (-\cos t, -\sin t)$, $F(0) = (0, 1)$, $F'(0) = (-1, 0)$;

(c) $F''(t) = (1 - 4t, 3 + 2t, 1 + 3t^2)$, $F(1) = (0, 0, 1)$, $F'(1) = (1, 4, 0)$.

9. (a) A particle moves on a curve in $\mathbb{R}^3$ in such a way that its acceleration is $F''(t) = (-4, 6t, -2 + 4t^2)$. If its initial position is $F(0) = (2, 0, 0)$ and its initial velocity is $F'(0) = (0, 0, 1)$, find its position at $t = 3$.

(b) Repeat Exercise 9(a) where $F''(t) = (e^t, 0, 1 + 6t)$.

10. (a) Let $F(t) = (1 - 7t, 5 + 9t)$, $G(t) = (3t, t^2 - 7)$, and $\varphi(t) = \langle F(t), G(t) \rangle$. Compute $\varphi'(t)$ in two ways: (i) using Theorem 2, and (ii) by substituting for F and G to find $\varphi(t)$, whose derivative is found as in elementary calculus.

(b) Repeat this for $F(t) = (3 - t^2, 1 + 4t, -9t)$, $G(t) = (1, e^t, 2/(1 + t^2))$.

11. Let $F(t)$ be the equation of a differentiable curve. If $F(t)$ is perpendicular to $F'(t)$ for all t, prove that $\| F(t) \|$ is a constant.

12. (a) If the speed of a particle is identically constant, prove that the acceleration vector is perpendicular to the velocity vector.

(b) Prove the converse: If the acceleration vector is perpendicular to the velocity vector, then the speed is constant.

13. (a) Let $F : \mathbb{R} \longrightarrow \mathbb{R}^3$ define a smooth curve that does not pass through the origin. If the point $X_0 = F(t_0)$ is the point on the curve closest to the origin, show that $F'(t_0) \perp F(t_0)$; that is, the velocity vector is orthogonal to the position vector. (Suggestion: Consider $\varphi(t) = \| F(t) \|^2$.)

(b) Apply this to prove anew the well-known fact that the radius vector to any point on a circle is perpendicular to the tangent vector at that point.

(c) What can you say about the point $X_1 = F(t_1)$ farthest from the origin?

14. (a) Sketch the curve $F(t) = (t(t^2 - 1), t^2 - 1)$, observing that it crosses itself at the origin, which corresponds to $t = \pm 1$.

(b) Find the equation of the tangent line for $t = +1$ and $t = -1$. Sketch these tangent lines.

15. (a) If $F(t)$ defines a smooth curve in $\mathbb{R}^3$, with the property that $F'(t) = O$ for all t, what can you conclude? Proof?

(b) What can you conclude if $F''(t) = O$ for all t? Proof?

16. If $F(t)$ defines a differentiable curve in $\mathbb{R}^3$, with the property that $F'(t) = V_0$ for all t, and $F(0) = X_0$, where X_0 and V_0 are fixed vectors, what can you conclude about $F(t)$? Proof?

17. Let $F \colon \mathbb{R} \longrightarrow \mathbb{R}^q$ define a smooth curve. If $F''(t) = C$, where C is a constant vector, prove that F has the form

$$F(t) = \tfrac{1}{2}Ct^2 + V_0 t + X_0,$$

where X_0 and V_0 are given fixed vectors. Show how X_0 and V_0 are determined by $F(0)$ and $F'(0)$. (This is precisely the situation for projectiles, such as baseballs, moving near the surface of the earth. Then the acceleration $F''(t)$ $= gN$, where N is a unit radius vector to the center of the earth and $g = 32$ ft/sec^2 is the acceleration due to gravity.)

18. A particle is traveling on a curve in outer space, so that its position at time t is given by $F(t) = (1 + t, t^2, -2t)$. At time $t = 2$, however, the particle flies off the curve and moves on the tangent line with constant velocity (we ignore gravity). Where is the particle at $t = 4$?

19. Let $F, G \colon \mathbb{R} \longrightarrow \mathbb{R}^q$ be differentiable maps and $h \colon \mathbb{R} \longrightarrow \mathbb{R}$ a differentiable function. Prove:

(a) $(F + G)' = F' + G'$.

(b) $\dfrac{d}{dt} \langle F(t), G(t) \rangle = ?$

(c) $\dfrac{d}{dt} (h(t)F(t)) = ?$

(d) $\dfrac{d}{dt} (F(t) \times G(t)) = ?$ (here $\times$ is the cross product)

20. Let $F \colon \mathbb{R} \longrightarrow \mathbb{R}^q$ describe a curve, and let $Z \in \mathbb{R}^q$ be a fixed vector. If $F'(t) \perp Z$ for all t and if $F(0) \perp Z$, prove that $F(t) \perp Z$ for all t. (Suggestion: Consider $\varphi(t) = \langle Z, F(t) \rangle$.)

21. Let $F \colon \mathbb{R} \longrightarrow \mathbb{R}^3$ describe the position of a particle, and let $Z \in \mathbb{R}^3$ be a fixed vector. If the acceleration $F''(t)$ is perpendicular to Z for all t and if the initial position and initial velocity vectors $F(0)$ and $F'(0)$ are perpendicular to Z, show that the position $F(t)$ is perpendicular to Z for all t.

22. Let $F \colon \mathbb{R} \longrightarrow \mathbb{R}^2$ be a continuously differentiable curve:

(a) If $\| F'(t) \| \geq 1$ for all t, then $F(0) \neq F(1)$. Proof or counterexample. You might want to think of $F(t)$ as the position of a particle.

(b) If $\| F'(t) \| \leq 1$ for all t, then $\| F(1) - F(0) \| \leq 1$. Proof or counterexample.

23. Let $F \colon \mathbb{R} \longrightarrow \mathbb{R}^3$ describe the position of a particle, and assume that $\| F''(t) \| \leq 1$ for all t, that $F(0) = 0$, and that $F'(0) = 0$. Show that

$$\| F(t) \| \leq \tfrac{1}{2}t^2.$$

In other words, given a bound on the acceleration, find a bound on the position. (Suggestion: Apply Taylor's Theorem with two terms to $\varphi(t) = \langle F(t), Z \rangle$, where $Z = F(s)$ and s is some fixed number. This gives $\| F(s) \| \le \frac{1}{2}s^2$.)

24. (a) Four spiders sit on the floor of a square room, each at a corner. They start walking simultaneously at the same rate, each moving steadily toward the spider on its right, who is also moving, and so on. Find the path of one spider, assuming that each wall is 10 ft long.
    (b) Generalize Exercise 24(a) to the case of N spiders, each at a corner of a regular N-gon (you may prefer to let $N = 3$ or 5).

25. Let A be a given $n \times n$ constant matrix, with the property that $\langle X, AX \rangle \ge 0$ for all $X \in \mathbb{R}^n$, and let $F : \mathbb{R} \longrightarrow \mathbb{R}^n$ satisfy
$$F'(t) + AF(t) = O, \qquad F(0) = X_0,$$
where X_0 is a given constant vector:
    (a) If
$$A = \begin{bmatrix} 2 & 0 \\ 0 & 3 \end{bmatrix},$$
    check that $\langle X, AX \rangle \ge 0$ for all $X \in \mathbb{R}^2$. Show that $F(t) = (ae^{-2t}, be^{-3t})$ satisfies $F' + AF = O$, with $F(0) = (a, b)$.
    (b) Let $E(t) = \| F(t) \|^2$. Prove that E is a decreasing function of time t; that is, $E'(t) \le 0$.
    (c) If $X_0 = O$, prove that $E(t) = 0$ for all $t \ge 0$, and then conclude that $F(t) = O$ for all $t \ge 0$.
    (d) If $F_1(t)$ and $F_2(t)$ both satisfy $F' + AF = O$, with $F_1(0) = X_0 = F_2(0)$, prove that $F_1(t) = F_2(t)$ for all $t \ge 0$. (Suggestion: Consider $F(t) = F_2(t) - F_1(t)$.) This problem shows that there is at most one solution of the differential equation $F' = -AF$, with $F(0) = X_0$. It is a uniqueness theorem.

26. (a) Let $F : \mathbb{R} \longrightarrow \mathbb{R}^n$ be a differentiable function, with derivative $F'(t_0)$ at $t = t_0$. Show that
$$\lim_{h \to 0} \frac{F(t_0 + h) - F(t_0)}{h} = F'(t_0).$$
    (b) Use the formula above to compute $F'(1)$, where $F(t) = (3t - t^2, 1 - 7t)$.

27. A particle moves on a straight line so that its position at time $t \ge 0$ is $F(t) = (\sqrt{t}, \sqrt{2} - \sqrt{t})$, until, at a certain time, it is tangent to the circle $x^2 + y^2 = 1$. It is then trapped into a circular orbit on that circle, with constant speed equal to the speed when it met the circle. When is the particle at the point $(0, 1)$ on the circle?

28. A particle moves so that its position $F(t)$ is a constant distance R from the origin. If you know that $F''(t) = -a^2 F(t)$, where a is a constant, prove that its speed is a constant. In fact, show that $\| F'(t) \| = aR$. Does this check with your intuition in the special case where $a = 0$?

7.2 GRAVITATIONAL MOTION:
AN EXCURSION

7.2a Newton's laws

Why is acceleration important? Suppose that we wish to study the motion of a particle, say a satellite, a photon of light, an electron, a base-ball. Its motion is described by some unknown function $X = X(t)$ (here we write $X(t)$, not $F(t)$ for the position, because we want to reserve the letter F for "force"). The genius of Newton is the pillar of our knowledge here. In particular, we invoke **Newton's Second Law of Motion**. It relates the acceleration to forces: "force equals mass times acceleration,"

$$mX'' = F,$$

where m is the mass of the particle involved and F is the net force on the particle.

Implicit in this wonderful formula is the assertion that the force F is a vector. This is not at all a priori obvious; in fact, it is a statement that requires experimental verification—and all experiments do confirm that force is a vector. Newton's Second tells us that forces "cause" accelera-tion and hence "cause" the motion. But in order for it to be useful, we must be able to say what the forces are (without using $F = mX''$, for this tautology gets us nowhere). Over the years, physics has accumulated theories that describe the forces in various situations.

One of the first and perhaps most widely known force laws is also due to Newton; it is the famous **Law of Gravitational Force**. Say that we have two particles, one of mass m at the point X and another of mass M at Y. See Fig. 7.18. Newton's Law states that the force acting on the particle at X is

$$F = -\gamma \frac{mM}{\|X - Y\|^2} e,$$

where e is a unit vector pointing from Y toward X and $\gamma > 0$ is a fixed con-stant, called the *universal gravitational constant*. The minus sign means that the force is an attractive force. If we observe that $e = (X - Y)/\|X - Y\|$, then the equation for the gravitational force on the particle at X due to the particle at Y is

$$F = -\gamma \frac{mM}{\|X - Y\|^3} (X - Y).$$

Combining Newton's Second with this equation for the force, we find that

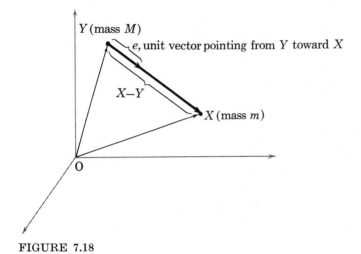

FIGURE 7.18

$$mX'' = -\gamma \frac{mM(X - Y)}{\|X - Y\|^3},$$

an equation that governs motion of all particles moving under the influence of gravity—the moon, spaceships, baseballs. Just one simple and profound formula. Incredible.

AFTERTHOUGHT: Is "incredible" the word? Probably not. It bespeaks a seventeenth-century mentality. As an outgrowth of Newton's work, the existence of such universal formulas may well be, in our culture, a basic article of belief. We take it in with our mother's milk. Such is the influence of Newton.

NOTE: We soon show that if the masses are not point masses, but rather are distributed in a spherical region, such as the earth or moon, approximately, then in applying the equation, it is possible to replace the mass distribution by a *single point* whose mass equals the total mass of the body but concentrated at the center of the body.

7.2b Three applications of the gravitational laws

The power and beauty of Newton's laws can be appreciated only by seeing them in action.

FIRST APPLICATION: motion near the earth's surface. We place the origin of our coordinate system at the center of the earth (mass M_e, radius

R_e). Let m denote the mass of a particle and $X(t)$ its position at time t. See Fig. 7.19. Thus the gravitational force due to the earth is

$$F = -\gamma \frac{mM_e}{\|X\|^3} X.$$

We are pleased to note that the minus sign here indicates that the force F acts in the opposite direction to the vector X from the earth to the particle. F *is* pulling the particle toward the earth. Also, we neglect the gravitational pull of the sun, moon, and other planets.

Let r denote the distance of the particle above the surface of the earth. Then $\|X\| = R_e + r$. Now if the particle remains near the earth, then r is very small compared with R_e. Thus we can approximate

$$\|X\| \approx R_e.$$

Consequently

$$\|F\| \approx \gamma \frac{mM_e}{R_e^2};$$

that is, the magnitude of the force is almost constant. Using Newton's Second, we conclude that the *magnitude of the acceleration near the earth's surface is essentially constant, independent of the mass of the object*:

$$\|X''\| \approx \gamma \frac{M_e}{R_e^2}.$$

The italicized statement but not the formula had been realized earlier by Galileo. Legend has it that he measured the magnitude, g, of this acceleration by dropping stones from the Leaning Tower of Pisa. He found that $g \approx 32$ ft/sec^2 (one must use objects like stones, rather than feathers, to allow us to minimize the influence of air resistance, a force we have ignored). You might try similar experiments yourself.

Armed with the numerical value of $g = \|X''\|$ near the earth's surface, and with the radius of the earth ($R_e \approx 4{,}000$ miles), Newton was able to make a further useful deduction from his formula just above,

$$g \approx \gamma \frac{M_e}{R_e^2};$$

that is, he found the value of the previously unknown constant γM_e,

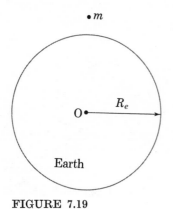

FIGURE 7.19

$$\gamma M_e \approx g R_e^2,$$

a fact we use shortly.

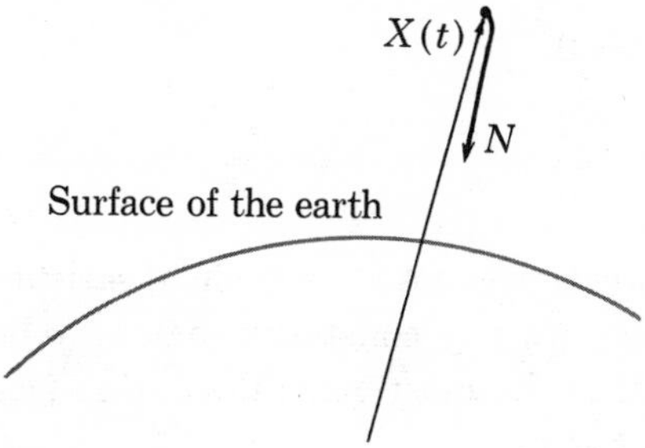

Combining Galileo's value of g with Newton's Second, we find that if $X(t)$ denotes the position of a particle moving near the earth's surface, then

$$X'' = gN,$$

FIGURE 7.20

where N is a unit vector from $X(t)$ toward the center of the earth (down). See Fig. 7.20. Assume that the particle does not move very far, so that we can let N be a constant vector. Then we can integrate this to find

$$X'(t) = gtN + V_0,$$

where $V_0 = X'(0)$ is the initial velocity of the particle. Integrating once again, we conclude that

$$X(t) = \tfrac{1}{2}gt^2N + tV_0 + X_0,$$

where $X_0 = X(0)$ is the initial position of the particle. The last equation governs the motion of all particles moving near the earth's surface under the influence of gravity only. Baseballs are an example. If friction is included, or if we allow the particle to move very far so that N is not constant, then our formula must be modified.

SECOND APPLICATION: motion of a satellite. Now we investigate the path $X(t)$ of a satellite moving about the earth. We *assume* that the satellite remains a constant distance R from the center of the earth. As in the first application, we find that

$$X'' = -\gamma\frac{M_e}{R^3}X,$$

where we again ignore the gravitational force of everything but the earth. An immediate consequence is that the motion does not depend on the mass of the satellite. At this point, we replace the somewhat inscrutable constant γM_e by the value found in the first application; also, for brevity, let $a^2 = \gamma M_e/R^3 = g R_e^2/R^3$, and note that a^2 is a constant independent of time. Then the equation of motion reads

$$X'' + a^2X = 0.$$

This is a second order (= involves second derivatives) differential equation for the unknown vector $X(t)$. In components $X(t) = (x(t), y(t), z(t))$ it splits into three separate equations:

$$x'' + a^2 x = 0,$$
$$y'' + a^2 y = 0,$$
$$z'' + a^2 z = 0.$$

Evidently, by direct substitution, the function $c_1 \cos at + c_2 \sin at$ satisfies each of these equations, where c_1 and c_2 are any constants that may be different for each of the equations. This is the most general solution (see Exercise 8). Thus the path $X(t)$ is

$$X(t) = A \cos at + B \sin at,$$

where the constant vectors A, B are found by observing that $A = X(0)$ is the position of the satellite at $t = 0$ and $aB = X'(0)$ is its velocity at $t = 0$. Consequently,

$$X(t) = X(0) \cos at + \frac{1}{a} X'(0) \sin at.$$

We can make several important deductions from this formula:

1. The satellite always moves in a fixed plane, the plane determined by the vectors $X(0)$ and $X'(0)$. Since $\| X(t) \| = R$, its path is a circle in this plane. See Fig. 7.21.

2. The satellite returns to its initial position at time $t = T$, where $aT = 2\pi$. Using the known value of a, we find that

$$T = \frac{2\pi}{a} = 2\pi \sqrt{\frac{R^3}{gR_e^2}}.$$

This is the *period* of the motion.

For example, say that a satellite orbits 4,000 miles above the earth's surface, so that $R = R_e + 4{,}000 = 8{,}000$ miles. Then the period is

$$T = 2\pi \sqrt{\frac{(8{,}000)^2}{g(4{,}000)^2}} = 2\pi \sqrt{\frac{32 \times 10^3}{32 \times 30^3/11}} \text{ hr} \approx 4 \text{ hr},$$

where we have used $g = 32 \text{ ft/sec}^2 = (32 \times 30^3/11) \text{ miles/hr}^2$.

Another example: Given the fact that the period of the moon (= lunar month) is about 27 days, one can also use the formula above to deduce that its average distance R from the earth is about 240,000 miles. We leave the easy but computationally tedious details to you.

3. The speed of the satellite is constant and is entirely determined by the radius of the orbit. There are two ways to see this. One can use the

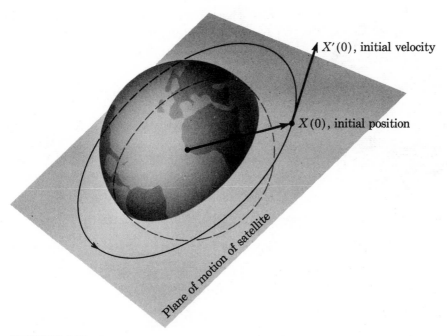

$X'(0)$, initial velocity

$X(0)$, initial position

Plane of motion of satellite

FIGURE 7.21

explicit formula for $X(t)$, but there is a more clever and less computational procedure. Since $\| X(t) \| = R$ for all t, by Corollary 3 we know that $\langle X(t),$ $X'(t) \rangle = 0$ for all t. On differentiating this with respect to t, we find that

$$\langle X'(t), X'(t) \rangle + \langle X(t), X''(t) \rangle = 0.$$

However, $X'' = -a^2 X$. Thus

$$\| X'(t) \|^2 = a^2 \| X(t) \|^2 = a^2 R^2,$$

so that, using the value of a, we conclude that

$$\| X'(t) \| = aR = R_e \sqrt{\frac{g}{R}}.$$

This formula gives the speed in terms of the radius R of the orbit. From it, we see that if a spaceship is in a circular orbit about the earth, then you cannot change its speed and still remain in the same orbit. Thus, if you want to catch up with something ahead of you, you must change the orbit. "Stepping on the gas" is a tricky maneuver in space travel.

THIRD APPLICATION: a hole through the earth. Imagine a hole drilled diametrically through the earth. If we drop a particle of mass m into the

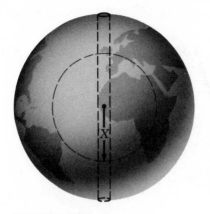

FIGURE 7.22

hole, what is the resulting motion $X(t)$ of the particle? See Fig. 7.22. We make the (unrealistic) assumption that the earth has constant mass density ρ, and so

$$M_e = \frac{4}{3}\pi R_e^3 \rho,$$

and use the fact, proved in the section in Chap. 10 entitled "Gravitational Force," that the portion of the earth farther than $\|X\|$ from the center (outside dotted ball) does not influence the motion. Then the force is

$$F = -\gamma \frac{mM(X)}{\|X\|^3} X,$$

where $M(X)$ is the mass of the earth at distance $\le \|X\|$ from the center of the earth, which we have taken to be the origin:

$$M(X) = \frac{4}{3}\pi \|X\|^3 \rho = \frac{M_e \|X\|^3}{R_e^3}.$$

Here we use the fact, proved in Chap. 9, but presumably known, that the volume of a sphere is $4\pi/3$ times the radius cubed.

Substituting this expression for the force into Newton's Second, we find that

$$X'' = \frac{-\gamma M_e}{R_e^3} X;$$

that is,

$$X'' + \alpha^2 X = 0,$$

where $\alpha^2 = \gamma M_e / R_e^3 = g/R_e$. This differential equation for $X(t)$ is exactly the one found in the second application. Thus its solution is

$$X(t) = A \cos \alpha t + B \sin \alpha t,$$

where $A = X(0)$ is the inital position and $\alpha B = X'(0)$ is the initial velocity. Since we assumed that the particle was dropped, not thrown, we have $X'(0) = 0$, and so $B = 0$. Therefore

$$X(t) = X(0) \cos \alpha t.$$

Consequently, the particle oscillates back and forth in a straight line through the diameter. See Fig. 7.23. Its period T is determined by $\alpha T = 2\pi$, so

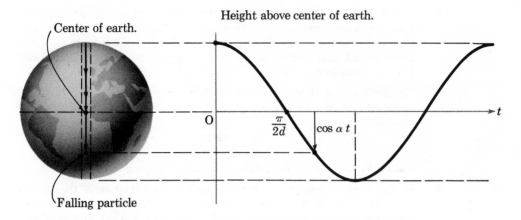

Showing periodic motion of falling particle.

FIGURE 7.23

$$T = \frac{2\pi}{\alpha} = 2\pi\sqrt{\frac{R_e}{g}}.$$

One can easily check that the particle is going fastest at the center of the earth, and its speed is zero at both ends of the diameter.

Exercises

1. Assume that the orbit of the moon is a circle about the earth. Deduce the radius of this orbit from the fact that the period of the moon is roughly 27 days.

2. The Syncom satellites are intended to hover over one point of the equator. Thus, their orbits are circular with period 24 hr in order to appear stationary. Find the radius of the orbit and the speed of the satellite.

3. Is it possible for a satellite, with its rockets turned off, to have a circular orbit that always remains over the Northern Hemisphere of the earth? Explain.

4. If $X(t)$ is the position of a particle of mass m moving under a gravitational force due to a mass M at the origin, prove that the energy

$$E(t) = \frac{m}{2}\,\|\,X'(t)\,\|^2 - \frac{\gamma mM}{\|\,X(t)\,\|}$$

is conserved; that is, $E'(t) = 0$, and so $E(t) \equiv$ constant.

5. A meteor falls toward earth. It is sighted 12,000 miles from the center of the earth, at which time its speed is observed to be 100 miles/hr. Neglecting air resistance, what is its speed when it hits the earth? (Suggestion: Use the result in Exercise 4.)

6. A rocket ship is fired vertically from the surface of the earth. At 6,000 miles from the earth's surface, its rockets turn off. Neglecting the presence of the sun, and so on, what must its speed be at that time in order to escape the gravitational attraction of the earth and "travel to infinity"? (Suggestion: Use the result in Exercise 4.)

7. The earth's orbit about the sun is roughly a circle of radius $R = 93,000,000$ miles, and its period is roughly $T = 365$ days. If M_s is the mass of the sun, show that

$$\frac{M_s}{M_e} = \frac{4\pi^2 R^3}{g T^2 R_e^2}$$

and evaluate the right side. Thus, you have "weighed" the sun.

8. In this exercise, you prove that the most general solution of the equation $x'' + a^2 x = 0$, $a \neq 0$, is $x(t) = c_1 \cos at + c_2 \sin at$, where c_1 and c_2 are any constants (this was used in the second application). Let $\varphi(t)$ be any solution, and let $x^*(t) = \varphi(0) \cos at + (\varphi'(0)/a) \sin at$. We claim that $\varphi(t) = x^*(t)$. This will prove our assertion, with $c_1 = \varphi(0)$, $c_2 = \varphi'(0)/a$:

(a) If $\psi(t) = \varphi(t) - x^*(t)$, show that $\psi'' + a^2\psi = 0$ and that $\psi(0) = 0$, $\psi'(0) = 0$.

(b) Let $E(t) = \psi'^2(t) + a^2\psi^2(t)$. Show that $E'(t) = 0$, so that E is a constant. Find the constant.

(c) Deduce that $\psi(t) \equiv 0$ and consequently that $\varphi(t) = x^*(t)$.

9. Say that a particle with position $X(t)$ moves in a **central force field** $F(X) = \varphi(X)X$, where $\varphi(X)$ is a scalar-valued function, so that the direction of the force F is the same as that of the position X:

(a) Prove that

$$\frac{d}{dt}(X(t) \times X'(t)) = 0 \qquad \times = \text{cross product.}$$

(b) If $Z = X(t) \times X'(t)$, deduce that the position and velocity vectors are both in the plane perpendicular to the constant vector Z. Thus, *central force motion always takes place in a plane*. The vector $mZ = X(t) \times mX'(t)$ is the **angular momentum** of a particle of mass m about the origin. The result of Exercise 9(a) asserts that it is constant in central force motion.

7.3 ARC LENGTH

7.3a Introduction

We comment first that the material in the preceding section is not used here.

Now let $F: \mathbb{R} \rightarrow \mathbb{R}^q$ be a mapping that traces out a smooth curve in $\mathbb{R}^q$. Suppose that $F(t)$ is defined for all t in some interval, $a \leq t \leq b$. How can we measure length along the curve from $F(a)$ to $F(b)$? See Fig. 7.24.

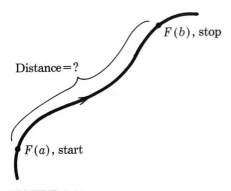

$F(b)$, stop

Distance=?

$F(a)$, start

FIGURE 7.24

This is the same as asking for the distance traveled by a particle whose position is $F(t)$ during the interval of time from $t = a$ to $t = b$.

If the curve is actually a straight line segment from a point X_0 to a point X, then we know its length; it is simply $\|X - X_0\|$. We have a precise formula for this, the Pythagorean formula; if $X_0 = (c_1, \ldots, c_q)$ and $X = (x_1, \ldots, x_q)$, then

$$\|X - X_0\| = \sqrt{(x_1 - c_1)^2 + \cdots + (x_q - c_q)^2}.$$

If the curve is *not* a straight line, then we have not even defined in a precise way what we mean by its length. And certainly we must define length before we can hope to discover a formula to compute it. Let us attend to this definition now.

7.3b The definition of arc length

We wish to define the length of the arc (piece of curve) traced out by $F(t)$ for $t = a$ to $t = b$, with $a \le b$. The idea is to approximate this arc by straight line segments, whose length we know. See Fig. 7.25.

We write $[a, b]$ for the interval $a \le t \le b$ in the t-axis. A **partition** of the interval $[a, b]$ is any finite set

$$\mathscr{P} = \{t_0, t_1, \ldots, t_m\}$$

consisting of numbers that satisfy

$$a = t_0 < t_1 < \cdots < t_m = b.$$

The points $F(a) = F(t_0)$, $F(t_1), \ldots, F(t_m) = F(b)$ are on the arc from $F(a)$ to $F(b)$. We connect $F(t_0)$ to $F(t_1)$ by a straight line segment, of length $\|F(t_1) - F(t_0)\|$; then connect $F(t_1)$ to $F(t_2)$ by another segment, of length $\|F(t_2) - F(t_1)\|$; then continue on to $F(t_3)$, and so on, until we reach the endpoint $F(t_m)$, that is, $F(b)$.

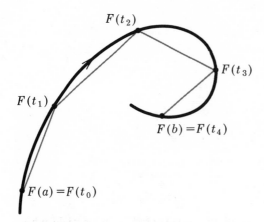

Approximating a curve by segments

FIGURE 7.25

The total distance traveled along these straight line segments from $F(t_0)$ to $F(t_1)$, $F(t_1)$ to $F(t_2)$, ..., $F(t_{m-1})$ to $F(t_m)$ is the finite sum

$$S(\mathscr{P}) = \| F(t_1) - F(t_0) \| + \cdots + \| F(t_m) - F(t_{m-1}) \|.$$

In Fig. 7.25, we have $\mathscr{P} = \{t_0, t_1, t_2, t_3, t_4\}$, and so

$$S(\mathscr{P}) = \| F(t_1) - F(t_0) \| + \| F(t_2) - F(t_1) \| + \cdots + \| F(t_4) - F(t_3) \|.$$

It would appear that this number $S(\mathscr{P})$ is somewhat less than the true length of the arc, because, in brief, a straight line gives the *shortest* distance between two points.

DEFINITION: The arc from $F(a)$ to $F(b)$ is **rectifiable** if and only if there is a number $s \geq 0$ such that:

1. $s \geq S(\mathscr{P})$ for all partitions $\mathscr{P}$ of $[a, b]$.
2. s is the smallest number having property 1.

If the arc is rectifiable, we define its **length**, denoted $\mathscr{L}(F; a, b)$, to be the number s above, $\mathscr{L}(F; a, b) = s$.

The rough idea here is, of course, that the more refined the partition $\mathscr{P}$ we select from the interval $[a, b]$, the closer the curve approximated by the path of straight line segments from $F(t_0)$ to $F(t_1)$, $F(t_1)$ to $F(t_2)$, and so on.

REMARKS: 1. You may check that if the arc from $F(a)$ to $F(b)$ is actually a straight-line segment, then the length as defined above is equal to

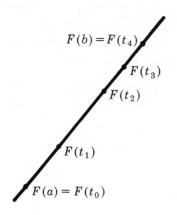

$F(b) = F(t_4)$

$F(t_3)$

$F(t_2)$

$F(t_1)$

$F(a) = F(t_0)$

Partition of a straight line

FIGURE 7.26

$\| F(b) - F(a) \|$. Note that, for instance, on the segment pictured in Fig. 7.26, $\mathscr{L}(F; a, b) = \| F(b) - F(a) \|$ does equal

$$\| F(t_1) - F(a) \| + \| F(t_2) - F(t_1) \| \\ + \| F(b) - F(t_3) \|.$$

2. If the arc is not rectifiable, then its length is infinite, by definition. It can be shown, however, that if $F(t)$ is a smooth curve, well-behaved near $t = a$ and $t = b$, then the arc from $F(a)$ to $F(b)$, with $-\infty < a \leq b < \infty$, is rectifiable. In this book we do not encounter arcs that are not rectifiable, except for an example in Exercise 5.

3. This definition of the length $\mathscr{L}(F; a, b)$ is geometrically convincing but not much good computationally. We certainly cannot form *all* partitions $\mathscr{P}$ of the interval $[a, b]$ and the corresponding sums $S(\mathscr{P})$; the number of such partitions is infinite. In the following section we derive a formula for the arc length $\mathscr{L}(F; a, b)$.

7.3c A formula for arc length

How may we compute the length of the arc from $F(a)$ to $F(b)$ on the curve $F(t)$? The answer is given in the following theorem.

THEOREM 4: Let $F(t)$ trace out a smooth curve in $\mathbb{R}^q$ for $t \in [a, b]$. Then the length of the arc from $F(a)$ to $F(b)$ is given by

$$\mathscr{L}(F; a, b) = \int_a^b \| F'(t) \| \, dt.$$

NOTE: $\| F'(t) \|$ is a nonnegative *scalar-valued* function of the real variable t. Hence the integral is the ordinary integral of calculus.

Outline of the Proof: For a partition $\mathscr{P}$ of $[a, b]$, a typical term in the approximate length $S(\mathscr{P})$ is of the form $\| F(t_{k+1}) - F(t_k) \|$. Let us write $\Delta_k t = t_{k+1} - t_k$. Then the typical term equals

$$\| F(t_k + \Delta_k t) - F(t_k) \| = \left\| \frac{F(t_k + \Delta_k t) - F(t_k)}{\Delta_k t} \right\| \Delta_k t.$$

Hence the approximating sum $S(\mathscr{P})$ has been put in the form

$$S(\mathscr{P}) = \sum_{k=0}^{m-1} \frac{\| F(t_k + \Delta_k t) - F(t_k) \|}{\Delta_k t} \Delta_k t.$$

Recall that we obtain the length $\mathscr{L}(F; a, b)$ from $S(\mathscr{P})$ by considering partitions $\mathscr{P}$ with more and more elements $t_0, t_1, \ldots, t_m$, that is, $m \to \infty$, and, moreover, insisting that *all* $\Delta_k t = t_{k+1} - t_k$ approach zero. If we do this, two things happen (these are plausible, but we omit details):

1. The expression $(F(t_k + \Delta_k t) - F(t_k))/\Delta_k t$ approaches $F'(t_k)$, since $\Delta_k t$ is approaching 0 and $F(t)$ is smooth. Hence $\|(F(t_k + \Delta_k t) - F(t_k))/\Delta_k t\|$ approaches $\|F'(t_k)\|$.

2. The sum

$$\sum_{k=0}^{m-1} \|F'(t_k)\| \, \Delta_k t$$

approaches the integral

$$\int_a^b \|F'(t)\| \, dt.$$

This integral exists as a finite number, because the integrand, $\|F'(t)\|$, is a continuous function. Thus, the limit of the approximating sums $S(\mathscr{P})$ over all $\mathscr{P}$ exists and equals the integral as claimed. $\ll$

REMARKS: 1. If $F(t) = (f_1(t), \ldots, f_q(t))$, then

$$\|F'(t)\| = \sqrt{f_1'(t)^2 + \cdots + f_q'(t)^2}.$$

In particular, if $F(t)$ traces out a curve in the plane, $F: \mathbb{R} \to \mathbb{R}^2$, then we sometimes write $x = f_1(t)$, $y = f_2(t)$. Thus $f_1'(t) = dx/dt$ or even $\dot{x}$, and $f_2'(t) = dy/dt = \dot{y}$. It follows that the arc length may therefore be written

$$\mathscr{L}(F: a, b) = \int_a^b \sqrt{\left(\frac{dx}{dt}\right)^2 + \left(\frac{dy}{dt}\right)^2} \, dt,$$

or

$$\mathscr{L}(F: a, b) = \int_a^b \sqrt{\dot{x}^2 + \dot{y}^2} \, dt.$$

2. If we think of $\|F'(t)\|$ as speed and dt as a measure of time, then the integrand in the formula is the product of speed and time, namely, distance or length, as expected.

EXAMPLES: 1. *The arc length of a circle.* Let $F(t) = (a \cos t, a \sin t)$. This traces out the circle of radius $a > 0$ centered at the origin as t varies in the interval $[0, 2\pi]$. What is the distance around this circle?

Since $F'(t) = (-a \sin t, a \cos t)$, we have $\|F'(t)\| = a$. Thus the length is

$$\mathcal{L}(F; 0, 2\pi) = \int_0^{2\pi} \| F'(t) \| \, dt = a \int_0^{2\pi} dt = 2\pi a,$$

as expected.

2. *The length of a graph.* The graph of $y = f(x)$ in the xy-plane is traced out by the mapping $F(t) = (t, f(t))$. We have $F'(t) = (1, f'(t))$, whence $\| F'(t) \| = \sqrt{1 + f'(t)^2}$. Using $x = t$, $y = f(x) = f(t)$, we conclude that the length $\mathcal{L}$ of the graph of $y = f(x)$ for $a \le x \le b$ is given by

$$\mathcal{L} = \int_a^b \sqrt{1 + f'(x)^2} \, dx,$$

a formula sometimes seen in ordinary calculus.

An explicit problem: Find the length of the graph $y = f(x) = x^{3/2}$ from $x = 0$ to $x = \frac{4}{3}$.

Solution: We have $f'(x) = \frac{3}{2}x^{1/2}$, whence the length is

$$\mathcal{L} = \int_0^{4/3} \sqrt{1 + \frac{9}{4}x} \, dx$$

$$= \frac{8}{27}\left(1 + \frac{9}{4}x\right)^{3/2}\Big]_{x=0}^{x=4/3} = \frac{56}{27}.$$

Thus $\mathcal{L} = \frac{56}{27}$, slightly more than 2 units long. See Fig. 7.27.

3. *The helix.* What is the length of one turn of the helix

$$F(\theta) = (\cos\theta, \sin\theta, \theta),$$

where $0 \le \theta \le 2\pi$.

We readily compute that $\| F'(\theta) \| = \sqrt{2}$, a constant, whence the length is given by

$$\mathcal{L}(F; 0, 2\pi) = \sqrt{2} \int_0^{2\pi} d\theta = 2\sqrt{2}\,\pi.$$

Note that in this example we used θ, not t, as the parameter. Needless to say, the concept of arc length does not depend on the name we give the parameter.

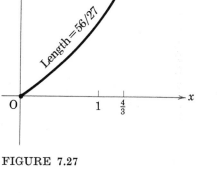

$y = f(x) = x^{3/2}$

Length $= 56/27$

FIGURE 7.27

4. *How to concoct some rectifiable plane curves.* The integrand $\sqrt{(dx/dt)^2 + (dy/dt)^2}$ may be formidable, in great part because of the square root. Here is a method for constructing curves $F(t) = (x(t), y(t))$ such that $(dx/dt)^2 + (dy/dt)^2$ is a perfect square. This will enable you to make up some reasonable practice problems.

If u, v are numbers, and if we define α, β, γ by

$$\alpha = u^2 - v^2, \qquad \beta = 2uv, \qquad \gamma = u^2 + v^2,$$

then it is easy to see that $\alpha^2 + \beta^2 = \gamma^2$. This is a classical method of generating right triangles—sides α, β, γ. Now let $u = u(t)$, $v = v(t)$ be ordinary functions such that the indefinite integrals

$$\int u^2(t)\, dt, \qquad \int v^2(t)\, dt, \qquad \int u(t) v(t)\, dt$$

are not too hard for you. For example, you might let $u(t)$ and $v(t)$ be any polynomials in t.

If we now define

$$x(t) = \int \{u^2(t) - v^2(t)\}\, dt, \qquad y(t) = 2 \int u(t) v(t)\, dt,$$

then we note, first, that each of these indefinite integrals can be done, and second, that

$$\left(\frac{dx}{dt}\right)^2 + \left(\frac{dy}{dt}\right)^2 = (u^2 + v^2)^2.$$

This is because $dx/dt = u^2 - v^2$ and $dy/dt = 2uv$ by the Fundamental Theorem of Calculus. Thus the problem of finding the length of the curve given by $F(t) = (x(t), y(t))$ as defined here reduces to the integration of $u^2(t) + v^2(t)$ (no square root), and we *chose* u, v so that this integration could be done.

You might pick two functions $u(t)$, $v(t)$ and try this procedure for yourself.

Exercises

1. Find the length of the following curves:
    (a) $F(t) = (1 - 3t, 5 + 4t)$, $\qquad -1 \le t \le 2$;
    (b) $G(r) = (r^3, 2 - 6r^2)$, $\qquad 0 \le r \le 3$;
    (c) $H(\tau) = (2 - \tau, 7 + 2\tau, -3 + 2\tau)$, $\qquad 0 \le \tau \le 4$;
    (d) $y = 7 + (x^2 + \tfrac{2}{3})^{3/2}$, $\qquad 1 \le x \le 2$;
    (e) $y = \dfrac{x^3}{6} + \dfrac{1}{2x}$, $\qquad 1 \le x \le 3$;
    (f) $F(\theta) = (\cos^3 \theta, 7, \sin^3 \theta)$, $\qquad 0 \le \theta \le \pi/2$;
    (g) $\Phi(t) = (e^{2t} \sin t, e^{2t} \cos t, e^{2t})$, $\qquad 0 \le t \le \log 5$;
    (h) $R(t) = (t \sin t, \sqrt{8}\,t, t \cos t, \tfrac{2}{3}\sqrt{6}\,t^{3/2})$, $\qquad 4 \le t \le 8$.

2. (a) The position, $R(t)$, of a particle at time t is given by
    $$R(t) = (5 + 3t^2, 2(1 + 2t)^{3/2}).$$
    How far does the particle travel between $t = 0$ and $t = 4$?
    (b) Repeat this for
    $$R(t) = (\tfrac{2}{3}t, 5 - t^2, 4 + t^3), \qquad 1 \le t \le 2.$$

3. Let P and Q be distinct points in $\mathbb{R}^n$, and let

$$F(t) = \frac{(t-a)P + (b-t)Q}{b-a}.$$

What is the arc length for $a \le t \le b$? Could you have guessed the answer?

4. The equations of curves are often expressed in polar coordinates by giving r, the distance from the origin, in terms of the angle θ, so that $r = r(\theta)$, where

$$x(\theta) = r(\theta) \cos \theta, \qquad y(\theta) = r(\theta) \sin \theta.$$

This is a parametric representation with the angle θ as parameter.

(a) Show that the length of a curve for $\alpha \le \theta \le \beta$ is

$$\mathcal{L} = \int_\alpha^\beta \sqrt{r'^2 + r^2}\, d\theta,$$

where $r' = dr/d\theta$.

(b) Find the arc length of a circle, $r(\theta) = 1$.

5. This is the standard example of a continuous curve that is not rectifiable. Let $y = x \sin (\pi/x)$ for $x \ne 0$, and $y = 0$ when $x = 0$.

(a) Show geometrically that the length of the portion of the curve for $1/(n+1) \le x \le 1/n$, that is, for one arch, is at least $2/(n + \frac{1}{2})$.

(b) Use this to show that the length of the curve from $x = 1/N$ to $x = 1/\pi$ tends to infinity as $N \longrightarrow \infty$ (you will need the fact that $1 + \frac{1}{2} + \frac{1}{3} + \cdots$

$$+ \frac{1}{N} \longrightarrow \infty \text{ as } N \longrightarrow \infty).$$

6. If $F(t) = (x_1(t), \ldots, x_n(t))$ is continuous, we define, as anticipated,

$$\int_a^b F(t)\, dt = \left(\int_a^b x_1(t)\, dt, \ldots, \int_a^b x_n(t)\, dt \right).$$

(a) Prove that

$$\left\| \int_a^b F(t)\, dt \right\| \le \int_a^b \| F(t) \|\, dt.$$

(b) If X_0 and X_1 are the end points of a smooth curve, show that

$$\text{length of curve} \ge \| X_1 - X_0 \|.$$

7. Let γ be a closed piecewise smooth curve in $\mathbb{R}^2$ that encloses a convex region D; that is, if P and Q are in D, so is every point on the straight line segment joining P and Q. See Fig. 7.28. We say that the curve γ_r is "parallel" to γ if every point on γ_r is outside D and has distance r from γ. Discover a formula relating the arc length γ_r to γ. (Suggestion: Look at the special cases in which γ is a circle, a rectangle, or a polygon.)

8. In Exercise 22 in Sec. 7.1, the spiders eventually meet in the center of the room. What is the total distance walked by each spider?

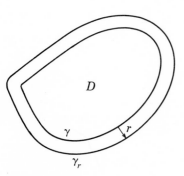

FIGURE 7.28

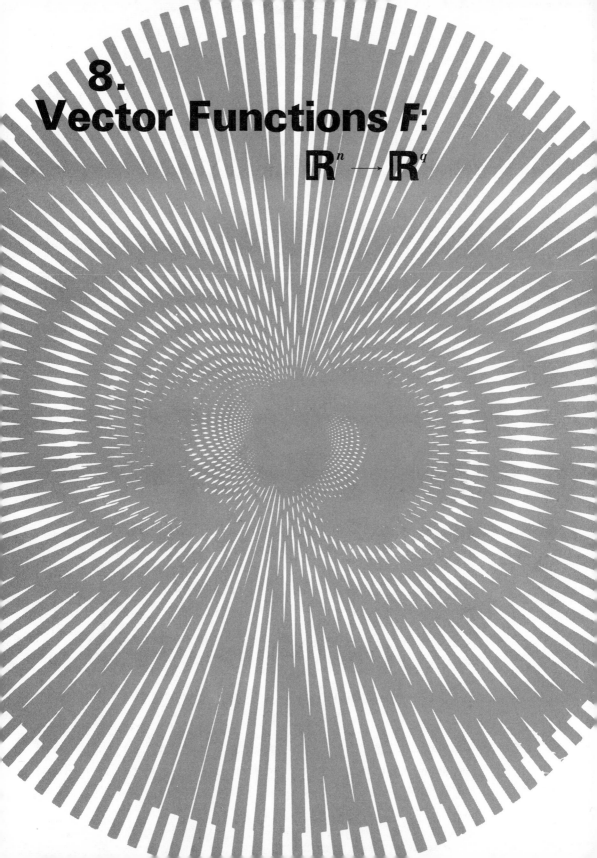

8.
Vector Functions F:
$$\mathbb{R}^n \longrightarrow \mathbb{R}^q$$

8.0 INTRODUCTION

Now we discuss the general case of $F: \mathbb{R}^n \to \mathbb{R}^q$, with n and q arbitrary positive integers. We have dealt with special cases before: fixing $n = q = 1$ yields ordinary one-variable calculus; n arbitrary, $q = 1$ yields scalar-valued functions of a vector; $n = 1$, q arbitrary yields curves in q-space.

Here, of course, the *function* or *map* F is a rule that assigns to each vector X in some subset $\mathscr{D}(F)$ of $\mathbb{R}^n$ a vector $Y = F(X)$ in $\mathbb{R}^q$. The subset $\mathscr{D}(F)$ is the *domain* of F; for us, $\mathscr{D}(F)$ is usually the entire space $\mathbb{R}^n$.

How are such functions specified? If $Y = F(X)$, with $X = (x_1, \ldots, x_n)$ and $Y = (y_1, \ldots, y_q)$, then each y_j depends on $x_1, \ldots, x_n$; that is,

$$y_j = f_j(x_1, \ldots, x_n) = f_j(X),$$

with $f_j: \mathbb{R}^n \to \mathbb{R}$, $j = 1, \ldots, q$. The scalar-valued functions $f_1, \ldots, f_q$ are the *coordinates* of F. It is sometimes instructive to write

$$F(X) = (f_1(X), \ldots, f_q(X)),$$

or even

$$F = (f_1, \ldots, f_q).$$

How do we interpret these maps? One standard interpretation is as geometric transformations of $\mathbb{R}^n$ into $\mathbb{R}^q$. We see this in the examples.

EXAMPLES: 1. Let $F: \mathbb{R}^2 \to \mathbb{R}^2$ be affine, $Y = F(X)$, given by

$$y_1 = f_1(x_1, x_2) = 2 + x_1 - 2x_2,$$
$$y_2 = f_2(x_1, x_2) = 1 + x_1 + x_2.$$

We think of this as a map from the $x_1 x_2$-plane to the $y_1 y_2$-plane. The origin $(0, 0)$ is mapped to $(2, 1)$; the x_1-axis, given by $x_2 = 0$, is mapped to the set of points $(y_1, y_2) = (2 + x_1, 1 + x_1)$; thus $y_1 - y_2 - 1 = 0$, and so a straight line in the $y_1 y_2$-plane is the image of the x_1-axis. The x_2-axis is mapped onto the straight line $y_1 + 2y_2 - 4 = 0$, as you should verify.

The shaded region in Fig. 8.1 indicates the image of the unit square

$$0 \le x_1 \le 1, \qquad 0 \le x_2 \le 1.$$

2. Now we embed a subset of the plane into 3-space. Let $G: \mathbb{R}^2 \to \mathbb{R}^3$ be given as follows: We let the coordinates in $\mathbb{R}^2$ be θ and φ. Then $G(\theta, \varphi) = Y = (y_1, y_2, y_3)$, where

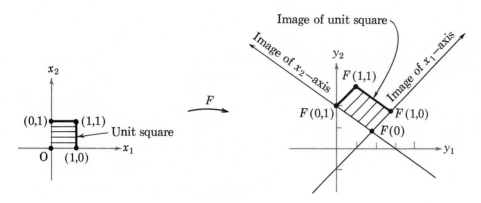

An affine map

FIGURE 8.1

$$y_1 = (a + b \cos \varphi) \cos \theta,$$

$$y_2 = (a + b \cos \varphi) \sin \theta,$$

$$y_3 = b \sin \varphi.$$

Here a, b are scalars, and $0 < b < a$.

We mentioned this function in Chap. 4, stating that the image of the square $0 \leq \theta \leq 2\pi$, $0 \leq \varphi \leq 2\pi$ under G is a torus in $\mathbb{R}^3$. To verify this, we examine where G maps vertical segments $\theta = $ constant, $0 \leq \varphi \leq 2\pi$, in the square. See Fig. 8.2.

First let $\theta = 0$. Then $G(0, \varphi)$ is a point in the vertical plane $y_2 = 0$. As φ varies from 0 to 2π, the point $G(0, \varphi)$ travels around a small circle of radius b centered at $(a, 0, 0)$.

Now let θ be arbitrary but fixed, $\theta = \theta_0$, $0 \leq \theta_0 \leq 2\pi$. Then $\cos \theta_0$, $\sin \theta_0$ are fixed scalars, and, if $(y_1, y_2, y_3) = G(\theta_0, \varphi)$, then it is easy to see that

$$(\sin \theta_0)y_1 - (\cos \theta_0)y_2 = 0.$$

This is the equation of a vertical plane containing the y_3-axis in $y_1y_2y_3$-space. Hence $G(\theta_0, \varphi)$ is a point in the plane for all φ. Moreover, as φ varies from 0 to 2π, the point $G(\theta_0, \varphi)$ travels around a small circle of radius b centered at the point $(a \cos \theta_0, a \sin \theta_0, 0)$.

Thus for each choice $\theta = \theta_0$, we get a small circle of radius b traced out by $G(\theta_0, \varphi)$. The centers of all these small circles lie on one large circle of radius a—recall that $0 < b < a$—in the horizontal plane $y_3 = 0$.

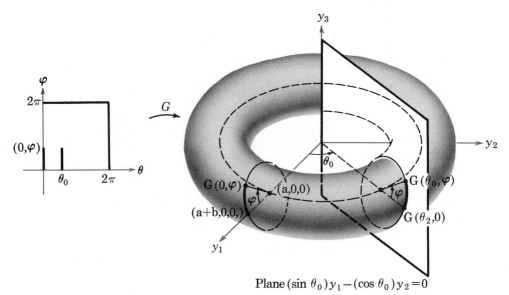

FIGURE 8.2

Hence as $\theta = \theta_0$ varies from 0 to 2π, the small circles trace out a torus (hollow doughnut) in 3-space.

Note that the periodicity of the sine and cosine imply that every point (θ, φ) is mapped by G onto this torus, not only those points in the square $0 \le \theta, \varphi \le 2\pi$. For instance, $G(0, 2\pi) = G(0, 4\pi) = G(0, 6\pi) = G(0, 0) = (a + b, 0, 0)$.

We urge you to work through this example with care. You will see the torus again in the following section. We dwell on it because it is geometrically interesting, easily visualized, and not beyond the powers of computation.

3. The graph of $z = f(x_1, x_2) = x_1^2 + x_2^2$ is a paraboloid in $\mathbb{R}^3$. We studied this in Chap. 5, since $f: \mathbb{R}^2 \to \mathbb{R}$. Now we claim that this graph may also be considered the *image* of a map $F: \mathbb{R}^2 \to \mathbb{R}^3$ by a useful artifice. Let $X = (x_1, x_2)$, $Y = (y_1, y_2, y_3)$, and let $Y = F(X) = (f_1(X), f_2(X), f_3(X))$ be determined by

$$y_1 = f_1(X) = x_1,$$
$$y_2 = f_2(X) = x_2,$$
$$y_3 = f_3(X) = f(X) = x_1^2 + x_2^2.$$

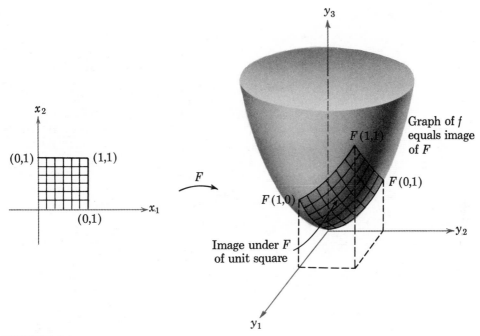

FIGURE 8.3

Hence F embeds the x_1x_2-plane into $y_1y_2y_3$-space as a surface. This surface is the image $F(\mathbb{R}^2) = \{F(X) \mid X \in \mathbb{R}^2\}$. See Fig. 8.3. The map F may be described more succinctly by

$$F(x_1, x_2) = (x_1, x_2, f(x_1, x_2)).$$

It should be clear that the graph of *any* scalar-valued function $f(x_1, x_2)$ may be embedded into $\mathbb{R}^3$ by such a map $F(x_1, x_2)$. This provides us with many more examples of maps from $\mathbb{R}^2$ to $\mathbb{R}^3$.

In the sections to come, we study the differential calculus of nonlinear maps $F: \mathbb{R}^n \rightarrow \mathbb{R}^q$. This is a program that should be familiar: To study F near the point $X_0 \in \mathbb{R}^n$, we construct its best affine approximation at X_0 and examine it. This is an affine map $T: \mathbb{R}^n \rightarrow \mathbb{R}^q$ that approximates F near X_0 in the sense that $T(X_0) = F(X_0)$ and $T(X)$ is close to $F(X)$, provided that X is close to X_0. Having defined the map T, we must learn how to compute it (partial derivatives again) and interpret it. Let us begin this now.

Exercises

1. Consider the maps $F: \mathbb{R}^2 \longrightarrow \mathbb{R}^2$, defined by:

    (a) $y_1 = \quad 1 + x_1 + x_2,$
    $y_2 = -2 + x_1 - x_2;$

    (b) $y_1 = x_1 - x_2,$
    $y_2 = x_1^2 + x_2^2,$

    (c) $y_1 = x_1 \cos \pi x_2,$
    $y_2 = x_1 \sin \pi x_2.$

    For each of these, draw sketches, as in Example 1, showing the square with vertices at $(1, 0)$, $(2, 0)$, $(2, 1)$, and $(1, 1)$ as well as the image of this square. (Suggestion: Find the image of the sides of the square.)

2. Consider the maps $F: \mathscr{D} \longrightarrow \mathbb{R}^3$, $\mathscr{D} \subset \mathbb{R}^2$, defined by:

    (a) $\mathscr{D} = \{x_1^2 + x_2^2 \le 1\},$
    $y_1 = x_1,$
    $y_2 = x_2,$
    $y_3 = \sqrt{4 - x_1^2 - x_2^2};$

    (b) $\mathscr{D} = \left\{ 0 \le x_1 \le \dfrac{\pi}{2}, \quad 0 \le x_2 \le \dfrac{\pi}{2} \right\},$
    $y_1 = \sin x_1 \cos x_2,$
    $y_2 = \sin x_1 \sin x_2,$
    $y_3 = \cos x_1;$

    (c) $\mathscr{D} = \{1 \le x_1 \le 2, \quad 0 \le x_2 \le 2\},$
    $y_1 = x_1 \cos \pi x_2,$
    $y_2 = x_1 \sin \pi x_2,$
    $y_3 = \dfrac{4}{x_1^2}.$

    For each of these, draw sketches of the image of $\mathscr{D}$.

8.1 THE BEST AFFINE APPROXIMATION AND FIRST DERIVATIVE

8.1a The basic definitions

We have $F: \mathbb{R}^n \to \mathbb{R}^q$ and suppose, for brevity, that $F(X)$ is defined for all $X \in \mathbb{R}^n$. Let X_0 be fixed in $\mathbb{R}^n$. We say that F is **differentiable** at X_0 if and only if there exists an affine map $T: \mathbb{R}^n \to \mathbb{R}^q$ satisfying:

1. $T(X_0) = F(X_0).$
2. $\displaystyle \lim_{X \to X_0} \frac{\| F(X) - T(X) \|}{\| X - X_0 \|} = 0.$

Note how this agrees with the definitions in Chaps. 5 and 7. The map T is called the **best affine approximation** to F at X_0. Since T is affine, it may be put in the form (see Chaps. 3 and 4)

$$T(X) = T(X_0) + L(X - X_0),$$

where $L: \mathbb{R}^n \to \mathbb{R}^q$ is a linear map determined by T (and hence by F).

This linear map L is called the **first derivative** or **total derivative** of F at X_0 and denoted $F'(X_0)$. Hence we have, using requirement 1 and the fact that $L = F'(X_0)$,

$$T(X) = F(X_0) + F'(X_0)(X - X_0).$$

This is a familiar expression.

8.1b The Main Theorem on Differentiable Maps

We suppose that we are given F in coordinate form and are told that it is differentiable at X_0. How do we compute the best affine approximation T?

We know $T(X) = F(X_0) + F'(X_0)(X - X_0)$. Given F, there is no problem computing the vector $F(X_0)$; it is the linear map $F'(X_0)$ that we must find. We recall, however, that since $F: \mathbb{R}^n \to \mathbb{R}^q$, then $F'(X_0)$ may be represented by a matrix $[F'(X_0)]$ of q rows and n columns, a q by n matrix.

THEOREM 1: (Main Theorem on Differentiable Maps.) Let $F: \mathbb{R}^n \to \mathbb{R}^q$, $Y = F(X) = (f_1(X), \ldots, f_q(X))$ and suppose F differentiable at X_0. Then:

1.   Each coordinate function f_i is differentiable at X_0 (that is, all partial derivatives $\partial f_i / \partial x_j (X_0)$ exist, $j = 1, \ldots, n$).

2.   The first derivative $F'(X_0)$ has the matrix

$$[F'(X)_0] = \left[\frac{\partial f_i}{\partial x_j}(X_0) \right] = \begin{bmatrix} \dfrac{\partial f_1}{\partial x_1}(X_0) & \cdots & \dfrac{\partial f_1}{\partial x_n}(X_0) \\ \vdots & & \vdots \\ \dfrac{\partial f_q}{\partial x_1}(X_0) & \cdots & \dfrac{\partial f_q}{\partial x_n}(X_0) \end{bmatrix}.$$

REMARK: This matrix is sometimes called the **Jacobian of** F, after the nineteenth-century mathematician C. J. G. Jacobi.

Proof: 1. Let us write $T(X) = (t_1(X), \ldots, t_q(X))$, where each t_i: $\mathbb{R}^n \to \mathbb{R}$ is affine and scalar-valued. Then we have, for each i, that

$$|f_i(X) - t_i(X)| \leq \|F(X) - T(X)\|.$$

Dividing by a positive quantity, we get

$$\frac{|f_i(X) - t_i(X)|}{\|X - X_0\|} \leq \frac{\|F(X) - T(X)\|}{\|X - X_0\|}.$$

Now if T is the best affine approximation to F at X_0, the right-hand expression approaches zero as X approaches X_0. Hence the left-hand expression approaches zero also. This means that f_i is differentiable at X_0 and that t_i is its best affine approximation there. Therefore statement 1 is proved.

2. Let us compute the first column of $[F'(X_0)]$. This is given by the vector $[F'(X_0)]e_1$ (for any map L, the first column of the matrix $[L]$ is $[L]e_1$). Write $X = X_0 + se_1$, where $s > 0$ is real. Then

$$\frac{\| F(X) - T(X) \|}{\| X - X_0 \|} = \frac{\| F(X_0 + se_1) - F(X_0) - F'(X_0)se_1 \|}{s}.$$

We let X approach X_0 by letting s approach zero. Now the left-hand side has limit zero, by hypothesis. This means that the vector

$$\frac{F(X_0 + se_1) - F(X_0)}{s}$$

approaches the vector $F'(X_0)e_1$ as s approaches zero. A typical coordinate of the quotient above is

$$\frac{f_i(X_0 + se_1) - f_i(X_0)}{s}.$$

This has the limit $\partial f_i / \partial x_1(X_0)$ as s approaches 0, because it is the quotient used to define the directional derivative in the direction of e_1. It follows, as claimed, that the first column of $F'(X_0)$ consists of partial derivatives with respect to x_1; that is,

$$\begin{bmatrix} \dfrac{\partial f_1}{\partial x_1}(X_0) \\ \cdot \\ \cdot \\ \cdot \\ \dfrac{\partial f_q}{\partial x_1}(X_0) \end{bmatrix}.$$

The proof for the remaining columns is entirely similar. This establishes statement 2 and the theorem. $\ll$

EXAMPLES OF THE BEST AFFINE APPROXIMATION: 1. Consider the affine map in Sec. 8.0, $Y = F(X)$, given by

$$y_1 = f_1(x_1, x_2) = 2 + x_1 - 2x_2,$$
$$y_2 = f_2(x_1, x_2) = 1 + x_1 + x_2.$$

We compute its best affine approximation $T(X)$ at $X_0 = O = (0, 0)$. To do this, we must (a) compute the vector $F(O)$, (b) compute the matrix of $F'(O)$, (c) use these to form $T(X)$:

a. Clearly $F(O) = (f_1(0, 0), f_2(0, 0)) = (2, 1)$.

b. According to the theorem, we compute the partial derivatives

$$\frac{\partial f_1}{\partial x_1}(X) = 1, \qquad \frac{\partial f_1}{\partial x_2}(X) = -2,$$

$$\frac{\partial f_2}{\partial x_1}(X) = 1, \qquad \frac{\partial f_2}{\partial x_2}(X) = 1.$$

Hence the matrix $[F'(O)]$ is

$$[F'(O)] = \begin{bmatrix} 1 & -2 \\ 1 & 1 \end{bmatrix}.$$

c. Thus $T(X) = F(O) + F'(O)X$ with $F(O)$ and $F'(O)$ given in (a) and (b) above. Let us write this in coordinates. We have $X = (x_1, x_2)$, so that

$$[T(x_1, x_2)] = \begin{bmatrix} 2 \\ 1 \end{bmatrix} + \begin{bmatrix} 1 & -2 \\ 1 & 1 \end{bmatrix} \begin{bmatrix} x_1 \\ x_2 \end{bmatrix} = \begin{bmatrix} 2 + x_1 - 2x_2 \\ 1 + x_1 + x_2 \end{bmatrix}.$$

But this is the same as

$$y_1 = 2 + x_1 - 2x_2,$$
$$y_2 = 1 + x_1 + x_2,$$

which is the original map F. What we have discovered here is the not very surprising fact that an affine map is its own best affine approximation.

In summary, if $F(X) = Y_0 + LX$ is any affine map, then $Y_0 = F(X_0)$, $L = F'(X_0)$, and the best affine approximation $T(X) = F(X)$ itself. In the example above, note that $Y_0 = (2, 1)$ and L has the matrix

$$\begin{bmatrix} 1 & -2 \\ 1 & 1 \end{bmatrix}.$$

2. We compute the best affine approximation at $X_0 = O$ to the torus map $G \colon \mathbb{R}^2 \to \mathbb{R}^3$, mentioned in Sec. 8.0. We recall that, for $X = (\theta, \varphi)$, this map is given by $G(X) = (g_1(X), g_2(X), g_3(X))$, where

$$g_1(X) = (a + b \cos \varphi) \cos \theta,$$
$$g_2(X) = (a + b \cos \varphi) \sin \theta,$$
$$g_3(X) = b \sin \varphi.$$

a. $G(O) = (a + b, 0, 0)$.

b. We compute, writing $g_{1\theta}$ for $\partial g_1/\partial \theta$, and so on,

$$g_{1\theta}(X) = -(a + b \cos \varphi) \sin \theta, \qquad g_{1\varphi}(X) = - b \sin \varphi \cos \theta,$$

$$g_{2\theta}(X) = (a + b \cos \varphi) \cos \theta, \qquad g_{2\varphi}(X) = -b \sin \varphi \sin \theta,$$

$$g_{3\theta}(X) = 0, \qquad\qquad\qquad g_{3\varphi}(X) = b \cos \varphi.$$

When $X = X_0 = (0, 0)$, we obtain the Jacobian matrix

$$[G'(O)] = \begin{bmatrix} 0 & 0 \\ a + b & 0 \\ 0 & b \end{bmatrix}.$$

c. Thus $T(X)$, when $X_0 = O$, is given by

$$T(X) = G(O) + G'(O)X = \begin{bmatrix} a + b \\ 0 \\ 0 \end{bmatrix} + \begin{bmatrix} 0 & 0 \\ a + b & 0 \\ 0 & b \end{bmatrix} \begin{bmatrix} \theta \\ \varphi \end{bmatrix} = \begin{bmatrix} a + b \\ (a + b)\theta \\ b\varphi \end{bmatrix},$$

that is, $T(X) = Y = (y_1, y_2, y_3)$, where

$$y_1 = a + b, \qquad y_2 = (a + b)\theta, \qquad y_3 = b\varphi.$$

d. What is the geometric image of T? We note first that $T(O) = G(O)$ $= (a + b, 0, 0)$, as expected. Moreover, we note that the y_1-coordinate of $T(X)$ is always $a + b$, independent of X. Using this and the expressions in (c) for y_2 and y_3, we see that $T(\mathbb{R}^2)$ is the vertical plane whose equation is $y_1 = a + b$. This is the tangent plane to the torus at the point $G(O)$ $= (a + b, 0, 0)$. See Fig. 8.4.

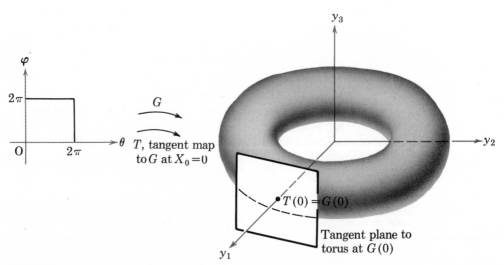

FIGURE 8.4

3. This time we approximate the same map G at a different point, namely, $X_1 = (\pi/4, \pi/4)$. Thus we must construct a new affine map $T_1(X)$, different from the map $T(X)$ obtained in Example 2, of the form

$$T_1(X) = G(X_1) + G'(X_1)(X - X_1).$$

  a. Using $\cos \pi/4 = \sin \pi/4 = \sqrt{2}/2$, we compute $G(X_1) = \frac{1}{2}(a\sqrt{2} + b, a\sqrt{2} + b, b\sqrt{2})$.

  b. In Example 2, we computed the partial derivatives $g_{1\theta}(X), \ldots,$ $g_{3\varphi}(X)$. Letting $X = X_1$, we obtain the 3 by 2 Jacobian matrix

$$[G'(X_1)] = \begin{bmatrix} -\dfrac{1}{2}(a\sqrt{2} + b) & -\dfrac{b}{2} \\[2mm] \dfrac{1}{2}(a\sqrt{2} + b) & -\dfrac{b}{2} \\[2mm] 0 & \dfrac{b\sqrt{2}}{2} \end{bmatrix}.$$

  c. Thus $T_1(X) = T_1(\theta, \varphi) = Y = (y_1, y_2, y_3)$ is given by

$$y_1 = \frac{1}{2}(a\sqrt{2} + b) - \frac{1}{2}(a\sqrt{2} + b)\left(\theta - \frac{\pi}{4}\right) - \frac{b}{2}\left(\varphi - \frac{\pi}{4}\right),$$

$$y_2 = \frac{1}{2}(a\sqrt{2} + b) + \frac{1}{2}(a\sqrt{2} + b)\left(\theta - \frac{\pi}{4}\right) - \frac{b}{2}\left(\varphi - \frac{\pi}{4}\right),$$

$$y_3 = \frac{b\sqrt{2}}{2} \qquad\qquad\qquad\qquad\qquad + \frac{b\sqrt{2}}{2}\left(\varphi - \frac{\pi}{4}\right).$$

Note the presence of $\theta - \pi/4$ and $\varphi - \pi/4$, the coordinates of the vector $X - X_1$.

The image $T_1(\mathbb{R}^2) = \{Y \in \mathbb{R}^3 \mid Y = T_1(X)\}$ is the plane tangent to the torus at the point $G(X_1)$. See Fig. 8.5.

  4. Let $f: \mathbb{R}^2 \to \mathbb{R}$ and define $F: \mathbb{R}^2 \to \mathbb{R}^3$, where $X = (x_1, x_2)$, by

$$F(X) = (x_1, x_2, f(x_1, x_2)).$$

As we pointed out in Example 3 in Sec. 8.0, the vector-valued map F embeds the plane $\mathbb{R}^2$ into $\mathbb{R}^3$ as the graph of the scalar-valued function f. Now we compute the best affine approximation T to F at a point X_0 and show

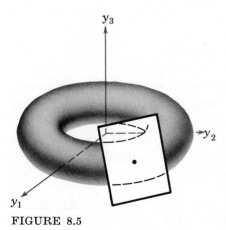

y_3

y_2

y_1

FIGURE 8.5

that the image, $T(\mathbb{R}^2)$, is the same as the tangent plane to the graph of f above X_0, that is, the *graph* of the best affine approximation to f at X_0.

a. Letting $X_0 = (c_1, c_2)$, we have $F(X_0) = (c_1, c_2, f(c_1, c_2))$.

b. You should be able to show that the Jacobian matrix is

$$[F'(X)] = \begin{bmatrix} 1 & 0 \\ 0 & 1 \\ f_{x_1}(x_1, x_2) & f_{x_2}(x_1, x_2) \end{bmatrix}.$$

c. As usual, $T(X)$ is given by $F(X_0) + F'(X_0)(X - X_0)$. The first-order term is obtained by matrix multiplication,

$$[F'(X_0)][X - X_0] = \begin{bmatrix} 1 & 0 \\ 0 & 1 \\ f_{x_1}(c_1, c_2) & f_{x_2}(c_1, c_2) \end{bmatrix} \begin{bmatrix} x_1 - c_1 \\ x_2 - c_2 \end{bmatrix}$$

$$= \begin{bmatrix} x_1 - c_2 \\ x_2 - c_2 \\ f_{x_1}(c_1, c_2)(x_1 - c_1) + f_{x_2}(c_1, c_2)(x_2 - c_2) \end{bmatrix}.$$

This column vector must be added to $[F(X_0)]$ to obtain $[T(X)]$. Let us note that $f(c_1, c_2) + f_{x_1}(c_1, c_2)(x_1 - c_1) + f_{x_2}(c_1, c_2)(x_2 - c_2)$ equals $f(X_0) + f'(X_0)(X - X_0)$, which is the best affine approximation to the function f at the point X_0. We denote it $t(X)$. We thereby may write, noting that $c_i + (x_i - c_i) = x_i$,

$$[T(X)] = \begin{bmatrix} x_1 \\ x_2 \\ t(X) \end{bmatrix};$$

that is, in coordinate form,

$$T(X) = (x_1, x_2, t(X)).$$

REMARK: The vector-valued F was constructed from the scalar-valued f. Hence it is not surprising that the map T that approximates F is constructed from the map t that approximates f.

Moreover, since $T(x_1, x_2) = (x_1, x_2, t(x_1, x_2))$, the image of $T(X)$ in $\mathbb{R}^3$ is the same as the graph of $t(X)$, that is, the tangent plane above X_0 to the graph (a surface) of the function f.

5. Let $f: \mathbb{R}^n \to \mathbb{R}$ be scalar-valued, as in Chap. 5. Then we may think of f' as a map from $\mathbb{R}^n$ to $\mathbb{R}^n$; to each $X_0 \in \mathbb{R}^n$, f' assigns the vector (1 by n matrix) given by $[f_{x_1}(X_0) \cdots f_{x_n}(X_0)]$. Since $f': \mathbb{R}^n \to \mathbb{R}^n$, *its* first derivative

at a point X_0 is a linear map $f''(X_0)\colon \mathbb{R}^n \to \mathbb{R}^n$. In fact, Theorem 1 implies that $[f''(X_0)] = [f_{x_i x_j}(X_0)]$, where $i, j = 1, \ldots, n$. Note that this is the way we *defined* $f''(X_0)$ in Chap. 6.

Exercises

1. Find the total derivative matrix of each of the following maps at the indicated points:

   (a) $y_1 = 1 + x_1 + x_2$,
   $y_2 = -2 + x_1 - x_2$,
   at $(3, -4)$;

   (b) $y_1 = x_1 - x_2$,
   $y_2 = x_1^2 + x_2^2$,
   at $(-1, 2)$;

   (c) $z_1 = y_1 \cos \pi y_2$,
   $z_2 = y_1 \sin \pi y_2$,
   at $(2, 0)$;

   (d) $y_1 = 3x_1 - 7$,
   $y_2 = x_1^2 + 4x_2^2$,
   $y_3 = x_2$,
   at $(1, -2)$;

   (e) $x = \sin \varphi \cos \theta$,
   $y = \sin \varphi \sin \theta$,
   $z = \cos \varphi$,
   at $(\theta, \varphi) = (\pi/4, \pi/4)$;

   (f) $u = e^x \cos y$,
   $v = e^x \sin y$,
   $w = 3z - 5$,
   at $(0, \pi, 2)$.

2. Find the best affine approximations to the maps in Exercise 1 at the given points.

3. (a) Show that the map in Exercise 1(e) has the property that $x^2 + y^2 + z^2 = 1$. Moreover, by interpreting θ and φ as in Fig. 8.6 (**spherical coordinates**), note that this map assigns to each point (θ, φ), $0 \le \theta \le 2\pi$, $0 \le \varphi \le \pi$ a unique point on the unit sphere $x^2 + y^2 + z^2 = 1$.

   (b) Find the best affine approximation to this map at $\theta = 0$, $\varphi = \pi/2$.

   (c) If T is the best affine approximation found in Exercise 3(b), draw a sketch indicating the image of T. Note that this is precisely the tangent plane to the sphere at the point corresponding to $\theta = 0$, $\varphi = \pi/2$.

   (d) Repeat Exercises 3(b) and (c) at $\theta = \pi/2$, $\varphi = \pi/4$.

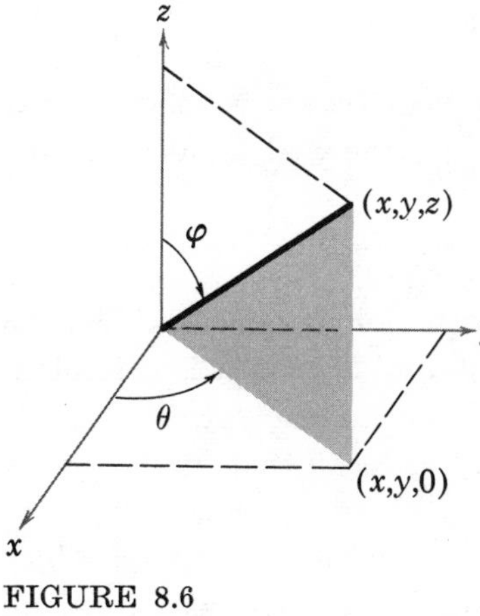

FIGURE 8.6

4. Find a map $F\colon \mathbb{R}^2 \to \mathbb{R}^3$ such that

$$F'(X) = \begin{bmatrix} 1 & -2 \\ 2 & 3 \\ -1 & 0 \end{bmatrix}$$

for all $X \in \mathbb{R}^2$, and $F(0, 1) = (1, 0, 2)$.

5. (a) Let $F\colon \mathbb{R}^2 \to \mathbb{R}^3$ have the property that $F'(X) = O$ for all $X \in \mathbb{R}^2$. What can you conclude about F? Proof?

(b) Let F and G map $\mathbb{R}^2$ to $\mathbb{R}^3$. If $F'(X) = G'(X)$ for all X, and $F(X_0) = G(X_0)$ for some $X_0 \in \mathbb{R}^2$, what can you conclude? Why?

(c) Let $F\colon \mathbb{R}^2 \longrightarrow \mathbb{R}^3$ have the property that $F'(X) = A$ for all $X \in \mathbb{R}^2$, where A is a constant matrix. If in addition $F(X_0) = Y_0$, where X_0 and Y_0 are fixed vectors, what can you conclude about F?

6. Let $F\colon \mathbb{R}^2 \longrightarrow \mathbb{R}^2$ be a differentiable map, and let γ_1 be the line $x_2 = b$ in parametric form, say $x_1 = t$, $x_2 = b$:

(a) If V_1 is the first column vector of $F'(a, b)$ and $F(\gamma_1)$ is the image of γ_1, prove that V_1 is tangent to the curve $F(\gamma_1)$ at $Y_0 = F(a, b)$. Draw a sketch that illustrates this.

(b) Obtain a similar result for γ_2, the line $x_1 = a$, $x_2 = t$.

(c) If $x_1 = a + \alpha t$, $x_2 = b + \beta t$ is the parametric equation of a line γ through (a, b), show that the tangent vector V to $F(\gamma)$ at Y_0 is $V = \alpha V_1 + \beta V_2$.

(d) Let γ and γ^* be any two straight lines through (a, b). If the angle between $F(\gamma)$ and $F(\gamma^*)$ at Y_0 equals the angle between γ and γ^*, prove that $V_1 \perp V_2$ and $\|V_1\| = \|V_2\|$. If we write $F(X) = (u(X), v(X))$, deduce that $u_{x_1} = v_{x_2}$ and $u_{x_2} = -v_{x_1}$. (An angle-preserving map is called **conformal**.)

8.2 THE CHAIN RULE

8.2a The general case

Let $\mathbb{R}^n \xrightarrow{F} \mathbb{R}^q \xrightarrow{G} \mathbb{R}^r$ be mappings. We suppose that G is defined for all vectors in $\mathbb{R}^q$, so that if $X \in \mathbb{R}^n$, then $G(F(X))$ is defined. Thus we have a composite map $G \circ F\colon \mathbb{R}^n \longrightarrow \mathbb{R}^r$ given by

$$(G \circ F)(X) = G(F(X)).$$

We discussed the composition of linear maps, for instance, in Chap. 3.

From knowledge of the derivative F' of F and G' of G it is possible to obtain the derivative of $G \circ F$ in a straightforward manner. This is the content of the Chain Rule. Before stating it, we recall that for each point $X_0 \in \mathbb{R}^n$ the first derivative $F'(X_0)$ may be given as a q by n matrix, and for each $Y_0 \in \mathbb{R}^q$ the first derivative $G'(Y_0)$ may be given as an r by q matrix. Finally, the first derivative $(G \circ F)'(X_0)$ is given as an r by n matrix.

THEOREM 2:   (Chain Rule.) Let F and G be differentiable maps $\mathbb{R}^n \xrightarrow{F} \mathbb{R}^q \xrightarrow{G} \mathbb{R}^r$. Let $T_F(X)$ be the best affine approximation to F at X_0, and let $T_G(Y)$ be the best affine approximation to G at the point $F(X_0)$. Then the best affine approximation to $G \circ F$ at X_0 is given by

$$T_{G \circ F}(X) = (T_G \circ T_F)(X);$$

that is,

$$T_{G \circ F}(X) = G(F(X_0)) + G'(F(X_0))F'(X_0)(X - X_0).$$

Thus, the first-derivative matrix for $G \circ F$ at X_0 is given by

$$[(G \circ F)'(X_0)] = [G'(F(X_0))][F'(X_0)].$$

REMARK: The derivative matrix of the composite map is the product of the derivative matrices of the components G and F. Thus, the Chain Rule is actually matrix multiplication. Imagine the monstrous formulas that would face us if we did not have linear maps and coordinate-free linear algebra.

Before discussing the proof of the Chain Rule, we illustrate its meaning.

EXAMPLE: Let $\mathbb{R}^2 \xrightarrow{F} \mathbb{R}^2 \xrightarrow{G} \mathbb{R}^3$ be mappings given by

$$F(X) = (f_1(X), f_2(X)) = (x_1^2 - x_2^2, 2x_1x_2),$$

$$G(Y) = (g_1(Y), g_2(Y), g_3(Y)) = (3 + y_1 + y_2, \quad y_1y_2, \quad y_1 - 2y_2).$$

We compute the matrix of $(G \circ F)'(X_0)$ at $X_0 = (1, 1)$:

1. We have $F(X_0) = (0, 2)$. To compute $F'(X_0)$ and $G'(F(X_0))$, we first need the partial derivatives related to F and G. Writing f_{11} for $\partial f_1/\partial x_1$, f_{12} for $\partial f_1/\partial x_2$, and so on, we see that

$$[F'(X)] = \begin{bmatrix} f_{11}(X) & f_{12}(X) \\ f_{21}(X) & f_{22}(X) \end{bmatrix} = \begin{bmatrix} 2x_1 & -2x_2 \\ 2x_2 & 2x_1 \end{bmatrix},$$

and

$$[G'(Y)] = \begin{bmatrix} g_{11}(Y) & g_{12}(Y) \\ g_{21}(Y) & g_{22}(Y) \\ g_{31}(Y) & g_{32}(Y) \end{bmatrix} = \begin{bmatrix} 1 & 1 \\ y_2 & y_1 \\ 1 & -2 \end{bmatrix}.$$

Since $F(X_0) = (0, 2)$, we obtain immediately

$$[F'(X_0)] = \begin{bmatrix} 2 & -2 \\ 2 & 2 \end{bmatrix}, \qquad [G'(F(X_0))] = \begin{bmatrix} 1 & 1 \\ 2 & 0 \\ 1 & -2 \end{bmatrix}.$$

These computations are entirely similar to those in Sec. 8.1.

2. Now, according to the Chain Rule, the first derivative matrix for the map $G \circ F$ at X_0 is given by a product:

$$[(G \circ F)'(X_0)] = \begin{bmatrix} 1 & 1 \\ 2 & 0 \\ 1 & -2 \end{bmatrix} \begin{bmatrix} 2 & -2 \\ 2 & 2 \end{bmatrix} = \begin{bmatrix} 4 & 0 \\ 4 & -4 \\ -2 & -6 \end{bmatrix}.$$

3. It is now easy to obtain the best affine approximation $T_{G \circ F}$ at X_0. We have $G(F(X_0)) = G(0, 2) = (5, 0, -4) \in \mathbb{R}^3$. Thus

$$[T_{G \circ F}(X)] = \begin{bmatrix} 5 \\ 0 \\ 4 \end{bmatrix} + \begin{bmatrix} 4 & 0 \\ 4 & -4 \\ -2 & -6 \end{bmatrix} \begin{bmatrix} x_1 - 1 \\ x_2 - 1 \end{bmatrix}$$

$$= \begin{bmatrix} 5 + 4(x_1 - 1) \\ 4(x_1 - 1) - 4(x_2 - 1) \\ 4 - 2(x_1 - 1) - 6(x_2 - 1) \end{bmatrix}.$$

Writing this in terms of coordinates, we have

$$T_{G \circ F}(X) = (5 + 4(x_1 - 1), 4(x_1 - 1) - 4(x_2 - 1),$$
$$4 - 2(x_1 - 1) - 6(x_2 - 1)).$$

Outline of the Proof: To show that $T_{G \circ F}(X) = T_G(T_F(X))$, we begin by observing that if F is differentiable at X_0, then the function $E(X, X_0)$ (E for error) defined by

$$E(X, X_0) = \frac{F(X) - T_F(X)}{\|X - X_0\|}$$

has the property that $E(X, X_0) \rightarrow 0$ as $X \rightarrow X_0$. Now we can write the equation above as

$$F(X) = T_F(X) + E(X, X_0) \|X - X_0\|,$$

which we think of as

$$F(X) = T_F(X) + \text{small},$$

where "small" $= E(X, X_0) \|X - X_0\|$ approaches 0 rapidly (faster than $\|X - X_0\|$) as $X \rightarrow X_0$. Similarly,

$$G(Y) = T_G(Y) + \text{small}.$$

Thus we have

$$G(F(X)) = T_G(T_F(X) + \text{small}) + \text{small} = T_G(T_F(X)) + \text{small}$$
$$= G(F(X_0)) + G'(F(X_0))F'(X_0)(X - X_0) + \text{small}$$

(the second equality requires a proof). Since $T_G(T_F(X))$ is affine, as you should check, we conclude that it is the best affine approximation to $G(F(X))$ at X_0, as claimed. $\ll$

It is worthwhile to recall the Chain Rule of ordinary calculus. If the composite function $g(f(x))$ is defined, then its derivative with respect to x is given by

$$g(f(x))' = g'(f(x))f'(x).$$

This is the case of $n = q = r = 1$ of Theorem 2. It is sometimes expressed differently. If we write $y = f(x)$, $z = g(y)$, then

$$\frac{dz}{dx} = \frac{dz}{dy}\frac{dy}{dx}.$$

Had we used the notation $Y = F(X)$, $Z = G(Y)$ and $dY/dX = dY/dX$ (X_0) for $F'(X_0)$, dZ/dY for $G'(Y_0)$, $Y_0 = F(X_0)$, and so on, then Theorem 2 would read

$$\frac{dZ}{dX} = \frac{dZ}{dY}\frac{dY}{dX}.$$

This is especially easy to remember, although somewhat more ambiguous than the notation we have used (and continue to use).

8.2b   An important special case

Let $\mathbb{R}^n \xrightarrow{F} \mathbb{R}^q \xrightarrow{g} \mathbb{R}$, so that the composite $g \circ F(X) = g(F(X))$ is scalar-valued. Call it $h(X)$. We recall from Chap. 5 that $h'(X_0)$ is a linear map with matrix $[h_{x_1}(X_0) \cdots h_{x_n}(X_0)]$, or, equivalently, the gradient is

$$\nabla h(X_0) = (h_{x_1}(X_0), \ldots, h_{x_n}(X_0)).$$

It is natural to ask for $h'(X_0)$, that is, the partial derivatives $h_{x_i}(X_0)$, in terms of the derivatives of g and F.

THEOREM 3:   (Chain Rule for Scalar-Valued Functions.) Let $h = g \circ F$ with $\mathbb{R}^n \xrightarrow{F} \mathbb{R}^q \xrightarrow{g} \mathbb{R}$. Writing $X = (x_1, \ldots, x_n)$ and $Y = (y_1, \ldots, y_q) = F(X)$, $Y_0 = F(X_0)$, we have the partial derivatives

$$\frac{\partial h}{\partial x_1}(X_0) = \frac{\partial g}{\partial y_1}(Y_0)\frac{\partial y_1}{\partial x_1}(X_0) + \cdots + \frac{\partial g}{\partial y_q}(Y_0)\frac{\partial y_q}{\partial x_1}(X_0)$$

$$\vdots \qquad\qquad \vdots \qquad\qquad\qquad \vdots$$

$$\frac{\partial h}{\partial x_n}(X_0) = \frac{\partial g}{\partial y_1}(Y_0)\frac{\partial y_1}{\partial x_n}(X_0) + \cdots + \frac{\partial g}{\partial y_q}(Y_0)\frac{\partial y_q}{\partial x_n}(X_0).$$

REMARK: h depends on the dependent variables $y_1, \ldots, y_q$, and we must include the rate of change of each of these with respect to x_1, that is, $\partial y_i/\partial x_1$, to obtain the rate of change of h with respect to x_1, that is, $\partial h/\partial x_1$.

   Proof: The Chain Rule gives $h'(X_0) = g'(Y_0)F'(X_0)$. The right-hand side here may be computed thus:

$$[g'(Y_0)][F'(X_0)] = \left[\frac{\partial g}{\partial y_1}(Y_0) \cdots \frac{\partial g}{\partial y_q}(Y_0) \right] \begin{bmatrix} \frac{\partial y_1}{\partial x_1}(X_0) & \cdots & \frac{\partial y_1}{\partial x_n}(X_0) \\ \cdot & & \cdot \\ \cdot & & \cdot \\ \cdot & & \cdot \\ \frac{\partial y_q}{\partial x_1}(X_0) & \cdots & \frac{\partial y_q}{\partial x_n}(X_0) \end{bmatrix}$$

(note here that $g: \mathbb{R}^q \to \mathbb{R}$, and so its total derivative matrix is a 1 by q row vector). Matrix multiplication gives the result. ≪

EXAMPLE: Let $\mathbb{R}^2 \xrightarrow{F} \mathbb{R}^2 \xrightarrow{g} \mathbb{R}$ be given by

$$F(X) = (x_1^2 - x_2^2, 2x_1 x_2), \qquad g(Y) = y_1 y_2.$$

We compute the partial derivative of $h(X) = g(F(X))$ with respect to x_1. We have, using Theorem 3,

$$\frac{\partial h}{\partial x_1}(X) = \frac{\partial g}{\partial y_1}(Y)\frac{\partial y_1}{\partial x_1}(X) + \frac{\partial g}{\partial y_2}(Y)\frac{\partial y_2}{\partial x_1}(X)$$

$$= y_2(2x_1) + y_1(2x_2)$$

$$= (2x_1 x_2)2x_1 + (x_1^2 - x_2^2)2x_2$$

$$= 6x_1^2 x_2 - 2x_2^3.$$

For instance, if $X_0 = (1, 1)$, then $(\partial h/\partial x_1)(X_0) = 4$.

Another way to obtain the expression for $(\partial h/\partial x_1)$ is to note that $h(X) = g(x_1^2 - x_2^2, 2x_1 x_2) = (x_1^2 - x_2^2)2x_1 x_2$ and then take the partial derivative in the usual way:

$$\frac{\partial h}{\partial x_1}(X) = \frac{\partial}{\partial x_1} 2(x_1^3 x_2 - x_1 x_2^3) = 6x_1^2 x_2 - 2x_2^3.$$

If the derivatives of F and g are already known, however, this second method may be wasteful; it may be simpler to apply the Chain Rule.

8.2c Another special case

A common situation is the following: We are given a scalar-valued function $f: \mathbb{R}^3 \to \mathbb{R}$, which might be the temperature function. A particle moves along a curve in $\mathbb{R}^3$ so that its position at time t is given by $X = \Phi(t)$. Then the temperature at the location of the particle is $h(t) = f(\Phi(t))$. We ask for the rate of change of temperature as seen by the particle; that is, we want $h'(t)$.

Thus, the situation is $\mathbb{R} \xrightarrow{\Phi} \mathbb{R}^n \xrightarrow{f} \mathbb{R}$, where we assume that f and $\Phi = (\varphi_1, \ldots, \varphi_n)$ are differentiable maps. Then $h = f \circ \Phi$, and so, by the Chain Rule,

$$h'(t) = f'(\Phi(t))\Phi'(t).$$

Since $\Phi \colon \mathbb{R} \to \mathbb{R}^n$, the matrix $[\Phi']$ is an n by 1 matrix; it looks like a column vector. Therefore, in coordinates

$$h'(t) = [f'][\Phi'] = \begin{bmatrix} \dfrac{\partial f}{\partial x_1} & \cdots & \dfrac{\partial f}{\partial x_n} \end{bmatrix} \begin{bmatrix} \varphi_1' \\ \vdots \\ \varphi_n' \end{bmatrix} = \frac{\partial f}{\partial x_1}\frac{d\varphi_1}{dt} + \cdots + \frac{\partial f}{\partial x_n}\frac{d\varphi_n}{dt}.$$

In terms of the gradient of f, we can also write this

$$h'(t) = \langle \nabla f(\Phi(t)), \Phi'(t) \rangle.$$

Note that Lemma 2 in Sec. 7.2 is a special case.

EXAMPLES: 1. Let $f(x, y) = x^2 y$, and $x = 1 + t^2$, $y = 3e^t$. If $h(t) = f(x(t), y(t))$, compute $h'(0)$.

We could substitute for x and y directly to obtain $h(t) = (1 + t^2)^2 3e^t$ and then differentiate, but this would not illustrate the Chain Rule (which is hardly needed in this simple situation). Now $\nabla f(X) = (2xy, x^2)$ and $\Phi(t) = (1 + t^2, 3e^t)$, and so $\Phi'(t) = (2t, 3e^t)$. At $t = 0$, $\Phi(0) = (1, 3)$, and so $\nabla f(\Phi(0)) = (6, 1)$ and $\Phi'(0) = (0, 3)$. Therefore

$$h'(0) = \langle \nabla f(\Phi(0)), \Phi'(0) \rangle = 6 \cdot 0 + 1 \cdot 3 = 3.$$

2. Let $f(x, y, t) = xt + 3y^2 - xy + t^3$ be the temperature at (x, y) at time t (so here, the temperature itself changes with time) and say that $x = 3t - 1$, $y = t^2$. We want $h'(1)$, where $h(t) = f(x(t), y(t), t)$. Thus $\Phi(t) = (3t - 1, t^2)$. Here $\Phi \colon \mathbb{R} \to \mathbb{R}^2$ and $f \colon \mathbb{R}^3 \to \mathbb{R}$, and so it is meaningless to speak of $f(\Phi(t))$. This is an artificial difficulty. We just observe that $t = t$ and introduce $\Phi^*(t) = (3t - 1, t^2, t)$, $\Phi^* \colon \mathbb{R} \to \mathbb{R}^3$. Now $f(\Phi^*(t))$ does make sense (if one kept calm, this was obvious). Then $\nabla f = (t - y, 6y - x, x + 3t^2)$, $\Phi^{*\prime}(t) = (3, 2t, 1)$, and since $\Phi^*(1) = (2, 1, 1)$,

$$h'(1) = \langle \nabla f(2, 1, 1), \Phi^{*\prime}(1) \rangle = \langle (0, 4, 5), (3, 2, 1) \rangle = 13.$$

By the same procedure as in this second example, we find that, given $f(x, y, t)$, where $x = \varphi(t)$, $y = \psi(t)$, the derivative of $h(t) = f(\varphi(t), \psi(t), t)$ is

$$h'(t) = \frac{\partial f}{\partial x}\frac{d\varphi}{dt} + \frac{\partial f}{\partial y}\frac{d\psi}{dt} + \frac{\partial f}{\partial t}.$$

One is sometimes given $f(x, y, s, t)$, where $x = \varphi(t)$, $y = \psi(t)$. If $h(s, t) = f(\varphi(t), \psi(t), s, t)$, then the computation above shows that

$$\frac{\partial h}{\partial t} = \frac{\partial f}{\partial x}\frac{d\varphi}{dt} + \frac{\partial f}{\partial y}\frac{\partial \psi}{\partial t} + \frac{\partial f}{dt},$$

and so there is no change, except that this time we are only computing a *partial* derivative of h.

Exercises

1. Given the following maps F, G, decide if $F \circ G$, or $G \circ F$, or both, or neither makes sense:

(a) $F: \mathbb{R}^2 \longrightarrow \mathbb{R}^2$, $G: \mathbb{R}^2 \longrightarrow \mathbb{R}^3$;
(b) $F: \mathbb{R}^2 \longrightarrow \mathbb{R}^2$, $G: \mathbb{R}^1 \longrightarrow \mathbb{R}^2$;
(c) $F: \mathbb{R}^1 \longrightarrow \mathbb{R}^1$, $G: \mathbb{R}^2 \longrightarrow \mathbb{R}^1$;
(d) $F: \mathbb{R}^2 \longrightarrow \mathbb{R}^3$, $G: \mathbb{R}^2 \longrightarrow \mathbb{R}^1$;
(e) $F: \mathbb{R}^2 \longrightarrow \mathbb{R}^3$, $G: \mathbb{R}^3 \longrightarrow \mathbb{R}^2$;
(f) $F: \mathbb{R}^3 \longrightarrow \mathbb{R}^1$, $G: \mathbb{R}^1 \longrightarrow \mathbb{R}^3$.

2. Let $F: \mathbb{R}^2 \longrightarrow \mathbb{R}^2$ and $G: \mathbb{R}^3 \longrightarrow \mathbb{R}^2$ be defined by $F(x, y) = (e^{x+y^2}, e^{y+x^2})$, $G(r, s, t) = (r + s^2 + t^2, s + t^2 + r^3)$:

(a) Compute F' and G'.
(b) If $G \circ F$ makes sense, compute $(G \circ F)'$ at $(-1, -1)$; if $F \circ G$ makes sense, compute $(F \circ G)'$ at $(-1, 0, 0)$.

3. Compute the partial derivatives with respect to x and y of:

(a) $f(u, v) = uv^2$, where $u = \sin x$, $v = y \cos x$ at $X = (\pi, 1)$;
(b) $g(r, s) = e^{rs}$, where $r = x^2 - y^2$, $s = xy$ at $X = (1, -1)$.

4. Let $F(X) = (x_2 - e^{x_1+2x_2}, x_1 x_2)$, $G(Y) = (y_2 + y_2 \sin y_1, (y_1 + y_2)^2)$:

(a) Compute F' at $X_0 = (-2, 1)$ and G' at $Y_0 = F(X_0)$.
(b) Let $H = G \circ F$. Compute H' at $X_0 = (-2, 1)$.

5. (a) Let $f(x, y) = xy - e^x$. If $x(t) = 1 - t^2$ and $y(t) = 2t - 3$, find the derivatives of $\varphi(t) = f(x(t), y(t))$ at $t = 1$.
(b) Repeat this for $f(x, y) = y \cos(x + y + 1)$.

6. Let $\varphi(x, y, t) = xy^2 - t \cos y$. If $x(t) = \sin \pi t$ and $y(t) = t^3 - 1$, let $h(t) = \varphi(x(t), y(t), t)$. Compute h' at $t = 1$.

7. Let $\varphi(x, s, t) = xs + xt + st$. If $x(t) = t^3 - 7$, compute the total derivative of $\varphi(x(t), s, t)$ at $t = 2$, $s = 3$.

8. (a) If $u(x, y) = \sin(x - y) + e^{2(x-y)}$, show that $u_x + u_y = 0$.
(b) If $v(x, y, t) = 4(x - t)^2 + \cos(t + y)$, show that $u_{xx} + u_{yy} = u_{tt}$.

9. Let $\varphi(t)$ be a suitably differentiable scalar-valued function.

(a) If $u(x, y) = \varphi(2x - 3y)$, show that $3u_x + 2u_y \equiv 0$.
(b) If $u(x, y) = \varphi(ax + by)$, show that $bu_x - au_y \equiv 0$.
(c) If $u(x, y) = \varphi(xy)$, show that $xu_x - yu_y \equiv 0$.
(d) If $u(x, y) = \varphi(x/y)$, show that (for $y \neq 0$) $xu_x + yu_y \equiv 0$.
(e) If $u(x, y) = \varphi(x^2 + y^2)$, and so u depends only on the distance to the origin, show that $yu_x - xu_y \equiv 0$.

(f) If $u(x, t) = \varphi(x + ct)$, show that $u_{tt} - c^2 u_{xx} \equiv 0$.

(g) If $u(x, y) = \varphi(ax + by)$, show that $u_{xx} u_{yy} - u_{xy}^2 \equiv 0$.

10. The temperature at a point $X = (x, y, z) \in \mathbb{R}^3$ is $f(X) = x^2 + y^2 + z^2$. A particle travels on the curve $X(t) = (\sin \pi t, \cos \pi t, t^2 - 2t + 2)$. What is the coolest point on the trajectory of the particle?

11. Let $X(t) \in \mathbb{R}^3$ be the position of a particle at time t. Compute the derivatives of $\| X(t) \|$ and $1/\| X(t) \|$, assuming that $X(t) \neq 0$.

12. A certain differentiable function $f: \mathbb{R}^2 \longrightarrow \mathbb{R}$ has the property that $f_x(1, 1) = 2, f_y(1, 1) = -3$. If $g(s, t) = f(2s - t, s^2)$, compute $g_s(1, 1)$ and $g_t(1, 1)$.

13. Say that $u(x, y)$ is a given differentiable function.

(a) If $v(s, t) = u(as + bt, cs + dt)$, show that
$$v_s = au_x + cu_y, \qquad v_t = bu_x + du_y.$$

(b) If $u_x - u_y = 0$, pick constants a, b, c, d so that $v_s = 0$ and the map $x = as + bt$, $y = cs + dt$ is invertible.

(c) Use this to show that the most general function satisfying $u_x - u_y = 0$ is $u(x, y) = \varphi(x + y)$, where φ is any differentiable function.

(d) If, in addition, you know that $u(x, 0) = x^2$, what must the function $u(x, y)$ in Exercise 13(c) be?

14. Apply the Chain Rule to $u(x, y) = f(x)g(y)$, with $x = t, y = t$, to deduce the standard formula for the derivative of $f(t)g(t)$.

15. Let $f: \mathbb{R} \longrightarrow \mathbb{R}$ be a differentiable function with the properties $f(1) = 7$ and $f'(1) > 0$. Consider the set of points $X = (x, y) \in \mathbb{R}^2$ such that $f(xy) = 7$. This gives a curve in $\mathbb{R}^2$. Show that the line $x + y = 2$ is tangent to the curve at $(1, 1)$ independent of the particular form of f. What is the equation of the tangent line to this curve at (a, b)?

16. (Euler) A function $f: \mathbb{R}^n \longrightarrow \mathbb{R}$ is **homogeneous of degree** λ if $f(tX) = t^\lambda f(X)$ for all $t > 0$ and all $0 \neq X \in \mathbb{R}^n$. Assume that f is differentiable.

(a) Show that $f(x, y) = x^2 y - 5y^3$ is homogeneous of degree 3 but that $f(x, y) = x^2 y - 5y^4$ is not homogeneous of any order.

(b) Show that $\langle X, \operatorname{grad} f(X) \rangle = \lambda f(X)$.

(c) Conversely, if f satisfies the equation in Exercise 16(b), show that f is homogeneous of degree λ.
(Suggestion: Show that $\partial(t^{-\lambda} f(tX))/\partial t = 0$.)

17. Let $u(x, y): \mathbb{R}^2 \longrightarrow \mathbb{R}$ be a twice-differentiable function that satisfies the *Laplace equation* $u_{xx} + u_{yy} = 0$. Assume that u depends only on the distance $r = \sqrt{x^2 + y^2}$ to the origin, and so $u(x, y) = \varphi(r)$ for some function φ. ,

(a) Show that $r\varphi'' + \varphi' = 0$.

(b) Deduce that $u(x, y) = a \log r + b$, where a and b are any constants.

18. Let $f: \mathbb{R}^2 \longrightarrow \mathbb{R}$ be a smooth function and $X = X(t)$ be a smooth curve in $\mathbb{R}^2$. The curve is a **level curve** of f if f is constant on the curve, so that $f(X(t)) \equiv a$ for some constant a. Prove that the gradient vector $\nabla f(X(t))$ is perpendicular to the tangent vector $X'(t)$ to the level curve. (Suggestion: Do something to $\varphi(t) = f(X(t))$.)

8.3 VECTORFIELDS

Now we discuss a new interpretation of functions $F: \mathbb{R}^n \to \mathbb{R}^n$.

Consider a thin layer of fluid (water, a gas) flowing across a flat plane surface. The fluid is shallow; we are not concerned with its depth. We may suppose that the plane surface is actually the xy-plane $\mathbb{R}^2$. At each point X_0 of the plane, we attach the tail of an arrow with the following properties:

    1. The arrow points in the direction of the flow.

    2. The length of the arrow is proportional to the speed of the particle at X_0, so that the longer the arrow, the faster the particle at X_0 is moving.

EXAMPLES OF VECTORFIELDS: 1. Here the arrows indicate that the flow is from left to right. See Fig. 8.7. Since all appear to have the same length, the speed at each point is the same.

2. Again the flow is from left to right, but now the arrows farther to the right are longer. See Fig. 8.8. As a particle flows to the right, it speeds up.

3. Now all particles flow with uniform speed toward a single point (called a **sink**, for obvious reasons; if the arrows all were reversed, that point would be a **source**). See Fig. 8.9.

It is customary to think of these arrows as vectors. If we denote by $F(X_0)$ the vector whose tail is attached to the point X_0, we are led to the

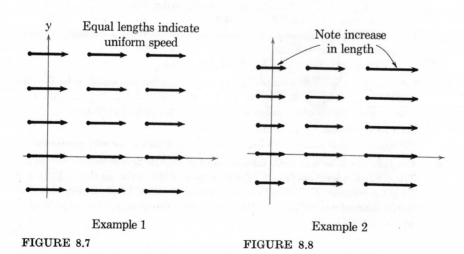

Example 1

FIGURE 8.7

Example 2

FIGURE 8.8

realization that all these vectors are given by a function $F: \mathbb{R}^2 \to \mathbb{R}^2$. For if F is such a function and X_0 a point in its domain, then $F(X_0)$ is some vector in $\mathbb{R}^2$. We think of this vector as an arrow and attach its tail to the point X_0.

We say that a **vectorfield** in $\mathbb{R}^2$ is a rule that assigns to each point $X \in \mathbb{R}^2$ a vector (arrow) $F(X) \in \mathbb{R}^2$. Hence a vectorfield is a function $F: \mathbb{R}^2 \to \mathbb{R}^2$ interpreted not as a geometric transformation of the plane but rather as a way of attaching an arrow $F(X)$ to each point X. The vectorfield associated with the fluid flows described above would be called **velocity fields**, because the length of the vector indicated speed.

Suppose that $F(X)$ is a vectorfield, $X \in \mathbb{R}^2$. If $X = (x, y)$, then $F(X) = F(x, y) = (p(x, y), q(x, y))$, where $p, q: \mathbb{R}^2 \to \mathbb{R}$ are the coordinate functions of F. See Fig. 8.10. Thus F is specified by giving the two functions $p(x, y), q(x, y)$. If F is a velocity field and $p(x, y) < 0$ at (x, y), then the particle is moving toward the left (and upward, horizontally, or downward, accordingly as $q(x, y) > 0, = 0,$ or < 0.)

EXAMPLES: Let us find analytic expressions for the vectorfields pictured above.

1. In this case the same vector is attached to each point (x, y). If we agree that its length is 1, then it is the vector $(1, 0)$, since it is horizontal

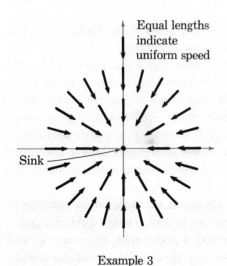

Sink

Example 3

FIGURE 8.9

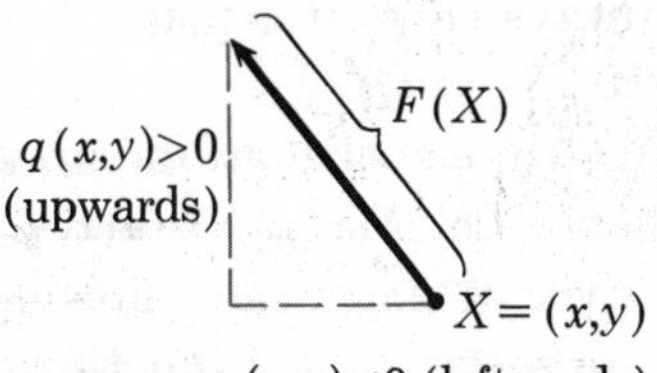

$q(x,y) > 0$
(upwards)

$F(X)$

$X = (x,y)$

$p(x,y) < 0$ (leftwards)

FIGURE 8.10

(so that $q = 0$) and points from left to right. Thus we conclude that $F(x, y)$ = $(1, 0)$, independent of the point (x, y). If these all had length $c > 0$, then, of course, $F(x, y) = (c, 0)$.

2. $F(x, y) = (p(x, y), q(x, y))$; $p(x, y) = ?$, $q(x, y) = ?$ Since each of these vectors is horizontal, its second coordinate is zero for all x, y, $q(x, y)$ = 0, just as in Example 1. Now $p(x, y)$ is positive and increasing as x increases, but does not vary with y, $p(x, y_1) = p(x, y_2)$. One candidate for such a function would be e^x (there are many others). Using this, we have

$$F(x, y) = (p(x, y), q(x, y)) = (e^x, 0).$$

3. Now $F(X)$ has constant length, which we suppose to be 1, and points away from X towards the origin. Such a vector is given by $F(X) =$ $-X/\|X\|$, as you should check (why the minus sign?). In coordinate notation (less elegant and less clear in this case) we have $F(x, y) = (p(x, y), q(x, y))$, where

$$p(x, y) = \frac{-x}{\sqrt{x^2 + y^2}}, \qquad q(x, y) = \frac{-y}{\sqrt{x^2 + y^2}}.$$

REMARKS: 1. The vectorfields above were introduced as velocity fields. There are several other situations in which vectorfields arise. A **force field** is given by attaching to each X a vector $F(X)$ that describes the push or pull that a particle at point X "feels." For example, the force of repulsion felt by a charged particle at X due to a particle of like charge at the origin O is pictured in Fig. 8.11. Note that the repellent force dies off (the arrows become shorter) as the distance from the particle at the origin increases.

2. We may think of vectorfields (velocity fields, force fields) in 3-space $\mathbb{R}^3$ also. We leave it to you to imagine some examples.

3. The vectorfields we gave above were "time-independent" in the sense that we assigned the same vector $F(X)$ to a fixed point X for all time. In practice, however, one often encounters "time-dependent" vectorfields $F(X, t)$, where t = time. A simple example is $F(X, t) = (e^{xy}, x^2 + \sin t)$. See Fig. 8.12. This assigns to each point X a vector that points toward the right but "wags" up and down as time passes. You might think of a weather vane, always fixed at a point, but changing direction with the wind. (In Fig. 8.12, $X_0 = O$).

We briefly mention an important notion in the differential calculus of vectorfields. Let $F(x, y) = (u(x, y), v(x, y))$ be the velocity field of a gas— a thin layer of smog, say—flowing across a plane area. Since $u(x, y)$ and $v(x, y)$ are scalar-valued functions, we may speak of their various partial

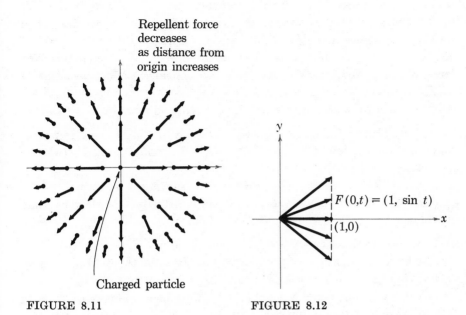

Repellent force
decreases
as distance from
origin increases

Charged particle

$F(0,t) = (1, \sin\ t)$

$(1,0)$

FIGURE 8.11   FIGURE 8.12

derivatives. The **divergence** of F at the point X_0 is defined to be the quantity

$$\text{div}\ F(X_0) = \frac{\partial u}{\partial x}(X_0) + \frac{\partial v}{\partial y}(X_0).$$

This is sometimes denoted $(\nabla \cdot F)(X_0)$, where ∇ is "gradient operator," $\nabla = (\partial/\partial x, \partial/\partial y)$, and $\nabla \cdot F$ (read "del dot F") denotes the symbolic inner product

$$\left\langle \left(\frac{\partial}{\partial x}, \frac{\partial}{\partial y}\right), (u(x, y), v(x, y)) \right\rangle = \frac{\partial}{\partial x} u(x, y) + \frac{\partial}{\partial y} v(x, y).$$

For example, if $F(x, y) = (x^2 y^2, -7ye^x)$, then div $F = 2xy^2 - 7e^x$.

The divergence, div $F(X_0)$, is a number. What does it signify? We note that $u(x, y)$ and $v(x, y)$ are, respectively, the horizontal and vertical velocities of a particle at (x, y). Hence the partials $u_x(X_0)$ and $v_y(X_0)$ measure the rates of change of these velocities with respect to changes in the x and y directions, respectively.

For instance, suppose that $u_x(X_0) > 0$. What does this say about particle motion at X_0? The answer is this: If $u(X_0) > 0$, then the particle is moving towards the right, and since $u_x(X_0) > 0$, then $u(X) > u(X_0) > 0$ for X slightly to the right of X_0. This means that the speed in the direction of horizontal motion (rightward) is increasing. On the other hand, if $u(X_0) < 0$, so that the particle at X_0 is moving toward the left, then $u_x(X_0) > 0$ means that $u(X) < u(X_0) < 0$ for X slightly to the *left* of X_0. Again the speed in the direction of the horizontal motion (this time leftward) is increasing. We may likewise consider the implications of $u_x(X_0) < 0$ and the sign of $v_y(X_0)$.

From such reasoning we derive the following summary:

1. div $F(X_0) > 0$. Then the particle at X_0 is tending to speed up as it flows in the direction of $F(X_0)$. The flow is "diverging" from X_0 (whence the name "divergence").

2. div $F(X_0) < 0$. Then the flow at X_0 is slowing or faltering.

3. div $F(X_0) = 0$. Then the flow at X_0 is "steady," neither speeding nor slowing.

For instance, let $F(x, y) = (e^x, 0)$, as in Example 2. At any $X_0 = (x_0, y_0)$, we have div $F(X_0) = e^{x_0} > 0$. This shows that as the fluid flows from left to right at each point X_0 it is tending to speed up.

In contrast, consider $G(x, y) = (e^{-x}, 0)$. This also produces a flow from left to right (horizontal, since $v = 0$). But now div $G(X_0) = -e^{-x_0} < 0$, reflecting the fact that the velocity in the direction of motion is tending to *decrease* at each point X_0.

We discuss vectorfields and divergence further in Chap. 10.

Exercises

1. Draw a sketch illustrating the given vectorfields as the velocity field of a fluid flow:

(a) $F(x, y) = (e^{-y}, 0)$;   (f) $F(x, y) = \left(\dfrac{y}{x^2 + y^2}, \dfrac{-x}{x^2 + y^2} \right)$, $(x, y) \neq 0$;

(b) $F(x, y) = (1, 2)$;   (g) $F(x, y) = (x, -y)$;

(c) $F(x, y) = (x, y)$;   (h) $F(x, y) = (2x, y)$;

(d) $F(x, y) = (y, 1)$;   (i) $F(x, y) = (x^2, 0)$.

(e) $F(x, y) = (y, -x)$;

2. Compute div F for each of the vectorfields in Exercise 1. Briefly interpret your results in terms of the sketch of the vectorfield.

3. A vectorfield $F = (p, q)$ is called **irrotational,** or **conservative,** if there is a scalar-valued function φ such that $F = \text{grad } \varphi$. Then φ is called a **potential function** for the field.

(a) Show that $F(x, y) = (2x, 2y)$ is irrotational.

(b) If a continuously differentiable vectorfield $F = (p, q)$ is irrotational, show that $p_y = q_x$. (Suggestion: If $\varphi(x, y)$ has two continuous derivatives, then $\varphi_{xy} = \varphi_{yx}$.)

(c) Show that the vectorfields in Exercises 1(a), (d), (e), are not irrotational, but the remainder are irrotational (for (f), consider arc tan (x/y) for $y \neq 0$).

4. Let $F: \mathbb{R}^3 \longrightarrow \mathbb{R}^3$ be $F(X) = -X/\|X\|^3$. This is essentially the gravitational force field. Draw a sketch of it. Show that div $F = 0$ and that it is conservative for $X \neq 0$. (Suggestion: Consider $\psi(X) = 1/\|X\|$.)

5. If $f, g: \mathbb{R}^2 \longrightarrow \mathbb{R}$ and $F: \mathbb{R}^2 \longrightarrow \mathbb{R}^2$ are sufficiently differentiable, show that:
    (a) div $(\operatorname{grad} f) = f_{xx} + f_{yy}$ (we write $f_{xx} + f_{yy}$ as $\nabla^2 f$ or as Δf and call $\nabla^2 = \Delta = \partial^2/\partial x^2 + \partial^2/\partial y^2$ the *Laplace operator*);
    (b) div $(fF) = \langle \operatorname{grad} f, F \rangle + f \operatorname{div} F$;
    (c) div $(f \operatorname{grad} g - g \operatorname{grad} f) = f\Delta g - g\Delta f$;
    (d) $\Delta(fg) = f\Delta g + 2\langle \operatorname{grad} f, \operatorname{grad} g \rangle + g\Delta f$.

6. Let $F(X): \mathbb{R}^3 \longrightarrow \mathbb{R}^3$ denote the force at a point X, and assume that the force is conservative, so that $F = -\operatorname{grad} \varphi$ for some scalar-valued function (we have adopted the usage in physics here and added the minus sign, which could have been absorbed into φ). If $X(t)$ denotes the position of a particle of mass m moving in the force field, we define its energy by

$$E(t) = \frac{m}{2}\|X'(t)\|^2 + \varphi(X(t)).$$

Use Newton's Second, $mX'' = F$, to show that energy is conserved; that is, $E'(t) = 0$ (this justifies the name "conservative force" for $F = -\operatorname{grad} \varphi$).

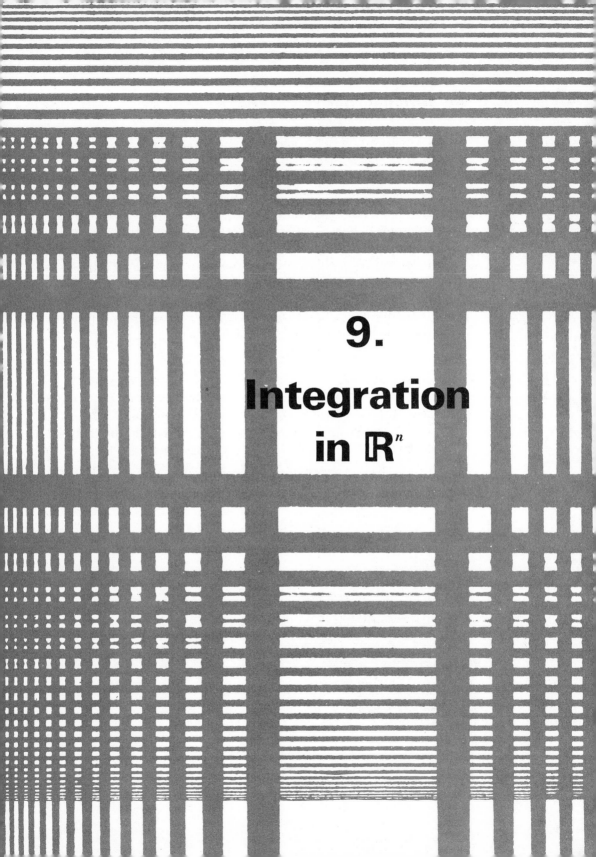

9.
Integration in $\mathbb{R}^n$

9.0 INTRODUCTION

Two elementary problems arise naturally and lead to the ideas of integration. The first is that of finding the area or volume of some set. The second, to be discussed later, is finding the mass of some region, given its density. Let us discuss volume now. Rather than construct a theoretical edifice, we work informally and rely on geometric intuition. The theory of integration involves some subtle and technical issues that would be out of place at the beginning.

Here is a typical problem. Say that we have a solid in $\mathbb{R}^3$ between a set B in the xy plane and the surface given by $z = f(x, y)$. See Fig. 9.1. Our problem: Find the volume, V, of this solid.

We note first that if the height $z = f(x, y)$ is constant, independent of (x, y), then the volume is given by

$$V = (\text{height})(\text{area of } B),$$

and so if we can compute the area of B, then we can compute V.

To deal with the general case where $f(x, y)$ varies with (x, y), we proceed as follows. First we partition the base B into smaller sets $B_1, B_2, \ldots, B_i,$ $\ldots, B_r$. We denote the volume above each B_i (and below the surface $z = f(x, y)$, of course) by ΔV_i (the ith little piece of V). See Fig. 9.2. Thus

$$V = \Sigma_i \, \Delta V_i = \Delta V_1 + \Delta V_2 + \cdots + \Delta V_i + \cdots + \Delta V_r.$$

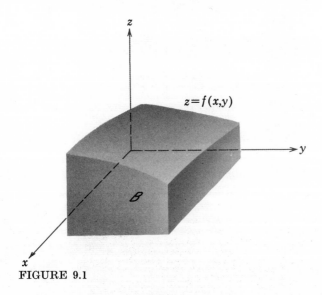

FIGURE 9.1

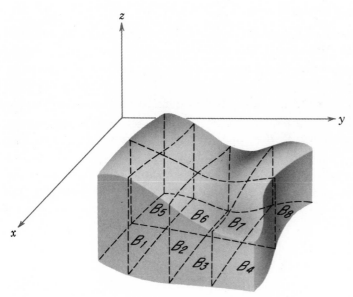

FIGURE 9.2

Here we use the notation Σ_i to mean "sum over all the indices i that occur in the present discussion." Since $i = 1, 2, \ldots, r$ now, we could be more formal and write $V = \sum_{i=1}^{r} \Delta V_i = \Delta V_1 + \cdots + \Delta V_r$. But we continue to be informal.

Second, we approximate V by approximating each ΔV_i. To do this, pick a point X_i in B_i, and form $f(X_i)$. The volume ΔV_i is approximately equal to the height $f(X_i)$ times the area of B_i, which we denote ΔA_i; that is,

$$\Delta V_i \approx f(X_i)\Delta A_i,$$

so that we have the approximation

$$V \approx \sum_i f(X_i)\Delta A_i.$$

The sum on the right is called an approximating *Riemann sum* to the integral, after the mathematician B. Riemann, who investigated the theory of the integral. If we break up B into smaller and smaller (consequently, more and more) pieces B_i, we expect that this Riemann sum approximation to V will become more accurate. It is this which motivates the notation

$$V = \int\int_B f \, dA$$

for the *exact* volume of the solid. Here the double integral signs should be thought of as an elongated *SS*, to stand for (double) sum. We refer to *f* as the **integrand**. The sum is "double" because the base *B*, the so-called **domain of integration**, is two-dimensional. We also say that *V* is given by a **double integral**.

Incidentally, if the height of a surface is negative, in the sense that the surface lies *below* the *xy*-plane, then we agree to call the resulting volume negative. For instance, if *B* is the square in the *xy*-plane ($z = 0$) with vertices at $(1, 1, 0), (1, -1, 0), (-1, 1, 0)$, and $(-1, -1, 0)$, and the surface is given by $z = f(x, y) = y$, then the net volume is zero, because the volume of the negative half, $y \leq 0$, exactly balances the positive half $y \geq 0$. See Fig. 9.3. The idea of negative volume may seem strange, but it is convenient.

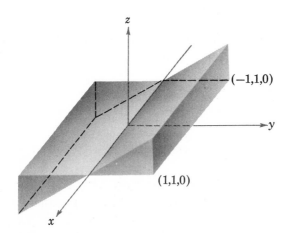

FIGURE 9.3

So far we have only raised a question ("find the volume") and introduced some notation. In Sec. 9.1, we approximate some volumes in the elementary way outlined above. Then, in Sec. 9.2, we learn how to calculate some volumes exactly. Finally, we discuss some of the theoretical aspects of integration in Sec. 9.3.

Exercises

1. On a sheet of graph paper, draw the triangle *A* with vertices at $(-2, 2)$, $(1, 1)$, and $(0, 0)$. By counting squares contained in *A* and containing *A*, obtain some estimates *m*, *M* for the area of *A*, $m \leq \text{area}(A) \leq M$. (By elementary geometry, area $(A) = 2$.)

2.   Find upper and lower bounds for the area bounded by the x-axis, the lines $x = -1$, $x = 2$, and the curve $y = 6(2 + x^2)^{-1}$. Please do not evaluate any integrals, but estimate the area naïvely by approximating the region by a simpler geometric configuration.

3.   Use a simple geometric argument to show that the volume V of a ball of radius 1 in $\mathbb{R}^3$ satisfies $8/3\sqrt{3} < V < 8$.

9.1   ESTIMATING THE VALUE OF INTEGRALS*

9.1a   Some numerical examples

Now we leave the descriptive introduction and begin the real work of obtaining numerical answers. Remember, the point is to rely on intuition gained from past experience. Before estimating a double integral, we recall a simpler example from elementary calculus.

EXAMPLE 1: Estimate the area (not volume now) between the curve $y = f(x) = 6/(2 + x^2)$ and the segment $0 \leq x \leq 2$ on the x-axis (see Fig. 9.4); that is, estimate the (single) integral

$$J = \int_0^2 \frac{6}{2 + x^2}\, dx.$$

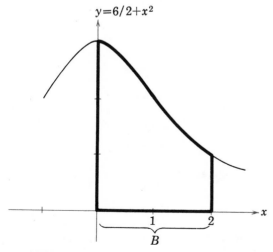

FIGURE 9.4

*This section provides a good opportunity to effectively use a computer.

If we denote the interval $0 \le x \le 2$ by B, then this integral could be written

$$J = \int_B f(x)\, dx,$$

which is closer to our notation for double integral. Since the maximum height is 3 ($= f(0)$) and the minimum is 1 ($= f(2)$), it is clear that

$$2 \cdot 1 < J < 2 \cdot 3.$$

Here $2 \cdot 3 = 6$ is the area of the large rectangle that contains the area we are approximating.

This first approximation, $2 < J < 6$, can be improved; that is, we can calculate numbers k_1, k_2 such that

$$2 < k_1 < J < k_2 < 6.$$

To do this, we break the area into two areas by breaking the base interval B into two intervals $0 \le x \le 1$ and $1 \le x \le 2$, denoted by B_1 and B_2 respectively. Then we estimate the areas above each interval separately, using inscribed and circumscribed rectangles as before. Let J_1, J_2 denote the exact areas above B_1, B_2 respectively. See Fig. 9.5. Then we see that

$$1 \cdot f(1) = 2 < J_1 < 1 \cdot f(0) = 3,$$

and

$$1 \cdot f(2) = 1 < J_2 < 1 \cdot f(1) = 2.$$

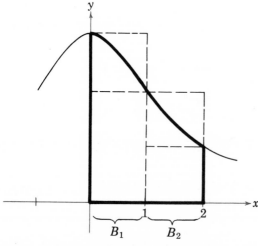

FIGURE 9.5

Since $J = J_1 + J_2$, we conclude that $2 + 1 < J < 3 + 2$; that is,

$$3 < J < 5.$$

Note that this new approximation sharpens the original estimate $2 < J < 6$. We now have a somewhat more refined idea of the size of the area J.

Even better accuracy can be had if we again subdivide B_1 and B_2. This process can be repeated until the area is estimated as closely as we desire.

Some might object, saying that this integral can be evaluated explicitly by the methods of basic calculus to give $J = (3/\sqrt{2})$ arctan $\sqrt{2}$, and so it is unnecessary to approximate. We point out, however, that one must consult tables to use this exact answer and might need only a rough estimate of the area, which we can now obtain without tables of the arctangent function. In addition, the procedure used here, or slight modifications of it, can be used easily by computers to evaluate numerically much more difficult integrals. To underscore this point, we mention that there is no "elementary" function $g(x)$, built up from polynomials, trigonometric, logarithmic, and exponential functions, such that $g'(x) = e^{-x^2}$. Hence

$$J = \int_{-1}^{1} e^{-x^2}\, dx$$

cannot be computed by searching for an anti-derivative $g(x)$ in the familiar manner of calculus. It is a fact of life that this integral, which is very important in probability and statistics, *must* be approximate.

Now we imitate the procedures of Example 1 to estimate volumes.

EXAMPLE 2: Consider the solid region in xyz-space $\mathbb{R}^3$ between the rectangle B with vertices at $(0, 0, 0)$, $(1, 0, 0)$, $(1, 2, 0)$, $(0, 2, 0)$ and the surface whose equation is $z = f(x, y) = 1 + x^2 + y^2$. See Fig. 9.6. As above, we denote the volume of this solid by

$$J = \iint_{B} (1 + x^2 + y^2)\, dA.$$

It is clear from Fig. 9.6 that

$1 \cdot 2 = (\text{min height})(\text{area of } B) < J < (\text{max height})(\text{area of } B) = 6 \cdot 2,$

that is,

$$2 < J < 12.$$

This gives a first, crude estimation of J.

It is easy to do better by regarding the solid region as built up from two regions obtained by breaking the base B into two rectangles, B_1 and

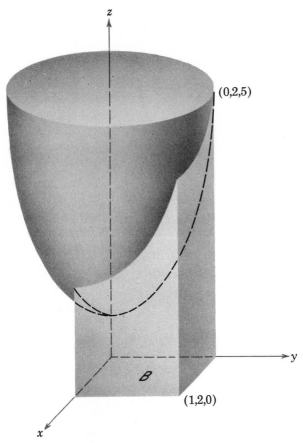

(0,2,5)

z

y

B

(1,2,0)

x

FIGURE 9.6

B_2. See Fig. 9.7. In the language of set theory, B is the "union" of B_1 and B_2, denoted $B = B_1 \cup B_2$. Then, in obvious notation,

$$J = \iint_B f(x, y) \, dA = \iint_{B_1} f(x, y) \, dA + \iint_{B_2} f(x, y) \, dA = J_1 + J_2,$$

and we can estimate the integrals J_1 and J_2 separately, just as we did J above.

For (x, y) in B_1 we see by inspection that $1 \leq 1 + x^2 + y^2 \leq 1 + 1 + 1 = 3$. Since the area of B_1 is 1, we know that $1 = 1 \cdot 1 \leq J_1 \leq 3 \cdot 1 = 3$. Similarly, in B_2 we have $2 \leq 1 + x^2 + y^2 \leq 6$, so that $2 = 2 \cdot 1 < J_2 < 6 \cdot 1 = 6$. Therefore, $1 + 2 < J_1 + J_2 < 3 + 6$, and since $J = J_1 + J_2$, we conclude that

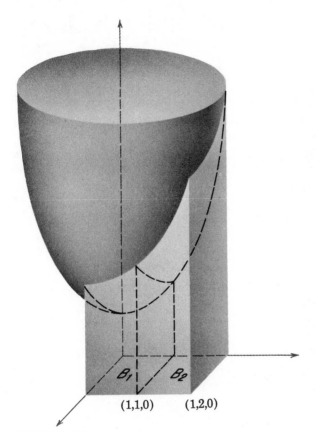

B_1 B_2

$(1,1,0)$ $(1,2,0)$

FIGURE 9.7

$$3 < J < 9,$$

which is a better estimate than $2 < J < 12$, found previously.

Let us do still better by partitioning B_2 into B_{21} and B_{22}, and writing $J_2 = J_{21} + J_{22}$. See Fig. 9.8. In B_{21} we find that

$$2 \le 1 + x^2 + y^2 \le 1 + 1 + \frac{9}{4} = \frac{17}{4}.$$

Since B_{21} has area $\frac{1}{2}$, we see that $1 < J_{21} < \frac{17}{8}$. Similarly, in B_{22},

$$1 + 0 + \frac{9}{4} = \frac{13}{4} \le 1 + x^2 + y^2 \le 1 + 4 + 1 = 6,$$

so that $\frac{13}{8} < J_{22} < 3$. Consequently, $\frac{21}{8} = 1 + \frac{13}{8} < J_2 < 3 + \frac{17}{8} = \frac{41}{8}$, which is sharper than the estimate $2 < J_2 < 6$, found above. Using this

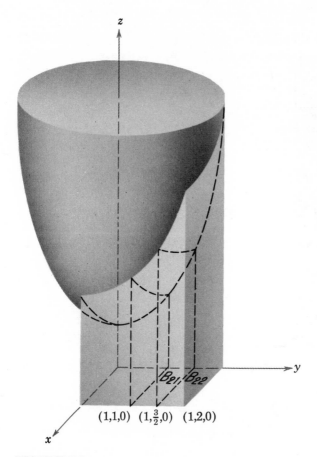

$(1,1,0)$ $(1,\tfrac{3}{2},0)$ $(1,2,0)$

FIGURE 9.8

new estimate for J_2 together with the old estimate $1 < J_1 < 3$, we have

$$\frac{29}{8} < J < \frac{65}{8},$$

which is better than our second estimate $3 < J < 9$.

By repeating this process and partitioning B_1 and B_2 even further, we presumably can estimate the exact volume J as closely as we desire. (This exact volume is $\tfrac{16}{3}$, by the way, and so we are not yet very close.) Although this process of estimation quickly loses its charm and becomes tedious, it has several important virtues. First, it is simpleminded and gets answers. Although we will, in Sec. 9.2, find easier ways of evaluating some integrals exactly, one should remember that in practice most integrals are actually evaluated by computers, using the method in Example 2. A second virtue is that the method leads us to a precise theoretical understanding of the integral; we will see more of this in Sec. 9.3.

Here is another problem we can attack by the same methods. Suppose that we are given a region B in the xy-plane and we want its area. See Fig. 9.9. We may regard B as the base of a solid whose height is constant and equal to 1. The volume of this cylinder is, in our notation,

$$J = \int\int_B 1 \, dA = \text{height} \cdot \text{area of } B = 1 \cdot \text{area of } B.$$

(Do not worry about units of measure here. The point is that the number of cubic feet in the three-dimensional solid equals the number of square feet in B.)

We conclude that the area of a plane set may be regarded as a very special case of double integration.

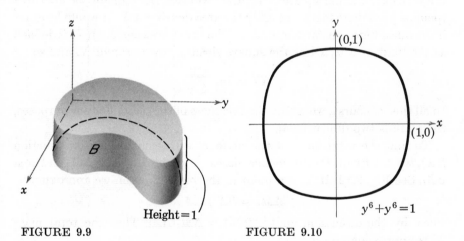

FIGURE 9.9   FIGURE 9.10

EXAMPLE 3: Estimate the area of the region B inside the curve $x^6 + y^6 = 1$. See Fig. 9.10. From the discussion above, the area of B is given by the double integral

$$J = \int\int_B 1 \, dA.$$

This number is estimated from below by inscribing a square inside B. Note that points (c, c), $(-c, c)$, $(-c, -c)$, and $(c, -c)$, where $c = (\frac{1}{2})^{1/6}$, are on the oval $x^6 + y^6 = 1$ and are the vertices of a square in the xy-plane. Now each side of this square has length $2 \cdot c = 2 \cdot (\frac{1}{2})^{1/6}$. Thus the area of this square is $(2 \cdot (\frac{1}{2})^{1/6})^2 = 4(\frac{1}{2})^{1/3} < J$.

By circumscribing another square (outside), we find easily that $J < 4$. These estimates of J, the area of B, can be refined by using many smaller such regions. In fact, one might estimate this area by drawing B on graph paepr and counting the number of squares inside and outside B.

9.1b Density

The second elementary problem that leads naturally to double integrals is that of finding the total mass (or weight) of a thin plate B in the xy-plane given its area density $f(X) = f(x, y)$, in pounds per square foot, say.

First, what do we mean by area density $f(X)$? We suppose that some parts of the plate are heavier than others of the same size situated at different locations in the plate. This might occur because the plate is cast from some inhomogeneous alloys. At the point X in question, we center a small square part of the plate. See Fig. 9.11. Let ΔM be the mass and ΔA the area of this small square. Then the ratio $\Delta M / \Delta A$ of mass per unit area (pounds per square foot) is roughly the area density at X. It would be exact if the mass were constant around X. The exact **area density** $f(X)$ is defined as the limit of $\Delta M / \Delta A$ as the square shrinks down around X, and so

$$f(X) = \lim_{\Delta A \to 0} \frac{\Delta M}{\Delta A}.$$

In all this, of course, we ignore the thickness of the plate. For our purposes, the plate is two-dimensional.

To find the total mass of the plate B with a given density function $f(X)$, we subdivide B into smaller pieces B_i, each with mass ΔM_i and area ΔA_i. See Fig. 9.12. If X_i is a point in the piece B_i, then we approximate

$$\Delta M_i \approx f(X_i)\, \Delta A_i,$$

since, by the discussion above, $f(X_i) \approx \Delta M_i / \Delta A_i$. Thus the total mass M is approximated by

$$M = \sum_i \Delta M_i \approx \sum_i f(X_i)\, \Delta A_i.$$

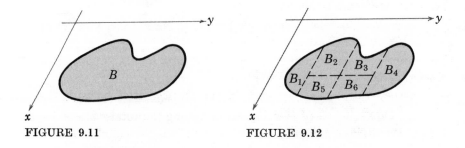

FIGURE 9.11   FIGURE 9.12

In the limit, using smaller and smaller subdivisions, we conclude that the total mass, like volume, is a double integral,

$$M = \int\int_B f \, dA.$$

EXAMPLE 4: Consider a rectangular plate B in $\mathbb{R}^2$ with corners at the points $(0, 0)$, $(1, 0)$, $(1, 2)$, $(0, 2)$ and whose density is $f(X) = f(x, y) = 1 + x^2 + y^2$. Hence the plate is "light" near $(0, 0)$ and "heavy" near $(1, 2)$. The total mass of this solid is

$$M = \int\int_B (1 + x^2 + y^2) \, dA.$$

True to form, we estimate the mass of this plate. Since the minimum and maximum densities are 1 and 6, respectively, that is, $1 \le 1 + x^2 + y^2 \le 6$ for all (x, y) in B, it is intuitively clear on physical grounds that

$$1 \cdot 2 = (\text{min density})(\text{area}) < M < (\text{max density})(\text{area}) = 6 \cdot 2,$$

that is,

$$2 < M < 12.$$

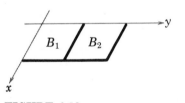

FIGURE 9.13

This estimates M crudely. By regarding the plate as the union of two plates obtained by partitioning B into two rectangles (see Fig. 9.13), we can use the same method to estimate the mass of the two parts separately and then add the result. Of course, this is a repetition of the arithmetic in Example 2. It yields

$$3 < M < 9.$$

From the point of view of computation, this density presents nothing new.

We remark, however, that contrary to volume there is a very natural situation in which the density $f(X)$ may be negative as well as positive. This is electric charge density, for charge may be $+$ or $-$. In this case $\int\int_B f \, dA$ is the total (net) electric charge on B. Many other kinds of problems lead to integrals with negative integrands.

9.1c   Triple integrals

All we have said concerning double integrals extends readily to triple integrals. One is led to triple integrals in finding the total mass of a solid B given its volume density $f(X) = f(x, y, z)$. See Fig. 9.14. We describe this now. If X is a point of the solid and ΔV is the volume of a small cube of mass ΔM centered at X, then the **volume density** at X is defined by

$$f(X) = \lim_{\Delta V \to 0} \frac{\Delta M}{\Delta V}.$$

We are generally given a solid B with volume density f and asked to find its total mass M. Subdividing B into smaller parts B_i, each with mass ΔM_i and volume ΔV_i, we have the by now familiar approximation to the mass

$$M = \sum_i \Delta M_i \approx \sum_i f(X_i)\, \Delta V_i,$$

where X_i is some point of B_i. In the limit as the B_i become small, we obtain the exact mass

$$M = \iiint_B f\, dV.$$

If the density f is identically 1 ($f \equiv 1$), then the mass of B is numerically equal to the volume V:

$$V = \iiint_B 1 \cdot dV.$$

We may consider the three-dimensional region B to carry an electric charge throughout. This leads to a charge density f that may be positive or negative. Of course, if the physical interpretation is alien to your mode of thought, then you are free to think of the triple integral $\iiint_B f\, dV$ in some other way.

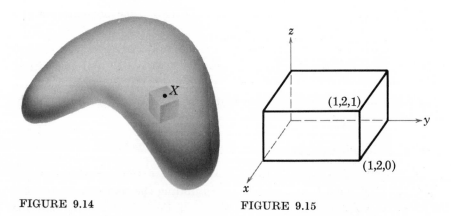

FIGURE 9.14   FIGURE 9.15

Let us show how to estimate the value of a triple integral. No new ideas are involved.

EXAMPLE 5: Let $B = \{(x, y, z) \in \mathbb{R}^3 \mid 0 \leq x \leq 1, \; 0 \leq y \leq 2, \; 0 \leq z \leq 1\}$, a box in $\mathbb{R}^3$. See Fig. 9.15. Let the density $f(x, y, z) = xz + y$. We are to estimate

$$J = \iiint_B f \, dV = \iiint_B (xz + y) \, dV.$$

A naïve upper bound for J results from observing that the maximum density in B occurs at the corner $(1, 2, 1)$, so that $xz + y \leq 3$. Since the volume of B is 2, we conclude that $J < 3 \cdot 2 = 6$. It should be clear that $0 < J$. Hence we have $0 < J < 6$.

A better estimate can be found by partitioning B into two parts, B_1, where $0 \leq y \leq 1$, and B_2, where $1 \leq y \leq 2$. Then $J = J_1 + J_2$, where J_1 and J_2 are the obvious triple integrals taken over B_1 and B_2. Now in B_1 we have $0 \leq xz + y \leq 2$, so that $0 < J_1 < 2$. Similarly, in B_2 we have $1 \leq xz + y \leq 3$, so that $1 < J_2 < 3$. Thus we estimate $1 < J < 5$, sharper than $0 < J < 6$.

You should be willing to believe that these same ideas immediately generalize to higher-dimensional multiple integrals, such as

$$\iiiint_B f \, dV,$$

where B is a set in $\mathbb{R}^4$, $f = f(x_1, x_2, x_3, x_4)$, and dV here signifies a tiny element of *four*-dimensional volume.

Exercises

1. Use a compass to draw a disk of radius 10 on a sheet of graph paper. Estimate the area of the disk from above and below by counting squares. Then use the result and the formula for the area of a disk to estimate the value of π.

2. Estimate the areas of the following sets in $\mathbb{R}^2$ by the methods in this section (do not use calculus of any kind):
   (a) $2 \leq x \leq 4, \; -1 \leq y \leq x^2$;
   (b) $x^2 + y^2 \leq 4$;
   (c) $0 \leq x \leq 1, \; x^2 \leq y \leq 1$;
   (d) $-1 \leq x \leq 2, \; -1 \leq y \leq (4 + x)^{-1}$;
   (e) $x^4 + y^4 \leq 16$;
   (f) $x^2 \leq y \leq 2 - x^2$.

3. Let B be the rectangle in $\mathbb{R}^2$ with corners at $(-1, 0)$, $(1, 0)$, $(1, 2)$, and $(-1, 2)$. Estimate each of the following integrals. You should first obtain a crude estimation, then a second, more refined estimation.

(a) $J = \iint_B (x^2 + y^2)\, dA$,

(f) $J = \iint_B \dfrac{1}{1 + x^2 + y}\, dA$,

(b) $J = \iint_B (1 + x^2 y)\, dA$,

(g) $J = \iint_B x(1 + y)\, dA$,

(c) $J = \iint_B (1 - 2x^2 - y^2)\, dA$,

(h) $J = \iint_B xy^2\, dA$,

(d) $J = \iint_B x(1 - 2y)\, dA$,

(i) $J = \iint_B e^{-xy-y}\, dA$,

(e) $J = \iint_B \dfrac{1}{5 + x + y}\, dA$,

(j) $J = \iint_B e^{-x^2-y^2}\, dA$.

4. Crudely estimate the integrals in Exercise 3, this time letting B be the set $-2 \le x \le 2$, $0 \le y \le 1 + x^2$.

5. Let $B \subset \mathbb{R}^3$ be the set $0 \le x \le 2$, $0 \le y \le 3$, $1 \le z \le 2$. Estimate the following integrals as in Exercise 3:

(a) $J = \iiint_B x\, dV$,

(c) $J = \iiint_B xyz\, dV$,

(b) $J = \iiint_B (x + 3y + z^2)\, dV$,

(d) $J = \iiint_B (x^2 y + y^2 z)\, dV$.

6. (a) Estimate the volume of the bounded region between the paraboloid $z = x^2 + y^2$ and the plane $z = 4$.
   (b) Estimate the volume of the bounded region between the paraboloids $z = x^2 + y^2$ and $z = 2 - x^2 - y^2$.
   (c) Estimate the volume of the region bounded from above by $z = 1 + x^2 + y^2$, from below by $z = -2x^2 - 2y^2$, and on the sides by the cylinder $x^2 + y^2 = 1$.

7. Let $y = f(x)$ define a smooth curve in $\mathbb{R}^2$. Find the exact area of the region bounded from below by $y = f(x)$ and from above by $y = f(x) + 7$ for $2 \le x \le 4$. (Hint: This needs a simple mental geometric argument.)

8. Let $z = f(x, y)$ define a smooth surface in $\mathbb{R}^3$. Find the exact volume of the solid region bounded from below by $z = f(x, y) - 1$ and bounded from above by $z = f(x, y) + 3$ for $1 \le x \le 5$, $-2 \le y \le 2$.

9. Let $B \subset \mathbb{R}^2$ be the set $y \ge 0$, $x^4 + y^4 \le 16$. Estimate:

(a) $J = \iint_B \dfrac{1}{1 + (x^4 + y^4)^{1/4}}\, dA$,

(b) $J = \iint_B \dfrac{3}{8 + \sqrt{x^4 + y^4}}\, dA$.

10. Let $B \subset \mathbb{R}^3$ be the set $x^4 + y^4 + z^4 \le 16$. Estimate:

(a) $J = \iiint_B (x^2 + y^2)z^2\, dV$,

(b) $J = \iiint_B (x^2 + y^2)z\, dV$.

11. (a) Give a brief intuitive proof that the integral in Exercise 3(h) must be zero.
    (b) Do the same for the integrals in Exercise 3(d), (g), Exercise 4(d), (g), (h), and Exercise 10(b).

(c) Let $B \subset \mathbb{R}^2$ be the set in Exercise 3. By inspection, evaluate

$$J = \iint_B (1 + 2x - xy^2)\, dA.$$

12. Let A be the area under the graph of $f(x) = 1/x$ for $1 \leq x \leq 2$ and B the area under the same graph for $2 \leq x \leq 4$. Give a geometric argument to show that $A = B$.

13. Let $B \subset \mathbb{R}^2$ have the property that $a \leq \text{Area}(B) \leq \alpha$, and let f be an integrable function with the property $m \leq f(X) \leq M$ for all $X \in B$. Give a brief intuitive proof that

$$ma \leq \iint_B f\, dA \leq M\alpha.$$

14. Let $B(t) \subset \mathbb{R}^2$ be the rectangle $\alpha \leq x \leq t, \; a \leq y \leq b$ whose width depends on t, and let $g(x)$ be a continuous increasing function depending on x only. Define $M(t)$ by

$$M(t) = \iint_{B(t)} g(x)\, dA.$$

(a) Give an intuitive argument to show that $M(\alpha) = 0$ and, for $h > 0$, that

$$g(t)(b - a) \leq \frac{M(t + h) - M(t)}{h} \leq g(t + h)(b - a).$$

(b) Deduce from this that $dM/dt = g(t)(b - a)$ and conclude that

$$\iint_{B(t)} g(x)\, dA = (b - a) \int_\alpha^t g(x)\, dx.$$

(c) Let $B \subset \mathbb{R}^2$ be the set $2 \leq x \leq 6, \; 1 \leq y \leq 3$. Use Exercise 14(b) to evaluate:

(i) $\iint_B (2x + x^3)\, dA,$ (iii) $\iint_B y^2\, dA,$

(ii) $\iint_B \sqrt{1 + 4x}\, dA,$ (iv) $\iint_B \log(xy^2)\, dA.$

15. Let $B(t) \subset \mathbb{R}^2$ be the rectangle $\alpha \leq x \leq t, \; a \leq y \leq b$, and let $f(x)$ and $g(y)$ be continuous increasing functions, with $g(y) \geq 0$. Define $S(t)$ by

$$S(t) = \iint_{B(t)} f(x)g(y)\, dA.$$

(a) Use the result of Exercise 14(b) to show that for $h > 0$,

$$f(t) \int_a^b g(y)\, dy \leq \frac{S(t + h) - S(t)}{h} \leq f(t + h) \int_a^b g(y)\, dy,$$

and deduce as in Exercise 14 that

$$\iint_{B(t)} f(x)g(y)\, dA = \left(\int_\alpha^t f(x)\, dx \right)\left(\int_a^b g(y)\, dy \right).$$

(b) Let $B \subset \mathbb{R}^2$ be the set $2 \leq x \leq 6, \; 1 \leq y \leq 3$. Evaluate:

(i) $\iint_B xy\, dA,$ (iii) $\iint_B (1 - 2xy)\, dA,$

(ii) $\iint_B (x + xy)\, dA,$ (iv) $\iint_B e^{2x + 3y}\, dA.$

9.2 COMPUTING INTEGRALS EXACTLY

9.2a Repeated or iterated integrals

We now have some practice in estimating the number $\iint_B f \, dA$; we would like to compute it exactly, if possible. Although the previous section might have been called "Evaluation of Integrals for Computers," this one is "Evaluation of Integrals for Humans." We present a method of computation that works, provided that the integrand f and the domain B of integration are not too complicated or pathological. The method reduces this computation of a double integral to the computation of two ordinary integrals from one-dimensional calculus. We assume that you are familiar with the standard method for computing a one-dimensional integral $\int_a^b f(x) \, dx$: If $F(x)$ can be found satisfying $F'(x) = f(x)$, then this integral has the value $F(b) - F(a)$. As always, we depend on intuition, pictures, and special examples to guide us.

For our first example we suppose that B is the rectangle in the xy-plane determined by $a \leq x \leq b, c \leq y \leq d$, and that we wish to evaluate $\iint_B f \, dA$. This represents the volume V of the solid in Fig. 9.16. We proceed as follows. Slice the solid into thin slabs parallel to the x-plane. The jth slab has width Δy_j and volume ΔV_j. It lies between the parallel planes $y = y_j$ and $y = y_{j+1} = y_j + \Delta y_j$. Its volume ΔV_j is approximately equal to $A(y_j) \, \Delta y_j$,

$$\Delta V_j \approx A(y_j) \, \Delta y_j,$$

where $A(y_j)$ is the exact cross-sectional area of the slice through y_j.

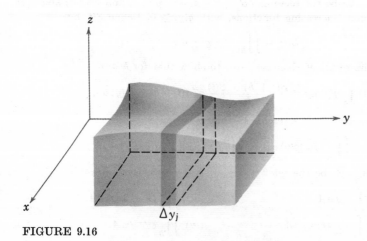

FIGURE 9.16

Now it is crucial to note that we can compute $A(y_j)$ exactly. By ordinary calculus, we have

$$A(y_j) = \int_a^b f(x, y_j)\, dx,$$

where y_j is kept fixed during this integration. Hence $A(y_j)$ is a number depending on y_j, a function of y_j.

Now we have the desired volume V approximated,

$$V = \sum_j \Delta V_j \approx \sum_j \left(\int_a^b f(x, y_j)\, dx \right) \Delta y_j.$$

If we allow the number of slabs to increase indefinitely and their widths Δy_j to tend to zero, this approximation tends to the exact volume V, and, moreover, it equals $\int_c^d (\int_a^b f(x, y)\, dx)\, dy$; that is,

$$V = \int\int_B f(x, y)\, dA = \int_c^d \left(\int_a^b f(x, y)\, dx \right) dy.$$

We call the integral on the right a **repeated** or **iterated integral**. Its virtue is that its evaluation involves two single integrals, each of which may be evaluated by the techniques of ordinary calculus.

Note that we could equally well have taken slices parallel to the yz-plane. This would have given

$$V = \int\int_B f(x, y)\, dA = \int_a^b \left(\int_c^d f(x, y)\, dy \right) dx,$$

in which we integrate first with respect to the variable y.

EXAMPLE: Find the volume of the solid between the rectangle B: $0 \le x \le 1$, $0 \le y \le 2$ in the xy-plane and the surface $z = f(x, y) = 1 + x^2 + y^2$. This is Example 2 in the previous section. The volume is given by the integral

$$V = J = \int\int_B (1 + x^2 + y^2)\, dA.$$

If we take slices parallel to the xz-plane, we have the repeated integral

$$J = \int_0^2 \left(\int_0^1 (1 + x^2 + y^2)\, dx \right) dy.$$

The inner integral is evaluated by treating y as a constant, and so

$$\int_0^1 (1 + x^2 + y^2)\, dx = x + \frac{x^3}{3} + y^2 x \Big]_{x=0}^{x=1} = 1 + \frac{1}{3} + y^2 = \frac{4}{3} + y^2.$$

This provides us with an integrand for the second integration,

$$J = \int_0^2 \left(\frac{4}{3} + y^2 \right) dy = \frac{4}{3}y + \frac{y^3}{3} \Big]_{y=0}^{y=2} = \frac{8}{3} + \frac{8}{3} = \frac{16}{3}.$$

Thus, the desired volume is $\frac{16}{3}$. How good were our estimates in Sec. 9.1? How might they have been improved?

Had we taken slices parallel to the yz-plane, we would have an integral with y as the first variable of integration,

$$\int_0^1 \left(\int_0^2 (1 + x^2 + y^2) \, dy \right) dx.$$

This should of course give the same volume $\frac{16}{3}$. We encourage you to verify that it does. It is easy.

9.2b Domains with curved boundaries

This same procedure can be applied to domains B more general than rectangles. For convenience in picturing, we draw only the set B in the xy-plane (a top view) and ask you to imagine the surface $z = f(x, y)$ above this plane. Suppose that B has the form

$$\alpha(y) \leq x \leq \beta(y), \qquad c \leq y \leq d,$$

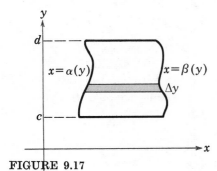

where $x = \alpha(y)$ and $x = \beta(y)$ are functions of y. Thus two sides of B are *curves*: the graphs of the functions α and β. It is natural in this case to consider slabs in three dimensions parallel to the xz-plane. The base of such a slab rests in the xy-plane (see Fig. 9.17.) A typical slab is determined by slicing B parallel to the x-axis through a typical y. This slice extends from $x = \alpha(y)$ to $x = \beta(y)$, so that the cross-sectional area of the three-dimensional solid there is

FIGURE 9.17

$$A(y) = \int_{\alpha(y)}^{\beta(y)} f(x, y) \, dx,$$

which rightly depends on y. Since y varies from $y = c$ to $y = d$, we have

$$\iint_B f \, dA = \int_c^d \left(\int_{\alpha(y)}^{\beta(y)} f(x, y) \, dx \right) dy.$$

This yields a number, because the second integral is taken from $y = c$ to $y = d$, and these limits are numbers. We will compute an example in a moment.

If, on the other hand, the base B has the form

$$a \leq x \leq b, \qquad \gamma(x) \leq y \leq \delta(x),$$

then it is natural to take slices parallel to the y-axis. See Fig. 9.18. This leads to

$$\iint_B f \, dA = \int_a^b \left(\int_{\gamma(x)}^{\delta(x)} f(x, y) \, dy \right) dx.$$

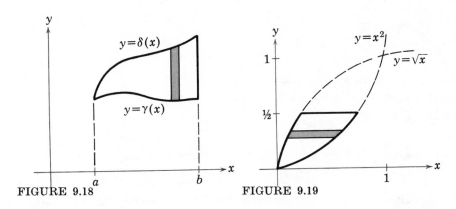

FIGURE 9.18   FIGURE 9.19

EXAMPLE: Evaluate $J = \iint_B x \, dA$, where B is the region given by

$$\sqrt{y} \leq x \leq y^2, \qquad 0 \leq y \leq \tfrac{1}{2}.$$

Solution: We slice parallel to the x-axis. See Fig. 9.19. Such a slice extends from $x = \alpha(y) = y^2$ to $x = \beta(y) = \sqrt{y}$. The values of y range from $y = 0$ up to $y = \tfrac{1}{2}$. Hence we must compute

$$J = \int_0^{1/2} \left(\int_{y^2}^{\sqrt{y}} x \, dx \right) dy.$$

The inner integral is

$$\int_{y^2}^{\sqrt{y}} x \, dx = \frac{x^2}{2} \Big]_{x=y^2}^{x=\sqrt{y}} = \frac{y}{2} - \frac{y^4}{2}.$$

Note that there is no longer any trace of the variable x; it has been "integrated out."

Now we obtain the value of J by integrating with respect to y,

$$J = \int_0^{1/2} \left(\frac{y}{2} - \frac{y^4}{2} \right) dy = \frac{y^2}{4} - \frac{y^5}{10} \Big]_0^{1/2} = \frac{19}{320}.$$

We have computed $\iint_B x \, dA$ exactly.

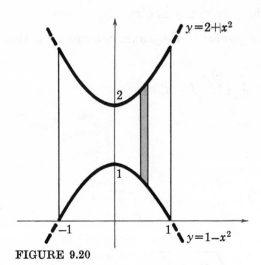

FIGURE 9.20

EXAMPLE: Compute $J = \iint_B xy \, dA$, where B is determined by
$$-1 \leq x \leq 1, \qquad 1 - x^2 \leq y \leq 2 + x^2.$$

Solution: We take slices parallel to the y-axis. See Fig. 9.20. The slice through a typical x extends in B from $y = 1 - x^2$ to $y = 2 + x^2$. Then x extends from -1 to 1. This leads to the iterated integral
$$J = \int_{-1}^{1} \left(\int_{1-x^2}^{2+x^2} xy \, dy \right) dx.$$

The first integral is
$$\int_{1-x^2}^{2+x^2} xy \, dy = \frac{xy^2}{2} \Big]_{y=1-x^2}^{y=2+x^2} = \frac{x}{2}(2 + x^2)^2 - \frac{x}{2}(1 - x^2)^2.$$

Thus the second integral becomes
$$\int_{-1}^{1} \left[\frac{x}{2}(2 + x^2)^2 - \frac{x}{2}(1 - x^2)^2 \right] dx = \frac{1}{12}(2 + x^2)^3 + \frac{1}{12}(1 - x^2)^3 \Big]_{-1}^{1}.$$

This is 0 by inspection, so that $J = 0$ (in fact, we could have deduced this by considering the behavior of the integrand xy in the domain B). Note that a certain amount of judgment was made in deciding to take slices parallel to the y-axis. If we had taken slices parallel to the x-axis, the computation would have been considerably more complicated, since for $y > 2$ and $y < 1$, each slice itself consists of two pieces. Sometimes it is equally reasonable to integrate with respect to either variable first, because of the nature of B. Consider

EXAMPLE: Evaluate $J = \iint_B (x - 2y)^2 \, dA$, where B is the triangular region bounded by the lines $x = 1$, $y = -2$, and $y + 2x = 4$.

First solution: We slice parallel to the x-axis through a typical y. See Fig. 9.21. This slice extends from $x = 1$ to $x = 2 - \frac{1}{2}y$. Then y extends from $y = -2$ up to $y = 2$. Thus we have

$$J = \int_{-2}^{2} \left(\int_{1}^{2-(1/2)y} (x - 2y)^2 \, dx \right) dy.$$

The first integration is

$$\int_{1}^{2-(1/2)y} (x - 2y)^2 \, dx = \frac{1}{3}(x - 2y)^3 \Big]_{x=1}^{x=2-(1/2)y}$$

$$= \frac{1}{3}\left(2 - \frac{5}{2}y\right)^3 - \frac{1}{3}(1 - 2y)^3.$$

Note that x has been integrated out.

Now we obtain J:

$$J = \frac{1}{3} \int_{-2}^{2} \left[\left(2 - \frac{5}{2}y\right)^3 - (1 - 2y)^3 \right] dy$$

$$= \frac{1}{12}\left[-\frac{2}{5}\left(2 - \frac{5}{2}y\right)^4 + \frac{1}{2}(1 - 2y)^4 \right]_{y=-2}^{y=2}$$

$$= \frac{164}{3}. \quad \text{Done.}$$

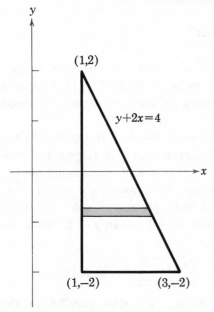

FIGURE 9.21

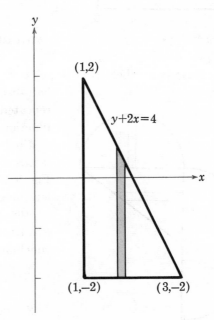

FIGURE 9.22

Second solution: This time we slice parallel to the y-axis and hence integrate first with respect to y. See Fig. 9.22. We note that y varies from $y = -2$ up to the hypotenuse $y = -2x + 4$. We get a slice at each x from $x = 1$ to $x = 3$. Thus

$$J = \int_1^3 \left(\int_{-2}^{-2x+4} (x - 2y)^2 \, dy \right) dx.$$

Note that we are left with a function of x alone after performing the first integration and evaluating at the designated limits; the variable y gets integrated out. You should verify that this procedure gives $J = \frac{164}{3}$ as computed earlier.

The emphasis of the next example is on determining the limits of integration.

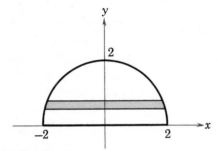

FIGURE 9.23

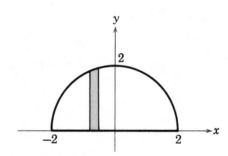

FIGURE 9.24

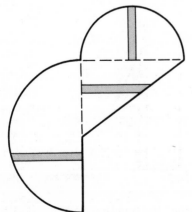

FIGURE 9.25

EXAMPLE: Let $B = \{(x, y) \in \mathbb{R}^2 \mid x^2 + y^2 \le 4, y \ge 0\}$ be the "upper half-disk" of radius 2. Write $J = \iint_B f \, dA$ as a repeated integral, where f is some continuous function.

First solution: We slice B parallel to the x-axis and integrate first with respect to x. See Fig. 9.23. A typical such slice extends from the left side of the circle, where $x = -\sqrt{4 - y^2}$, to the right side, where $x = \sqrt{4 - y^2}$. Then y varies from $y = 0$ up to $y = 2$. Thus we conclude that

$$J = \int_0^2 \left(\int_{-\sqrt{4-y^2}}^{\sqrt{4-y^2}} f(x, y) \, dx \right) dy.$$

Second solution: We slice parallel to the y-axis first. See Fig. 9.24. Now y varies from

$y = 0$ to $y = \sqrt{4 - x^2}$ for a given x. The value of x varies from $x = -2$ to $x = 2$. This leads to the iterated integral

$$J = \int_{-2}^{2} \left(\int_{0}^{\sqrt{4-x^2}} f(x, \acute{y}) \, dy \right) dx$$

This integral and the one obtained above yield the same number for J, of course. Depending on the nature of $f(x, y)$, one of the two might be easier to compute, however.

Domains somewhat more complicated than the ones just treated may often be decomposed into unions of simpler kinds that may be handled by the methods above. See Fig. 9.25.

9.2c Integration on disks; polar coordinates

There is a particular kind of double integral that is worthy of special mention. The problem is this. Compute $\iint_{B(a)} f \, dA$, where $B(a)$ is the disk of radius $a > 0$ centered at the origin in the xy-plane. Very often, moreover, the integrand f is given in terms of coordinates x, y, but it can be seen to have some kind of symmetry with respect to the origin. For example, consider $\iint_{B(a)} (x^2 + y^2) \, dA$. Here the integrand $f(x, y)$ attains the same value at two different points (x_1, y_1) and (x_2, y_2) if and only if they are equidistant from the origin. To capitalize on these circular symmetries in the disk $B(a)$ and in the integrand $f(x, y)$, we transform the double integral into polar coordinates.

Suppose that (x, y) denotes a typical point in the plane different from the origin. Then this point is also determined by the two numbers r and θ, where r is the distance from the origin and θ is the counterclockwise angle, in radians, from the x-axis to the segment, or arrow, from the origin to the point. Thus $r > 0$, and we may take $0 \leq \theta < 2\pi$, since θ and $\theta + 2\pi$ refer to the same angle in the plane. The pair (r, θ) gives the **polar coordinates** of the point. The polar and cartesian coordinates are related by trigonometry (see Fig. 9.26):

$$x = r \cos \theta, \qquad y = r \sin \theta,$$

whence

$$r = \sqrt{x^2 + y^2}, \qquad \theta = \arctan \frac{y}{x}.$$

We note that the function given by $f(x, y) = x^2 + y^2$ in cartesian coordinates has the very

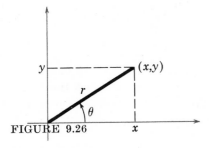

FIGURE 9.26

simple form r^2 in polar coordinates. The fact that the coordinate θ does not appear in this polar representation of f shows vividly that f depends only on distance from the origin.

Here is an example of some notation we use. Given $f(x, y) = x^2 + y^2$ again, we use the formulas for x and y above to write

$$f(x, y) = f(r \cos \theta, r \sin \theta) = (r \cos \theta)^2 + (r \sin \theta)^2 = r^2,$$

since $\cos^2 \theta + \sin^2 \theta = 1$. The moral is this: Given f in terms of x, y, we may rewrite it in terms of r, θ, using

$$f(x, y) = f(r \cos \theta, r \sin \theta).$$

We might call the last expression $\hat{f}(r, \theta)$, the hat above the f reminding us that r, θ are polar and not cartesian coordinates. Thus, if $f(x, y) = x^2 + y^2$ again, then $\hat{f}(r, \theta) = r^2$; if, on the other hand, we wrote $f(r, \theta)$ for this f, we would commit the sin of ambiguity: $f(r, \theta)$ might stand for $r^2 + \theta^2$ as well as r^2. You should make your own private peace with this issue.

Now we state a result that enables us to integrate functions over disks by iterating ordinary integrals.

THEOREM: Let $B(a)$ denote the disk of radius a centered at the origin. Then for any continuous function f,

$$\iint_{B(a)} f(x, y) \, dA = \int_0^{2\pi} \int_0^a f(r \cos \theta, r \sin \theta) r \, dr \, d\theta.$$

NOTE: On the right side, r varies from 0 to a and θ varies from 0 to 2π.

Before attempting to justify this theorem, let us apply it in a case already familiar. We know that

$$\iint_{B(a)} 1 \, dA = \text{area (disk of radius } a) = \pi a^2.$$

Here the integrand $f(x, y) = 1$ identically. The theorem tells us that

$$\iint_{B(a)} 1 \, dA = \int_0^{2\pi} \left(\int_0^a 1 \cdot r \, dr \right) d\theta = \int_0^{2\pi} d\theta \cdot \int_0^a r \, dr = \theta \Big]_0^{2\pi} \cdot \frac{r^2}{2} \Big]_0^a$$

$$= 2\pi \cdot \frac{a^2}{2} = \pi a^2,$$

as desired. This is reassuring. Note that we were able to break the iterated double integral into a product of two independent integrals, one in θ and one in r. This is because the integrand was independent of θ.

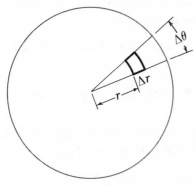

FIGURE 9.27

Idea of Proof of Theorem: Recall that $\iint_{B(a)} f(x, y) \, dA$ is obtained by breaking up the domain of integration—in this case $B(a)$—into small pieces of area ΔA_i and considering sums of the form $\sum_i f(X_i) \, \Delta A_i$, where X_i is a point in the small piece ΔA_i. Since $B(a)$ is a disk, it is natural to partition it into small "curved rectangles" as pictured in Fig. 9.27. The straight sides of this curved rectangle have length Δr, and the curved side closest to the origin has length $r \, \Delta \theta$. This is because an entire circle of radius r has length $2\pi r$ and the curved side is only a fraction of the entire circle, $\Delta\theta/2\pi$, to be exact. It follows that the area of the "curved rectangle" is approximately $r \, \Delta r \, \Delta \theta$.

Also, since $x = r \cos \theta$, $y = r \sin \theta$, the value $f(x, y)$ equals $f(r \cos \theta, r \sin \theta)$. Hence the integral $\iint_{B(a)} f(x, y) \, dA$ may be obtained by considering sums of the form

$$\sum f(r \cos \theta, r \sin \theta) r \, \Delta r \, d\theta.$$

If we let the "curved rectangles" that pave the disk shrink in size and increase in number and consider the limit of these sums as r varies from 0 to a and θ varies from 0 to 2π, that is, r and θ vary over the disk $B(a)$, then we obtain the iterated integral

$$\int_0^{2\pi} \int_0^a f(r \cos \theta, r \sin \theta) r \, dr \, d\theta,$$

as claimed in the theorem. ≪

We mention once again that in many cases the formidable expression $f(r \cos \theta, r \sin \theta)$ readily reduces to a function of r alone.

EXAMPLE: Compute $\iint_{B(a)} (1 + x^2 + y^2)^{1/3} \, dA$.

Solution: By the theorem this is equal to

$$\int_0^{2\pi} \left(\int_0^a (1 + r^2)^{1/3} r \, dr \right) d\theta = \int_0^{2\pi} d\theta \cdot \int_0^a (1 + r^2)^{1/3} r \, dr$$

$$= \theta \Big]_0^{2\pi} \cdot \frac{3}{8} (1 + r^2)^{4/3} \Big]_0^a$$

$$= \frac{3\pi}{4} \cdot ((1 + a^2)^{4/3} - 1).$$

We remark that this would have been much more difficult if we had reduced the original integral to cartesian coordinates x, y.

Now we observe that polar coordinates may be used on appropriate subsections of the disk as well.

EXAMPLE: Compute $\iint_B 2xy\,dA$, where B is the "piece of pie" shown in Fig. 9.28.

Solution: Here the radius $a = 1$, and the angle θ varies from $\theta = 0$ to $\theta = \pi/3$ (*not* to $\theta = 2\pi$, as in the full disk). Hence we have

$$
\iint_B 2xy\,dA = \int_0^{\pi/3}\left(\int_0^1 (2r^2\cos\theta\sin\theta)r\,dr\right)d\theta
$$
$$
= \int_0^{\pi/3} 2\sin\theta\cos\theta\,d\theta \cdot \int_0^1 r^3\,dr
$$
$$
= \sin^2\theta\Big]_0^{\pi/3}\frac{r^4}{4}\Big]_0^1 = \frac{3}{4}\cdot\frac{1}{4} = \frac{3}{16}.
$$

Finally, we apply polar coordinates to a famous computation, that of computing the area under the bell-shaped curve, the graph of $f(x) = e^{-x^2}$. See Fig. 9.29. This function is very important in probability and statistics.

EXAMPLE: Compute $J = \int_{-\infty}^{\infty} e^{-x^2}\,dx$.

Solution: This involves a clever trick. We note that $J^2 = \int_{-\infty}^{\infty} e^{-x^2}dx$ $\cdot \int_{-\infty}^{\infty} e^{-y^2}\,dy$. Since x and y are independent, we convert the product of integrals into an iterated integral:

$$
J^2 = \int_{-\infty}^{\infty}\int_{-\infty}^{\infty} e^{-(x^2+y^2)}\,dx\,dy.
$$

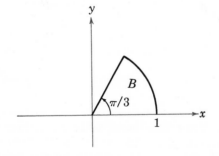

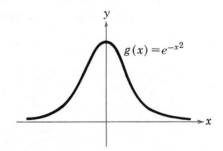

FIGURE 9.28   FIGURE 9.29

This is an integral over the entire xy-plane. Hence we may replace x, y by polar coordinates r, θ, with $0 \leq r < \infty$, $0 \leq \theta < 2\pi$. Carrying this out, we obtain

$$J^2 = \int_0^{2\pi} d\theta \cdot \int_0^{\infty} e^{-r^2}\, r\, dr = 2\pi \lim_{b \to \infty} \int_0^b e^{-r^2} r\, dr$$

$$= -\pi \lim_{b \to \infty} e^{-r^2} \Big]_0^b = \pi,$$

since $\lim_{b \to \infty} e^{-b^2} = 0$. Thus

$$J = \int_{-\infty}^{\infty} e^{-x^2}\, dx = \sqrt{\pi}. \quad \text{Done.}$$

9.2d Triple integrals

How do we compute $\iiint_B f\, dV$? It is not surprising, but nevertheless reassuring, that the approach used in Sec. 9.2b may be extended to triple integrals as well. We now compute some of these integrals as repeated integrals. The most difficult aspect of these problems is very likely the determination of the limits of integration. Unfortunately, there is no recipe. Only practice and experience, along with geometric visualization, can guide you.

We have already given an informal definition of $\iiint_B f\, dV$, in terms of limits of sums $\sum_i f(X_i)\, \Delta V_i$, and an interpretation as well: We spoke of the triple integral as the total mass of the solid B, provided that $f = f(x, y, z)$ was the density at the point (x, y, z). In case $f = 1$ identically, the triple integral yields the volume of B. Let us discuss volume first.

EXAMPLE: Let B be the solid (three-dimensional) ball of radius $a > 0$. Compute the volume of B (it depends on a, of course) as a repeated integral. See Fig. 9.30.

Solution: We know that the ball B is the set of all points (x, y, z) satisfying $\sqrt{x^2 + y^2 + z^2} \leq a$; that is,

$$x^2 + y^2 + z^2 \leq a^2.$$

Just as 2-space breaks up into four quadrants, 3-space breaks up into eight octants. It is clear that the volume of B is eight times the volume of that part B_+ in the first octant. B_+ is determined by the conditions

$$x^2 + y^2 + z^2 \leq a^2 \quad \text{and} \quad x \geq 0, y \geq 0, z \geq 0.$$

It is easier to draw and deal with B_+.

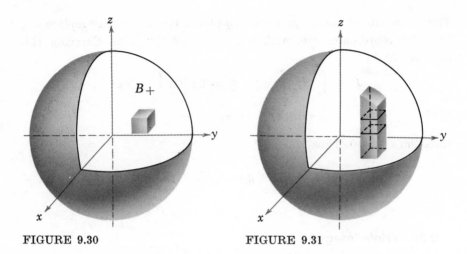

FIGURE 9.30   FIGURE 9.31

Let us compute the volume V_+ of B_+. You will recall that this volume may be approximated by stuffing B_+ with "bricks" or "boxes" whose edges are parallel to the coordinate axes and then adding up the volumes of all these boxes. Now we apply ordinary calculus to this approach.

Let (x, y, z) be one corner of a small box Δx by Δy by Δz inside B_+. We may construct a vertical column or stack of boxes, each having a base of the same dimensions, Δx by Δy. The height of such a column is, of course, obtained by adding the heights of the various boxes. See Fig. 9.31. The column extends from the xy-plane $z = 0$ up to the surface of the ball, given by $z = \sqrt{a^2 - x^2 - y^2}$. It follows that

$$\text{Height of column} = \int_0^{\sqrt{a^2-x^2-y^2}} dz,$$

and so

$$\text{Volume of column} = \left(\int_0^{\sqrt{a^2-x^2-y^2}} dz \right) \Delta x \, \Delta y.$$

Now we consider a slab parallel to the x-axis and constructed from vertical columns. The volume of such a slab is obtained by adding the volumes of the various columns. See Fig. 9.32. A typical slab has a base in the xy-plane extending from $x = 0$ to $x = \sqrt{a^2 - y^2}$. We suppose that the thickness of each column, and hence of the slab, is Δy. Adding the volumes of the columns in the manner of ordinary calculus gives

$$\text{Volume of slab} = \left(\int_0^{\sqrt{a^2-y^2}} \left(\int_0^{\sqrt{a^2-x^2-y^2}} dz \right) dx \right) \Delta y.$$

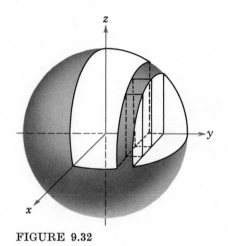

FIGURE 9.32

It should be clear what we do now. We obtain the volume V_+ of B_+ by adding up the volumes of the various slabs. These slabs extend from $y = 0$ to $y = a$, so that

$$V_+ = \int_0^a \left(\int_0^{\sqrt{a^2-y^2}} \left(\int_0^{\sqrt{a^2-x^2-y^2}} dz \right) dx \right) dy.$$

Thus we have expressed $V_+ = \iiint_{B_+} dV$ as a repeated triple integral. Note that we integrate first with respect to z, then x, then y. Note, further, that we could have carried things out in a different order. For example, we have also

$$V_+ = \int_0^a \left(\int_0^{\sqrt{a^2-z^2}} \left(\int_0^{\sqrt{a^2-y^2-z^2}} dx \right) dy \right) dz.$$

We are not finished yet. Now let us use ordinary calculus to get a formula for the volume of the ball. We discuss the integral a few lines above, integrating in the order z, x, y. The first integration yields

$$V_+ = \int_0^a \left(\int_0^{\sqrt{a^2-y^2}} \sqrt{a^2 - x^2 - y^2} \, dx \right) dy.$$

This inner integral in x is best attacked by making the trigonometric substitution

$$\rho^2 = a^2 - y^2, \qquad x = \rho \sin \theta, \qquad dx = \rho \cos \theta \, d\theta,$$

where we regard y, and hence ρ, as being fixed. You can compute that this inner integral in x gives $(\pi/4)\rho^2$, so that

$$V_+ = \frac{\pi}{4} \int_0^a \rho^2 \, dy;$$

that is,

$$V_+ = \frac{\pi}{4} \int_0^a (a^2 - y^2) \, dy.$$

This yields $V_+ = (\pi/6)a^3$. Since the volume of the ball is eight times V_+, we have the famous formula

$$\text{Volume of ball} = \frac{4}{3}\pi a^3.$$

Now we integrate a nonconstant function over a pyramid in $\mathbb{R}^3$.

EXAMPLE: Compute $J = \iiint_P (x - yz) \, dV$, where P is the pyramid bounded by the planes $x = 0$, $y = 0$, $z = 0$ and $x + y + z = 1$. See Fig. 9.33.

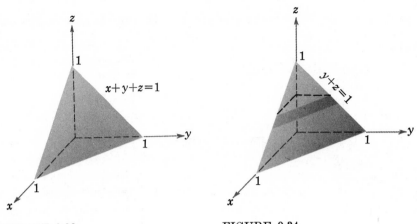

FIGURE 9.33   FIGURE 9.34

Solution: We compute J as an iterated triple integral:

$$J = \int_{z_0}^{z_1} \int_{y_0}^{y_1} \int_{x_0}^{x_1} (x - yz)\, dx\, dy\, dz.$$

Here the order of integration we have chosen is clearly first x, then y, then z, and the limits of integration x_0, x_1, y_1, and so on, must be found. To obtain x_0 and x_1, observe that a horizontal column built of small bricks of the Δx by Δy by Δz variety extends from $x = x_0 = 0$ to the slant face $x = x_1 = 1 - y - z$; that is,

$$x_0 = 0, \qquad x_1 = 1 - y - z.$$

Since we chose to integrate next with respect to y, we now build a horizontal slab from these columns as in Fig. 9.34. A typical slab extends from $y = y_0 = 0$ to $y = y_1 = 1 - z$. Finally, these horizontal slabs are piled from $z = z_0 = 0$ to $z = z_1 = 1$. Thus we conclude that

$$y_0 = 0, \qquad y_1 = 1 - z, \qquad z_0 = 0, \qquad z_1 = 1.$$

Now we compute

$$
\begin{aligned}
J &= \int_0^1 \int_0^{1-z} \int_0^{1-y-z} (x - yz)\, dx\, dy\, dz \\
&= \int_0^1 \int_0^{1-z} \left[\frac{x^2}{2} - xyz \, \Big|_{x=0}^{x=1-y-z} \right] dy\, dz \\
&= \frac{1}{2} \int_0^1 \int_0^{1-z} [(1 - y - z)^2 - 2(1 - y - z)yz]\, dy\, dz \\
&= \frac{1}{6} \int_0^1 \left[-(1 - y - z)^3 - 6 \Big(\frac{1}{2}(1 - z)y^2 z - \frac{1}{3} y^3 z \Big) \right] \Big|_{y=0}^{y=1-z} dz \\
&= \frac{1}{6} \int_0^1 (1 - z)^4\, dz = \frac{1}{30}.
\end{aligned}
$$

We conclude our discussion of triple integrals with a very useful observation.

EXAMPLE: Compute $\iiint_B x^2 y^2 z \, dV$, where B is the ball of radius a centered at the origin in $\mathbb{R}^3$.

Solution: The integral has the value 0. We see this without any computation. For the domain B is symmetric about the plane $z = 0$. Moreover, $f(x, y, z) = x^2 y^2 z$ is *odd* with respect to the variable z:

$$f(x, y, -z) = -f(x, y, z).$$

If we denote the northern hemisphere of the ball (points with $z \geq 0$) by B_1 and the southern hemisphere by B_2, then

$$\iiint_B x^2 y^2 z \, dV = \iiint_{B_1} x^2 y^2 z \, dV + \iiint_{B_2} x^2 y^2 z \, dV$$

$$= \iiint_{B_1} x^2 y^2 z \, dV - \iiint_{B_1} x^2 y^2 z \, dV = 0.$$

No need to perform any integrations. The moral here is "Look for symmetries and odd functions. Don't compute blindly."

Is it clear to you now that if B is the ball of unit radius in $\mathbb{R}^3$, then

$$\iiint_B (3 + (x^2 + y^2 + z^2) \sin z) \, dV = 4\pi?$$

Exercises

1. For each of the following, sketch the domain of integration and evaluate the integral:

   (a) $\int_3^4 \left(\int_{-1}^2 (x + x^2 y - 1) \, dx \right) dy,$

   (c) $\int_0^2 \left(\int_0^{\sqrt{4-x^2}} 6xy^2 \, dy \right) dx,$

   (b) $\int_0^1 \left(\int_y^{2y} e^{x-y} \, dx \right) dy,$

   (d) $\int_0^{2\pi} \left(\int_{-\sin x}^x y^2 \, dy \right) dx.$

2. For each of the following regions B, use the methods in Sec. 9.1 to estimate, roughly,

   $$J = \iint_B xy \, dA.$$

   Then write it as a repeated integral in two different ways, as

   $$\int \left(\int xy \, dy \right) dx \quad \text{and as} \quad \int \left(\int xy \, dx \right) dy,$$

   and evaluate *both* repeated integrals:

   (a) B is the rectangle with vertices at $(1, 1)$, $(1, 5)$, $(3, 1)$, and $(3, 5)$.

   (b) B is the triangle with vertices at $(1, 1)$, $(3, 1)$, and $(3, 5)$.

   (c) B is the curvilinear triangle enclosed by the lines $x = 1$, $y = 8$, and the curve $y = x^3$.

   (d) B is the region enclosed by the curves $y = x^2$ and $y = \sqrt{x}$.

(e) B is the triangle with vertices at $(-1, 1)$, $(0, 0)$, and $(2, 1)$.

(f) B is the region enclosed by the curves $y = x^2$ and $y = 2 - x^2$.

(g) B is the parallelogram with vertices at $(0, 0)$, $(2, 1)$, $(1, 2)$, and $(3, 3)$. Don't evaluate this one. It's too messy.

3. Evaluate

$$\iint_B \sin \pi(2x + y) \, dA,$$

where B is the triangle bounded by the lines $x = 1$, $y = 2$, and $x - y = 5$.

4. Determine the limits of integration but do *not* evaluate

$$\iint_B \sqrt{2x + y} \, dA,$$

where B is the region defined by:

(a) $x + y \geq 1$, $x + y \leq 2$, $x \geq 0$, $x \leq 1$;

(b) $x + y \geq 1$, $x + y \leq 2$, $x \geq 0$, $y \geq 0$.

5. Evaluate:

(a) $\int_1^2 \left(\int_{-1}^1 \left(\int_{-2}^1 2xy + z^2 \, dz \right) dx \right) dy,$ (c) $\int_0^1 \left(\int_0^{\sqrt{1-y^2}} \left(\int_{x^2+y^2}^{2-x^2-y^2} y \, dz \right) dx \right) dy,$

(b) $\int_1^2 \left(\int_x^2 \left(\int_{x-y}^{x+y} 4x + 3dz \right) dy \right) dx,$ (d) $\int_1^2 \left(\int_1^z \left(\int_{z+y}^{z+2y} 2z - 6 \, dx \right) dy \right) dz.$

6. For each of the following regions B, use the methods in Sec. 9.1 to estimate, roughly,

$$J = \iiint_B 6z \, dV.$$

Then write this as some repeated integral and evaluate if:

(a) B is the parallelepiped $0 \leq x \leq 1$, $-1 \leq y \leq 3$, $1 \leq z \leq 2$.

(b) B is the pyramid bounded by the planes $x = 0$, $y = 0$, $z = 0$ and $2x + y + z = 2$.

(c) B is the solid enclosed by the paraboloid $y = x^2 + z^2$ and the plane $y = 4$.

(d) B is the ball $x^2 + y^2 + z^2 \leq a^2$.

(e) B is the solid between the paraboloids $z = x^2 + y^2$ and $z = 2 - x^2 - y^2$.

(f) B is the region bounded by the cone $x^2 = y^2 + z^2$ and between the planes $x = 1$ and $x = 3$.

7. Use polar coordinates to evaluate the following integral:

(a) $\iint_B 8(1 - x^2 - y^2)^7 \, dA,$ B is the disk $x^2 + y^2 \leq 1$;

(b) $\iint_B \dfrac{x}{\sqrt{x^2 + y^2}} \, dA,$ B is the ring $1 \leq x^2 + y^2 \leq 4$;

(c) $\iint_B xy(x^2 + y^2)^{7/2} \, dA,$ B is the half-disk $x^2 + y^2 \leq 9$, $y \geq 0$;

(d) $\iint_B \dfrac{y}{x^2 + y^2} \, dA,$ B is the portion of the ring $1 \leq x^2 + y^2 \leq 4$ in the second quadrant (where $y \geq 0$, $x \leq 0$);

(e) $\iint_B \exp(2x^2 + 2y^2) \, dA,$ B is the region bounded by $x^2 + y^2 = 4$, $x^2 + y^2 = 25$, $y = x$, $x = 0$, $y > 0$;

(f) $\iint_B \sin(x^2 + y^2) \, dA,$ B is the disk $x^2 + y^2 \leq 4$.

8. Evaluate $\iint_B x^y \, dA$, where B is the square $0 \leq x \leq 1$, $1 \leq y \leq 2$.

9. Evaluate $\iint_B x^2 y^2 \, dA$, where B is the disk $x^2 + y^2 \leq 1$.

10. Evaluate $\iint_B \sqrt{25 - x^2 - y^2} \, dA$, where B is the portion of the disk $x^2 + y^2 \leq 25$ in the right half-plane $x \geq 0$.

11. Evaluate by first interchanging the order of integration:

    (a) $\int_0^1 \left(\int_y^1 y e^{x^3} \, dx \right) dy$,

    (b) $\int_0^{\pi/2} \left(\int_x^{\pi/2} \frac{\sin y}{y} \, dy \right) dx$.

12. In computing the mass M of a plate of density f, it was found that

$$M = \int_{-3}^0 \left(\int_0^{3-2x-x^2} f(x, y) \, dy \right) dx + \int_{-1}^3 \left(\int_0^{3-y} f(x, y) \, dx \right) dy.$$

Sketch the domain of integration.

13. Find the volume of the ellipsoid $x^2/a^2 + y^2/b^2 + z^2/c^2 \leq 1$.

14. If B is the rectangle $a \leq x \leq b$, $c \leq y \leq d$, show that

$$\iint_B f(x) g(y) \, dA = \left(\int_a^b f(x) \, dx \right) \left(\int_c^d g(y) \, dy \right).$$

15. (a) By interchanging the order of integration, show that if $x \geq 0$,

$$\int_0^x \left(\int_0^s f(t) \, dt \right) ds = \int_0^x (x - t) f(t) \, dt.$$

    (b) $\int_0^x \left(\int_0^r \left(\int_0^s f(t) \, dt \right) ds \right) dr = ?$, where $x \geq 0$.

16. (a) Let B be a rectangle with vertices at $P = (a, c)$, $Q = (b, c)$, $R = (b, d)$, and $S = (a, d)$. If f is twice continuously differentiable, show that

$$\iint_B f_{xy}(x, y) \, dA = f(P) + f(R) - f(Q) - f(S).$$

    (b) Use this to again evaluate the result in Exercise 2(a).

    (c) If $u(x, y)$ satisfies $u_{xy}(x, y) = 0$ for $0 < y < x$, $u(x, x) = 0$, and $u(x, 0) = x \sin x$, find $u(x, y)$ for all points (x, y) in the wedge $0 < y < x$. See Fig. 9.35.

17. Find the volume of the ball $x_1^2 + x_2^2 + x_3^2 + x_4^2 \leq a^2$ of radius a in $\mathbb{R}^4$. Generalize.

18. Let B be the rectangle $a \leq x \leq b$, $c \leq y \leq d$. Evaluate

$$\iint_B f_x(x, y) - g_y(x, y) \, dA,$$

where f and g are given continuously differentiable functions. Note that the result is, in some sense, an integral around the boundary of B.

19. Let $f(x, y)$ be a given continuous function (perhaps the height or temperature)

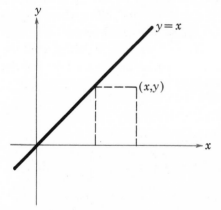

FIGURE 9.35

defined on a region B in $\mathbb{R}^2$. Then its *average value*, or *mean value*, is defined to be

$$f_{av} = \frac{1}{\text{Area }(B)} \iint_B f \, dA.$$

If B is the rectangle $-2 \leq x \leq 2$, $0 \leq y \leq 3$, compute the average values of

(a) $f(x, y) = 3x^2y$,   (b) $f(x, y) = y \sin x$.

20.   Jill waits on a certain corner for a bus every day. There are two different bus lines that stop at this corner, each going to her destination (the dentist). Each line runs buses every 10 min but beginning at varying times every day. What is the *average* time Jill must wait before a bus comes?

Say that service on the route is improved so that three bus lines stop there, each on 10 min schedules. What is Jill's average waiting time?

9.3 THEORY OF THE INTEGRAL

9.3a Introduction

It has come to this: We have an intuitive understanding of integrals that is sufficient to estimate them (Sec. 9.1), and we can use repeated integration to evaluate many simple integrals exactly (Sec. 9.2). But we do not have a precise definition of the integral, nor do we have precise proofs of the assertions made so far in this chapter. In fact, filling these gaps is an ambitious task that is reserved to a more advanced course.

Our intention here is more humble. We just want to mention what the problem is, why it is difficult, and to state some results. As usual, we restrict our attention mainly to double integrals.

9.3b Properties of the integral

The following properties of the integral are basic:

1. Additivity: $\iint_B (f + g) \, dA = \iint_B f \, dA + \iint_B g \, dA;$

2. Homogeneity: $\iint_B cf \, dA = c \iint_B f \, dA$, c a constant;

3. Positivity: $f \geq 0$ in $B \Rightarrow \iint_B f \, dA \geq 0;$

4. Normalization: $\iint_R dA = $ area of R,   where R is a rectangle in the plane.

In all these formulas, we naturally assume that the given integrals exist. They should be very plausible in view of the interpretation of the integral

$$\iint_B f \, dA$$

as the volume of a solid with base B and height $z = f(x, y)$.

We cannot prove these, since any such proof must be founded on a precise definition of the integral, which we do not give here. But there is another attitude one can take. One can feel that any concept of integration in $\mathbb{R}^2$ must have the four properties above. From this viewpoint, these four properties become the *axioms for the integral*. It turns out that these few axioms are all that are needed for the integral; no additional axioms are required. In other words, *if one uses only Properties 1 to 4, then one can deduce all other properties of the integral.*

It is instructive to deduce some other properties; they are both useful and intuitive. In order not to repeat ourselves, we assume once and for all that all the integrals we write do exist (as finite numbers).

CLAIM : $f \geq g$ in $B \Rightarrow \iint_B f \, dA \geq \iint_B g \, dA.$

Proof: By Properties 3, 1, and 2 (in this order), $f \geq g$ implies that $f - g \geq 0$, which implies that

$$\iint_B (f - g) \, dA \geq 0,$$

that is, $$\iint_B f \, dA - \iint_B g \, dA \geq 0.$$

CLAIM : $\left| \iint_B f \, dA \right| \leq \iint_B |f| \, dA.$

Proof: Note that $-|f| \leq f \leq |f|$. Therefore, by the first claim

$$-\iint_B |f| \, dA \leq \iint_B f \, dA \leq \iint_B |f| \, dA.$$

This is just another way of writing the desired conclusion.

CLAIM : If $|f| \leq M$ in a rectangle R, then

$$\iint_R |f| \, dA \leq MA,$$

where A is the area of R.

Proof: By the first claim, since $|f| \leq M$ in R, we have

$$\iint_R |f| \, dA \leq \iint_R M \, dA = M \iint_R dA = MA.$$

Note that Property 4 was needed to prove this.

There is one further result from advanced calculus that we quote, since it is often needed. It is a theorem about single integrals.

THEOREM: Let $f(x, y)$ be a differentiable function of x, with $f(x, y)$ and $f_x(x, y)$ continuous for all x, y. If

$$\varphi(x) = \int_a^b f(x, y) \, dy,$$

then we can compute the derivative of φ as

$$\varphi'(x) = \int_a^b f_x(x, y) \, dy.$$

More generally, if a and b depend on x also, and are differentiable, then

$$\varphi'(x) = \int_{a(x)}^{b(x)} f_x(x, y) \, dy + f(x, b(x))b'(x) - f(x, a(x))a'(x).$$

EXAMPLE: Let

$$\varphi(x) = \int_0^{1+x^2} \frac{e^{xy}}{1 + y^2} \, dy.$$

Then $$\varphi'(x) = \int_0^{1+x^2} \frac{ye^{xy}}{1 + y^2} \, dy + \frac{e^{x+x^3}}{1 + (1 + x^2)^2} \, 2x.$$

9.3c   Existence of the integral: the issues

Let us begin with the special case where the integrand is identically equal to 1. If the integral

$$\iint_B dA$$

exists, we agree to call the resulting number the area of B. Actually, one works in the opposite direction and say that if a set B has a well-defined number that is its "area," then the integral above exists and equals the area of B.

PROBLEM: Under what conditions does the area of a set $B \subset \mathbb{R}^2$ exist?

This problem is so subtle that most people would answer immediately that every set has area. But they just have no idea of how bad a set can be. Let Q denote the unit square $0 \leq x \leq 1$, $0 \leq y \leq 1$. Let B denote the set of all points (x, y) in Q, where x and y are both ratio-

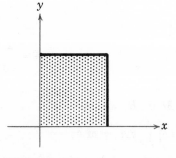

FIGURE 9.36

nal numbers. See Fig. 9.36. Thus we "throw away" an infinite number of points from Q. What is the area of B? One can even construct wilder sets. It should now be clear that the idea of the area of a set is not at all so simple as one might naïvely believe.

Step 1. Now in Sec. 9.1 we gave a numerical procedure, using appropriate Riemann sums, that approximates the area of most reasonable sets. Thus, we agree to say that a set B has area if the given numerical procedure *converges* to some number α as one takes partitions of B into smaller and smaller rectangles. Then we *define* the area of B to be the resulting number α. (It turns out that, using this definition of area, which is adequate for many purposes, the area of the set B defined above does *not* exist; the partition process does not converge. It does exist, however, and equals zero, if one uses a more sophisticated definition of area: Lebesgue area.)

Area exists for this set

FIGURE 9.37

Step 2. Having informally defined the concept of area, we ask for a workable criterion to determine if the area of a set exists. An example of such a criterion is: *The area of a bounded set exists if its boundary is a finite collection of differentiable curves.* See Fig. 9.37.

Step 3. Next, one extends these ideas to

$$J = \iint_B f \, dA.$$

Again one agrees to say that this integral exists if the numerical procedure in Sec. 9.1—approximate Riemann sums—converges to some number α as one takes a partition of B into smaller and smaller rectangles; and we define the value of the integral to be this number α.

Step 4. With this definition, a workable *criterion for the integral to exist is that the set B be bounded, that its boundary ∂B consist of a finite collection of differentiable curves, and that f be continuous on the union $B \cup \partial B$.*

Step 5. Finally, one justifies that the techniques of integration presented in Sec. 9.2 (repeated integrals) give the same values to the integral that one obtains by using the definition, that is, by taking the limiting value of the Riemann sum approximations found by using the methods in Sec.

9.1. Whenever one has two different methods for computing something, one must prove that both methods always give the same result; for if the methods give different results, then. . . .

Exercises

1. Use only the four properties of the integral stated in Sec. 9.3b to evaluate

$$J = \iint_B 3 \, dA,$$

where B is the rectangle $-1 \le x \le 5$, $2 \le y \le 4$.

2. Let $f(X)$ be a function equal to 2 on the rectangle $-1 \le x \le 1$, $0 \le y \le 3$, equal to -1 on the rectangle $2 \le x \le 3$, $-4 \le y \le -2$, and equal to zero elsewhere. Use only the four properties of the integral in Sec. 9.3b to evaluate

$$\iint f \, dA.$$

3. Let B be the rectangle $|x| \le 2$, $|y| \le 1$ in $\mathbb{R}^2$. Show that

$$\iint_B \frac{x^2 + y^2}{2 + \sin (xy)} \, dA \le 40.$$

4. (a) If $f(t) = \int_1^4 e^{tx^2} \, dx$, find $f'(t)$.

(b) If $g(t) = \int_t^{t^2} e^{tx^2} \, dx$, find $g'(t)$.

(c) If $u(x) = \int_0^x (x - t)^2 \sin t \, dt$, find u' and u''.

(d) If $u(x) = \int_0^x (x - t)e^{2(x-t)}f(t) \, dt$, where f is continuous, show that $u'' - 4u' + 4u = f$.

(e) If $h(s) = \int_0^1 \int_0^s f(x, y) \, dx \, dy$, where f is continuous, find h'.

5. If $v(x, t) = \int_{2x-t}^{2x+t} g(s) \, ds$, where g is a given continuous function, find v_x and v_t. Further, assuming that g is differentiable, show that $v_{xx} = 4v_{tt}$.

6. Let $w(x, y) = \int_0^y f(x + s - y, s) \, ds$, where $f(r, s)$ is a given differentiable function. Show that $w_x + w_y = f$.

7. If $u(x, y) = \int_0^1 (x + y^2 + t)^{15} \sin t \, dt$, show that $u_y = 2yu_x$.

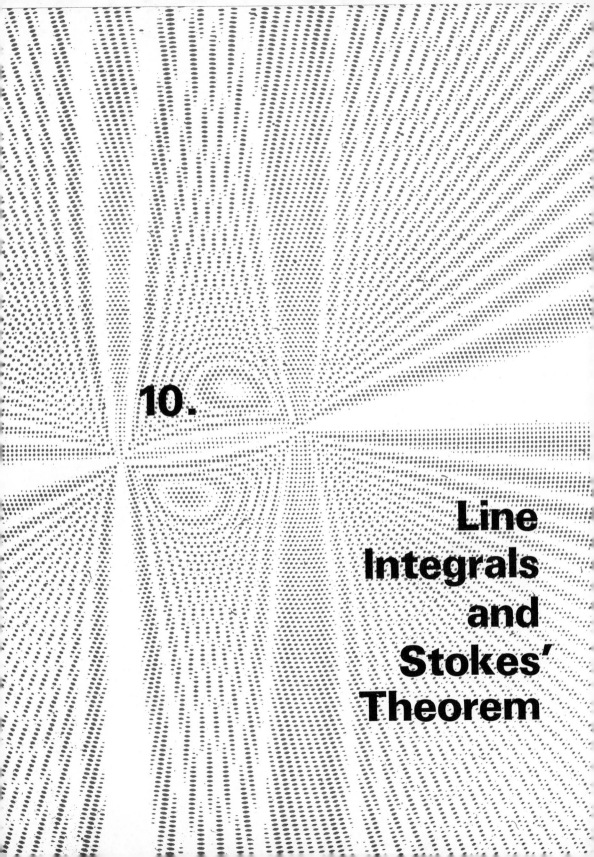

10.

Line Integrals and Stokes' Theorem

10.0 INTRODUCTION

Let C be a smooth curve in the plane, and let $p(x, y)$ and $q(x, y)$ be given scalar-valued functions. See Fig. 10.1. We want to define the line integral

$$\int_C [p(x, y)\, dx + q(x, y)\, dy]$$

along the curve C. Line integrals arise often in physics and engineering as well as in mathematics. In the special case where $p(x, y) = p(x)$, $q(x, y) = 0$, and C is the segment $a \le x \le b$ on the x-axis, the line integral reduces to the ordinary integral

$$\int_a^b p(x)\, dx$$

of elementary calculus.

One of the main results of this chapter is a beautiful and useful generalization of a version of the Fundamental Theorem of Calculus:

$$\int_a^b f'(x)\, dx = f(b) - f(a).$$

To describe it, let $F(x, y) = (p(x, y), q(x, y))$ be a given vectorfield on $\mathbb{R}^2$, and let $B \subset \mathbb{R}^2$ be a given region with boundary ∂B, a curve or curves. See Fig. 10.2. The generalization mentioned above states that

$$\iint_B \left(\frac{\partial q}{\partial x} - \frac{\partial p}{\partial y}\right) dA = \int_{\partial B} [p\, dx + q\, dy],$$

where the integral on the right is a line integral around the boundary of B.

In order to make the analogy with the one-variable case more plausible, we rewrite the one-variable case. Let $B = \{a \le x \le b\}$, and observe that

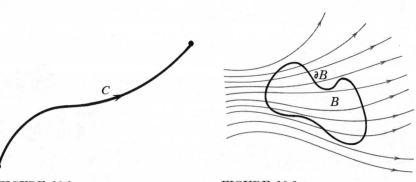

FIGURE 10.1   FIGURE 10.2

380

the boundary ∂B consists of the two points $x = a$, $x = b$. With this notation, the one-variable case reads

$$\int_B \frac{df}{dx}\, dx = f|_{\partial B},$$

where we agree to the convention $f|_{\partial B} = f(b) - f(a)$; that is, we attach a negative sign to the left boundary point $x = a$.

There are several useful forms of the generalized "fundamental theorem," attributed to Gauss, Green, Ostrogradsky, and Stokes. Still another version is called the Divergence Theorem. It has become customary to lump them collectively under the name Stokes' Theorem, although the attribution to Stokes violates historically valid claims of prior discovery.

When discussing a new topic, it is useful to make certain assumptions that avoid technicalities and allow one to concentrate on the new ideas at hand. In this spirit, we assume that all functions and vectorfields have infinitely many continuous derivatives. Such functions and vectorfields are called *smooth* (take care: some authors have a slightly different definition of "smooth"). If the reader so desires, he can examine the proofs for himself and determine how much differentiability was actually used; it turns out that three continuous derivatives always suffice for us.

Similarly, we recall the definition of a *smooth* (parametrized) *curve*. A curve C in the plane is a map $\alpha : \mathscr{D} \to \mathbb{R}^2$, where $\mathscr{D} \subset \mathbb{R}$ is an interval, say $\mathscr{D} = \{a \leq t \leq b\}$. We think of α as describing the position of a particle at time t, $C : \alpha(t) = (\phi(t), \psi(t))$, and so the coordinates describing the motion are $x = \phi(t)$, $y = \psi(t)$. The curve C is *smooth* if (1) the functions ϕ, ψ are smooth (infinitely differentiable) and (2) the corresponding velocity vector is never zero: $\alpha'(t) = (\phi'(t), \psi'(t)) \neq O$. In the absence of this last and perhaps peculiar requirement, a curve with sharp corners could be smooth. One can see this by imagining a particle traversing the curve; in order to have a continuous velocity and also get around the corner, the particle would have to stop at the corner momentarily (see Example 2 in Sec. 7.1e). Without mentioning it repeatedly, we assume that our parametrizations are smooth. We also assume that the interval $a \leq t \leq b$ in a parametrization is bounded. The upshot of the last several paragraphs is that we intend to avoid unessential difficulties so that we can concentrate on the business at hand.

One more concept: that of a *closed curve*. A curve $C : \alpha(t)$, $a \leq t \leq b$, is *closed* if $\alpha(a) = \alpha(b)$, that is, if the initial and terminal points of the curve coincide. In coordinates, $\alpha(t) = (\phi(t), \psi(t))$, this reads $\phi(a) = \phi(b)$ and $\psi(a) = \psi(b)$.

10.1 LINE INTEGRALS

10.1a The definition

Let $F(x, y) = (p(x, y), q(x, y))$ be a smooth vectorfield, and let C be a smooth curve defined by the equations

$$x = \phi(t), \qquad y = \psi(t), \qquad a \leq t \leq b.$$

We define the line integral of $F = (p, q)$ along C, written

$$\int_C p\, dx + q\, dy,$$

by

$$\int_C p\, dx + q\, dy = \int_a^b \left[p(\phi(t), \psi(t)) \frac{d\phi}{dt} + q(\phi(t), \psi(t)) \frac{d\psi}{dt} \right] dt,$$

which is often abbreviated

$$\int_C p\, dx + q\, dy = \int_a^b \left(p \frac{dx}{dt} + q \frac{dy}{dt} \right) dt.$$

Note that the integral on the right is an ordinary single integral. As a *mnemonic* device, one can think of the curve C as $X = \alpha(t)$, where $X = (x, y)$, $\alpha(t) = (\phi(t), \psi(t))$, and write, symbolically, $dX = (dx, dy)$. Then we have (using a dot for the inner product of vectors; that is, we write $X \cdot Y$ instead of $\langle X, Y \rangle$)

$$p\, dx + q\, dy = F(X) \cdot dX, \qquad \frac{d\alpha(t)}{dt} = \alpha'(t) = (\phi'(t), \psi'(t))$$

and the definition of the line integral reads

(A) $$\int_C F(X) \cdot dX = \int_a^b F(\alpha(t)) \cdot \alpha'(t)\, dt.$$

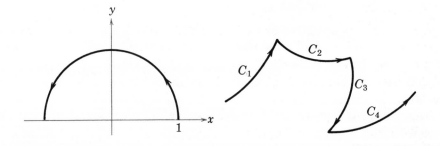

FIGURE 10.3   FIGURE 10.4

EXAMPLE 1: Let C be the curve $\alpha(t) = (\cos 2t, \sin 2t)$, $0 \leq t \leq \pi/2$. See Fig. 10.3. (The image of α is the upper half-unit circle traversed counterclockwise.) We evaluate

$$\int_C y\, dx - x\, dy.$$

Now $x = \cos 2t$, $y = \sin 2t$ on C. Thus, by definition,

$$\int_C y\, dx - x\, dy = \int_0^{\pi/2} [(\sin 2t)(-2\sin 2t) - (\cos 2t)(2\cos 2t)]\, dt$$

$$= -2\int_0^{\pi/2} [\sin^2 2t + \cos^2 2t]\, dt$$

$$= -2\int_0^{\pi/2} dt = -\pi.$$

There is an obvious way to extend the definition of the line integral to the useful case where the curve C is only *piecewise smooth*, that is, where C consists of a finite number of smooth curves. See Fig. 10.4. Thus, if $C = C_1 + C_2$, in obvious notation, then we define

$$\int_C F(X)\cdot dX = \int_{C_1} F(X)\cdot dX + \int_{C_2} F(X)\cdot dX.$$

In words, each term picks up the integration where the other left off. Incidentally, the reason for our increased generality to consider piecewise smooth curves is not obscure. We merely want to include triangles and squares.

EXAMPLE 2: Let $C = C_1 + C_2$, where C_1 is defined by $\alpha_1(t) = (2, 3-t)$, $0 \leq t \leq 2$, and C_2 by $\alpha_2(t) = (4-t, 1)$, $2 \leq t \leq 3$, and so C is composed of two straight-line segments. See Fig. 10.5. Then, because

$$\int_C = \int_{C_1} + \int_{C_2},$$

we evaluate

$$\int_{C_1} x^2\, dx - xy\, dy = \int_0^2 2(3-t)\, dt = 8,$$

and

$$\int_{C_2} x^2\, dx - xy\, dy = \int_2^3 (4-t)^2(-dt) = -\frac{7}{3}.$$

Consequently,

$$\int_C x^2\, dx - xy\, dy = 8 - \frac{7}{3} = \frac{17}{3}.$$

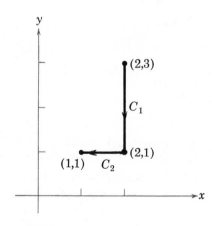

FIGURE 10.5

10.1b Properties of line integrals

The first and most basic property of a line integral is its linearity; that is,

$$\int_C [F(X) + G(X)] \cdot dX = \int_C F(X) \cdot dX + \int_C G(X) \cdot dX,$$

and
$$\int_C kF(X) \cdot dX = k \int_C F(X) \cdot dX,$$

where F and G are smooth vectorfields, k is a constant, and C is a smooth or piecewise smooth curve. A proof of these follows immediately from the definition.

The next property is an inequality. Assume that $\| F(X) \| \leq M$ for all X on the curve C. We claim that

$$\left| \int_C F(X) \cdot dX \right| \leq ML,$$

where L is the arc length of C. As one should expect, the proof begins with the definition. Given a curve C defined by $\alpha(t)$ for $a \leq t \leq b$, then

$$\left| \int_C F(X) \cdot dX \right| = \left| \int_a^b F(\alpha(t)) \cdot \alpha'(t) \, dt \right| \leq \int_a^b | F(\alpha(t)) \cdot \alpha'(t) | \, dt.$$

By the Schwarz inequality, however, we have

$$| F(\alpha(t)) \cdot \alpha'(t) | \leq \| F(\alpha(t)) \| \, \| \alpha'(t) \| \leq M \, \| \alpha'(t) \|.$$

Therefore,

$$\left| \int_C F(X) \cdot dX \right| \leq M \int_a^b \| \alpha'(t) \| \, dt = ML.$$

In order to state other properties of line integrals, we must attempt to cope with the fact that, although we privately think of curves in $\mathbb{R}^2$ as geometric point sets in $\mathbb{R}^2$, the official definition of a curve C is as a *parametrized* curve, that is, as a map $\alpha : \mathscr{D} \to \mathbb{R}^2$, where $\mathscr{D} = \{a \leq t \leq b\}$. Our immediate goal is to discuss different parametrizations of a curve and investigate their effect on line integrals.

Let $C : \alpha(t)$, $a \leq t \leq b$, be a smooth curve and let $t = h(\tau)$, $c \leq \tau \leq d$, where h is a smooth real-valued function with $h'(\tau) \neq 0$. Then the function $\hat{C} : \beta = \alpha \circ h$, that is, $\beta(\tau) = \alpha(h(\tau))$, is a *reparametrization* of C. For example, if $\alpha(t) = (1 + t, t^2)$, $-1 \leq t \leq 2$, and $t = h(\tau) = 2 - \tau$, then $\beta(\tau) = \alpha(h(\tau)) = (1 + 2 - \tau, (2 - \tau)^2) = (3 - \tau, 4 - 4\tau + \tau^2)$ is a reparametrization of C, where $0 \leq \tau \leq 3$ ($t = -1$ gives $\tau = 3$, and $t = 2$ gives $\tau = 0$). We say that the curves $C : \alpha(t)$ and $\hat{C} : \beta(\tau)$ are *equivalent* if there is a function h of the kind above such that $\beta = \alpha \circ h$.

The assumption $h' \neq 0$ implies that β is smooth. Indeed, by the Chain Rule

$$\frac{d\beta}{d\tau} = \alpha'(h(\tau))\frac{dh}{d\tau},$$

which is not zero, since $\alpha' \neq 0$ and $dh/d\tau \neq 0$.

One can think of β as describing the path of a particle traversing the same set of points in $\mathbb{R}^2$ as does α, but with different speeds—possibly going backward; that is, if $\alpha(a) = X_1$ and $\alpha(b) = X_2$, then β may begin at X_2 and go backward to X_1. See Fig. 10.6. (This was the case in the example above.) If $h'(\tau) > 0$, then, of course, $C: \alpha$ and $\hat{C}: \beta = \alpha \circ h$ have the *same orientation*—they both begin at $X_1 = \alpha(a) = \beta(c)$ and end at $X_2 = \alpha(b) = \beta(d)$, because as τ increases from c to d, then $t = h(\tau)$ also increases. On the other hand, if $h'(\tau) < 0$, then C and $\hat{C}$ have the *opposite orientation*, with $a = h(d)$ and $b = h(c)$. In particular, if $C: \alpha(t), a \leq t \leq b$, is a given curve, it is customary to denote the specific curve $\beta(\tau) = \alpha(a + b - \tau), a \leq \tau \leq b$, by $-C$, since it is the "same curve run backward."

As an example, consider the half-circle $C: \alpha(t) = (\cos t, \sin t), \ 0 \leq t \leq \pi$. See Fig. 10.7. Then $-C$ is the curve with parametrization

$$\beta(\tau) = (\cos(\pi - \tau), \sin(\pi - \tau))$$

$$= (-\cos \tau, \sin \tau), \qquad 0 \leq \tau \leq \pi.$$

Note here that $t = h(\tau) = \pi - \tau$ and $h'(\tau) = -1 < 0$.

Implicit in the discussion above is the assertion that any two equivalent curves $C: \alpha$ and $\hat{C}: \beta = \alpha \circ h$ have the same image in $\mathbb{R}^2$. To prove this, let $X_0 \in \mathbb{R}^2$ be in the image of $\hat{C}$. Then $X_0 = \beta(\tau_0)$ for some τ_0, and so

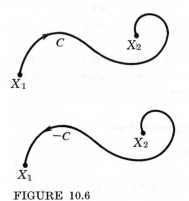

FIGURE 10.6

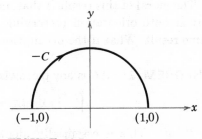

FIGURE 10.7

$X_0 = \beta(\tau_0) = \alpha(h(\tau_0)) = \alpha(t_0)$, where $t_0 = h(\tau_0)$. This shows that X_0 is also in the image of C. In the other direction, let X_0 be in the image of C, and so $X_0 = \alpha(t_0)$ for some t_0. The assumption $h'(\tau) \neq 0$ implies, by a harder theorem of calculus, that given t_0, there is a unique τ_0 such that $t_0 = h(\tau_0)$. Consequently $X_0 = \alpha(t_0) = \alpha(h(\tau_0)) = \beta(\tau_0)$, and so X_0 is also in the image of $\hat{C}$.

The following question now arises: How does the value of a line integral change if the parametrization of the curve is changed? We do not expect it to change very much, since, after all, any two parametrizations do have the same image in $\mathbb{R}^2$. Our expectation is essentially affirmed in

THEOREM 1: Let C and $\hat{C}$ be equivalent piecewise smooth curves with the same orientation. Then

$$\int_C F(X) \cdot dX = \int_{\hat{C}} F(X) \cdot dX.$$

Proof: Since any piecewise smooth curve can be decomposed into a finite number of smooth parts, we can assume that the given curve $C : \alpha(t)$, $a \leq t \leq b$, is smooth. Let $\hat{C} : \beta(\tau) = \alpha(h(\tau))$, $c \leq \tau \leq d$, be an orientation preserving reparametrization of C, so that $a = h(c)$ and $b = h(d)$. Now

$$\int_C F(X) \cdot dX = \int_a^b F(\alpha(t)) \cdot \alpha'(t) \, dt.$$

Make the change of variable $t = h(\tau)$ in the last integral. We find that

$$\int_C F(X) \cdot dX = \int_c^d F(\alpha(h(\tau))) \cdot \alpha'(h(\tau)) h'(\tau) \, d\tau$$

$$= \int_c^d F(\beta(\tau)) \cdot \beta'(\tau) \, d\tau$$

$$= \int_{\hat{C}} F(X) \cdot dX. \ll$$

The moral of this result is that, in evaluating a line integral, we can use any smooth orientation preserving parametrization, because all give the same result. What if the orientation is reversed?

THEOREM 2: If C is any piecewise smooth curve, then

$$\int_{-C} F(X) \cdot dX = -\int_C F(X) \cdot dX.$$

Proof: This is essentially identical to Theorem 1, except that now $t = h(\tau) = a + b - \tau$, $a \leq \tau \leq b$, and $-C$ is parametrized as $\beta = \alpha \circ h$.

The key change is in the limits of integration. Here $a = h(d)$ and $b = h(c)$. Thus

$$\int_C F(X) \cdot dX = \int_d^c F(\beta(\tau)) \cdot \beta'(\tau)\, d\tau$$

$$= -\int_c^d F(\beta(\tau)) \cdot \beta'(\tau)\, d\tau$$

$$= -\int_{-C} F(X) \cdot dX. \ll$$

FURTHER EXAMPLES: 3. The straight line segment from $(2, 3)$ to $(1, 1)$ can be parametrized by many curves. See Fig. 10.8. One such is $C : \alpha(t) = (2 - t, 3 - 2t)$, $0 \leq t \leq 1$ (note that the words *"from (2,3) to (1, 1)"* indicate the orientation). Then

$$\int_C x^2\, dx - xy\, dy = \int_0^1 [-(2 - t)^2 + 2(2 - t)(3 - 2t)]\, dt = 4.$$

By Theorem 1, we would have found the same result had we used any other parametrization of this segment.

4. Evaluate

$$\int_C (2y - 3)\, dx + x^2\, dy,$$

from $(-1, 1)$ to $(1, 1)$ along the parabola $y = x^2$. Here we have not given

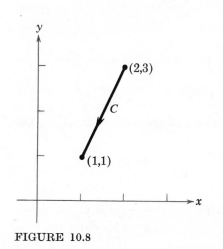

FIGURE 10.8

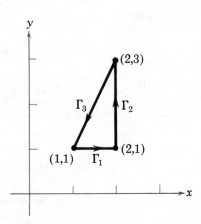

FIGURE 10.9

C as a parametrized curve; but the problem makes sense, since, by Theorem 1, we get the same result for any parametrized curve. One parametrization is $\alpha(r) = (r, r^2)$, $-1 \leq r \leq 1$ (for variety, we use r instead of t as the name of the parameter):

$$\int_C (2y - 3)\, dx + x^2\, dy = \int_{-1}^{1} [(2r^2 - 3) + 2r^3]\, dr = -\frac{14}{3}.$$

5. Let Γ denote the boundary of the triangle with vertices at $(1, 1)$, $(2, 1)$, and $(2, 3)$ traversed once counterclockwise. See Fig. 10.9. We compute

$$\int_\Gamma x^2\, dx - xy\, dy.$$

Again, this makes sense with any parametrization for Γ. Now $\Gamma = \Gamma_1 + \Gamma_2 + \Gamma_3$, where Γ_1 is the segment from $(1, 1)$ to $(2, 1)$, Γ_2 from $(2, 1)$ to $(2, 3)$, and Γ_3 from $(2, 3)$ to $(1, 1)$. Thus $\Gamma_1 = -C_2$ and $\Gamma_2 = -C_1$ in Example 2, and $\Gamma_3 = C$ in Example 3. Thus, by Theorem 2,

$$\int_\Gamma = \int_{\Gamma_1} + \int_{\Gamma_2} + \int_{\Gamma_3}$$

$$= \int_{-C_2} + \int_{-C_1} + \int_C = -\int_{C_2} - \int_{C_1} + \int_C$$

$$= \frac{7}{3} - 8 + 4 = -\frac{5}{3}.$$

This example illustrates the usefulness of our symbolism. It also shows that *the line integral between two points* $X_1 = (1, 1)$ *and* $X_2 = (2, 3)$ *may depend on the path taken* (but not on the parametrization of a particular path), since

$$-\frac{17}{3} = \int_{\Gamma_1 + \Gamma_2} x^2\, dx - xy\, dy \neq \int_{-\Gamma_3} x^2\, dx - xy\, dy = -4.$$

The two paths from X_1 to X_2 are $\Gamma_1 + \Gamma_2$ and $-\Gamma_3$.

6. Let C denote the straight line segment from $(2, 1)$ to $(1, 1)$. One convenient parametrization is $\alpha(s) = (-s, 1)$, $-2 \leq s \leq -1$. Thus

$$\int_C x^2\, dx - xy\, dy = \int_{-2}^{-1} -s^2\, ds = -\frac{7}{3}.$$

Of course, this should have been anticipated, because α is just a reparametrization of the curve C_2 in Example 2. Another method for the same problem is to observe that $-C$ is somewhat simpler to parametrize: $\beta(t) = (t, 1)$, $1 \le t \le 2$. Then we find that

$$\int_C = -\int_{-C} = -\int_1^2 t^2 \, dt = -\frac{7}{3}.$$

10.1c Work

If the vectorfield F is a force field so that $F(X)$ is the force "felt" at the point $X = (x, y)$, then the line integral

$$\int_C F(X) \cdot dX$$

is defined in physics to be the *work done by the force* in moving a unit mass along the curve C. If $C : \alpha(t)$, $a \le t \le b$, is the given curve, then

$$\text{work} = \int_C F(X) \cdot dX = \int_a^b F(\alpha(t)) \cdot \alpha'(t) \, dt.$$

But recall that the vector $\alpha'(t)$ is tangent to the curve. We conclude that only the component of the vectorfield F *tangent* to the curve influences the value of the line integral, and hence the work done by the force. In particular, if a force field is perpendicular to the path C of a moving particle, then this force does zero work.

For example, if the force field, like gravity, is vertical, so that $F = (0, q)$, then this force does no work in horizontal motion of the object. Work done by this force is nonzero only when moving an object up or down. It is friction, not gravity directly, that causes us to expend work in moving objects horizontally. On smooth ice, where sliding friction is quite small, it is very easy to move objects horizontally.

Since the velocity $\alpha'(t)$ appears in the formula for work, one might be led to believe that work depends on the velocity, and so if $\beta(\tau)$ describes the motion of another particle moving along the same path, then the work done by the force is different for the two particle motions $\alpha(t)$ and $\beta(\tau)$. Not true. *The work does not depend on the particular motions but only on the path.* More precisely, we are just reasserting Theorem 1 in a physical context. The statement that $\alpha(t)$ and $\beta(\tau)$ describe the motion of two particles along the same path is defined, in physics and mathematics, to mean that α and β are equivalent curves with the same orientation. We also can interpret Theorem 2 as stating that if a force F does work W in moving a particle along a curve C, then in running the particle backward, that is, along $-C$, the force does work $-W$.

Exercises

1. Compute $\int_C 2x\, dx + 6(x - y)\, dy$ for each of the following curves:

(a) C_1: $\alpha(t) = (t, t),\ 0 \le t \le 1$;

(b) C_2: $\beta(s) = (s, s^2),\ 0 \le s \le 1$;

(c) C_3: $\gamma(\tau) = (\sin \tau, \tau),\ 0 \le \tau \le 2\pi$;

(d) C_4: $\Phi(t) = (e^t, 1),\ 0 \le t \le 1$;

(e) C_5: $\alpha(\theta) = (1 - \cos \theta, \sin \theta),\ 0 \le \theta \le \pi/2$;

(f) C_6: $\psi(r) = (e^r, e^{-r}),\ -1 \le r \le 1$.

For the remainder, you will first have to write the set C as a parametrized curve:

(g) C_7: $y = 2x,\ 1 \le x \le 2$;

(h) C_8: the straight line segment beginning at $(-1, 1)$ and ending at $(2, -2)$;

(i) C_9: the two legs of the triangle with vertices at $(0, 0)$, $(1, 0)$, and $(1, 1)$, oriented to begin at $(0, 0)$ and end at $(1, 1)$ (suggestion: evaluate two separate line integrals);

(j) C_{10}: the two legs of the triangle with vertices at $(0, 0)$, $(0, 1)$, and $(1, 1)$, oriented to begin at $(0, 0)$ and end at $(1, 1)$;

(k) C_{11}: the parabola $x = y^2$, from $(1, -1)$ to $(1, 1)$;

(l) C_{12}: the larger arc of the circle $(x - 1)^2 + y^2 = 1$ from $(0, 0)$ to $(1, 1)$.

2. Compute $\int_C (6x - y^2)\, dx - 2xy\, dy$ for the curves C in Exercise 1(a), (b), (j). The result may surprise you. (Challenge: Find a reason.)

3. (a) Evaluate $\int_C y^2\, dx + 3x^2\, dy$ for each of the following curves:

(i) C_1 is a parametrization of the straight line segment from $(-1, 1)$ to $(2, -2)$,

(ii) C_2 is a parametrization of the straight line segment from $(2, -2)$ to $(-1, 1)$.

(b) How could Theorem 2 have helped you find the result in Exercise 3(a)(ii) after having done 3(a)(i)?

4. Let A and B be the answers to Exercises 1(a) and (j), respectively, and let C be the triangle with vertices at $(0, 0)$, $(0, 1)$, and $(1, 1)$, oriented so that it is traversed counterclockwise. Show that

$$\int_C 2x\, dx + 6(x - y)\, dy = A - B.$$

5. (a) Let C be a parametrization of the entire circle $(x - 1)^2 + y^2 = 1$ (this is a closed curve) oriented counterclockwise. Evaluate

$$J = \int_C 2x\, dx + 6(x - y)\, dy.$$

(b) Let E and L be the answers to Exercises 1(e) and (l), respectively. Express the number J in terms of E and L.

6. (a) Evaluate

$$\int_{C_1} e^x \sin y\, dx + e^x \cos y\, dy,$$

where C_1 is the portion of the circle $x^2 + y^2 = 1$ with $x \geq 0$, oriented so that it begins at $(0, -1)$ and ends at $(0, 1)$.

(b) Evaluate the same integral, but replace C_1 by the curve C_2, which is C_1 with its orientation reversed, so that it begins at $(0, 1)$ and ends at $(0, -1)$.

7.   Evaluate $\int_C (x^2 + y^2)\, dx + 2xy\, dy$, where C is the curve $x = 1 - t^2$, $y = 3 - 2t$, $1 \leq t \leq 2$.

8.   Let a force field be $F(x, y) = (xy, x^2 - 3y^2)$. Compute the work done by this force to move a particle of unit mass from $(0, 0)$ to $(1, 1)$ along each of the paths in Exercise 1(a), (b), (h), (i), (j).

9.   If a force field is $F(x, y) = (ye^{xy}, xe^{xy})$, find the work done by this force to move a particle of mass 1 counterclockwise around the boundary of the square $|x| \leq 1, |y| \leq 1$.

10.   Proof or counterexample. Here p and q are smooth functions.

(a) If C is a vertical line segment, then $\int_C p(x, y)\, dx = 0$.
(b) If C is a horizontal line segment, then $\int_C q(x, y)\, dy = 0$.
(c) If $p \geq 0$, $q \geq 0$, then for any smooth curve C,

$$\int_C p\, dx + q\, dy \geq 0.$$

11.   Let $F(X)$ be a smooth vectorfield and $\| F(X) \| \leq M \| X \|$ for all X. If C: $\alpha(t)$ is a smooth curve of length L, with $\| \alpha(t) \| \leq R$ for all t, show that

$$\left| \int_C F(X) \cdot dX \right| \leq MRL.$$

12.   Let $F(X)$ be a smooth force field in the plane. Say that a particle of mass m moves so that its position $X(t)$—here we use $X(t)$ instead of $X = \alpha(t)$—is determined by Newton's Second Law:

$$m\frac{d^2 X}{dt^2} = F(X)$$

Show that the work done by the force F for $t_1 \leq t \leq t_2$ is equal to the change in kinetic energy of the particle:

$$\int_{t_1}^{t_2} F \cdot X'\, dt = \frac{m}{2} \| X'(t_2) \|^2 - \frac{m}{2} \| X'(t_1) \|^2.$$

(Suggestion: First show that $\frac{1}{2}m\, (d/dt) \| X' \|^2 = F \cdot X'$.)

13.   The given definition (A) [on p. 382] of a line integral extends immediately to curves C in $\mathbb{R}^3$; just let C: $\alpha(t) = (\varphi_1(t), \varphi_2(t), \varphi_3(t))$ be a curve in $\mathbb{R}^3$. Evaluate

$$\int_C (1 + x)\, dx - xy\, dy + (y + z)\, dz,$$

where C is the helix $x = \cos \pi t$, $y = \sin \pi t$, $z = t$, $-1 \leq t \leq 2$.

14.   Evaluate the line integral in Exercise 13 if:

(a) C is the straight line from $(0, 0, 0)$ to $(1, 2, -3)$ (one parametrization of this line is $x = t$, $y = 2t$, $z = -3t$, $0 \leq t \leq 1$);
(b) C is the path consisting of three straight line segments: C_1 from $(0, 0, 0)$ to $(1, 0, 0)$, C_2 from $(1, 0, 0)$ to $(1, 2, 0)$, and C_3 from $(1, 2, 0)$ to $(1, 2, -3)$;

(c) C is the curve $x = t^2$, $y = 2t^2$, $z = -3t$ from $(0, 0, 0)$ to $(1, 2, -3)$;

(d) C is the closed circle $x^2 + z^2 = 1$, $y = 1$, oriented counterclockwise when viewed from the point $(0, 2, 0)$.

10.2 STOKES' THEOREM

10.2a Stokes' Theorem for a rectangle

The time has come to generalize the Fundamental Theorem of Calculus to multiple integrals. Here is the simplest case.

THEOREM 3r: Let R denote the rectangle $a < x < b, c < y < d$ and $F = (p, q)$ a smooth vectorfield. Then

$$\iint_R \left(\frac{\partial q}{\partial x} - \frac{\partial p}{\partial y} \right) dA = \int_{\partial R} p \, dx + q \, dy,$$

where the boundary ∂R is oriented counterclockwise.

Proof: We treat the p and q terms separately. Now

$$\iint_R \frac{\partial q}{\partial x} dA = \int_c^d \left(\int_a^b \frac{\partial q\,(x, y)}{\partial x} \, dx \right) dy = \int_c^d [q(b, y) - q(a, y)] \, dy.$$

In the notation of Fig. 10.10, however, we have

$$\int_c^d q(b, y) \, dy = \int_{C_2} q \, dy$$

and

$$-\int_c^d q(a, y) \, dy = -\int_{-C_4} q \, dy = \int_{C_4} q \, dy,$$

and since $y \equiv$ constant on C_1 and C_3,

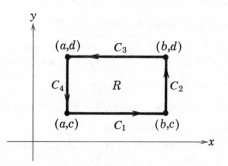

FIGURE 10.10

$$\int_{C_1} q\,dy = 0 \quad \text{and} \quad \int_{C_3} q\,dy = 0.$$

Therefore

$$\iint_R \frac{\partial q}{\partial x}\,dA = \int_{\partial R} q\,dy.$$

Similar considerations give

$$\iint_R \frac{\partial p}{\partial y}\,dA = -\int_{\partial R} p\,dx.$$

Adding these two equations, we arrive at the result. ≪

10.2b More general regions

Precisely the same result holds for more general regions than rectangles. We prove this in two steps. First is the case where a region $B \subset \mathbb{R}^2$ can be written both as

$$\varphi(x) \le y \le \psi(x), \qquad a \le x \le b,$$

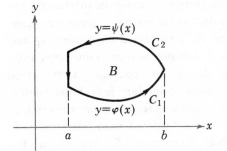

so that φ and ψ are the lower and upper boundary curves, respectively, and as

$$\Phi(y) \le x \le \Psi(y), \qquad c \le y \le d,$$

so that Φ and Ψ are the left and right boundary curves, respectively. See Fig. 10.11. We also suppose that the boundary curves are piecewise smooth; that is, the boundary ∂B consists of a finite number of smooth curves. Such regions are called *simple regions*.

THEOREM 3s: Let $B \subset \mathbb{R}^2$ be a simple region and $F = (p, q)$ a smooth vectorfield. Then

$$\iint_B \left(\frac{\partial q}{\partial x} - \frac{\partial p}{\partial y} \right) dA = \int_{\partial B} p\,dx + q\,dy,$$

where the boundary ∂B is oriented counterclockwise.

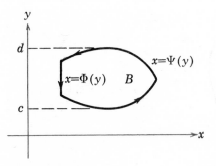

Proof: We discuss only the p term. The q term is treated similarly. Now

FIGURE 10.11

$$\iint_B \frac{\partial p}{\partial y}\, dA = \int_a^b \left(\int_{\varphi(x)}^{\psi(x)} \frac{\partial p\,(x, y)}{\partial y}\, dy \right) dx = \int_a^b [p(x, \psi(x)) - p(x, \varphi(x))]\, dx.$$

Let C_1 denote the bottom curve of B, and C_2 the top curve, oriented so that ∂B is oriented counterclockwise. Then

$$\int_a^b p(x, \psi(x))\, dx = \int_{-C_2} p\, dx, \qquad \int_a^b p(x, \varphi(x))\, dx = \int_{C_1} p\, dx.$$

Thus, as claimed,

$$\iint_B \frac{\partial p}{\partial y}\, dA = -\int_{C_1 + C_2} p\, dx = -\int_{\partial B} p\, dx$$

(observe that $\int p\, dx = 0$ over any vertical line segment). ≪

Finally, we extend this to more general regions, possibly containing holes. The idea is to dissect the region into simple regions and apply Theorem 3s to these simple regions separately. An example makes this clear. Let $B \subset \mathbb{R}^2$ be the indicated region. We have dissected B into a number of simple regions. See Fig. 10.12. If one applies Stokes' Theorem to each of these simple regions separately, then line integrals along the interior dotted lines appear. The integrals along the dotted lines cancel, however, since, for example, the line shared by B_1 and B_2 is traversed once going up and once going down. By Theorem 2, the net result is zero. Thus, only integration along ∂B remains, that along the "outer" boundary is traversed counterclockwise and that along the "inner" boundary is traversed clockwise. There is a mnemonic device that enables us to remember the orientation of the boundary: "If you walk along ∂B so that your left hand is in B, then you are walking in the 'positive' direction of ∂B." We call this the *positive orientation* of ∂B.

The discussion above has established

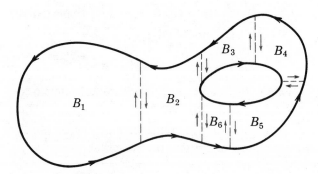

FIGURE 10.12

THEOREM 3: (Stokes' Theorem) If $B \subset \mathbb{R}^2$ is decomposable into a finite number of simple regions with piecewise smooth boundaries, then for any smooth vectorfield $F = (p, q)$,

$$\iint_B \left(\frac{\partial q}{\partial x} - \frac{\partial p}{\partial y} \right) dA = \int_{\partial B} p \, dx + q \, dy,$$

where ∂B has positive orientation.

EXAMPLES: 1. Let B denote the triangle with vertices at $(0, 0)$, $(1, 0)$, and $(0, 1)$, with positively oriented boundary. We evaluate

$$\int_{\partial B} (e^x + y - 2x) \, dx + (7x - \sin y) \, dy.$$

Let $p = e^y + y - 2x$ and $q = 7x - \sin y$. Then, by Stokes' Theorem,

$$\int_{\partial B} (e^x + y - 2x) \, dx + (7x - \sin y) \, dy = \iint_B 6 \, dA.$$

But

$$\iint_B dA = \text{area } (B) = \frac{1}{2}.$$

Thus

$$\int_{\partial B} (e^y + y - 2x) \, dx + (7x - \sin y) \, dy = 3.$$

The theorem greatly simplified the amount of computation.

2. Let B denote the disk of radius a centered at the origin and let $p(r) = p(x, y)$, $q(r) = q(x, y)$ be smooth functions that depend only on the distance r from the origin. Then

$$\iint_B \left(\frac{\partial q}{\partial x} - \frac{\partial p}{\partial y} \right) dA = \int_{\partial B} p \, dx + q \, dy.$$

But on $\partial B =$ circle of radius a, $p = p(a)$, $q = q(a)$ are constant. Therefore

$$\iint_B (q_x - p_y) \, dA = \int_{\partial B} p(a) \, dx + q(a) \, dy.$$

To evaluate the line integrals, we could introduce a parametrization and go through the easy computation. Instead, we are sneaky and again apply Stokes' Theorem, using the fact that $p(a)$ and $q(a)$ are constants. A moment's thought (do not write anything) reveals that the value of the integral is zero; that is, if p and q are constant on the circle

$$\iint_B (q_x - p_y) \, dA = 0.$$

10.2c Other versions

In practice, people often use several different variants of Stokes' Theorem. For the first variant we need the idea of the *outer normal* vector to the boundary of a region B. Let $X = \alpha(t) = (\varphi(t), \psi(t))$ be a parametrization of ∂B, oriented positively. Then $\alpha'(t) = (\varphi'(t), \psi'(t))$ is tangent to the boundary at the point $\alpha(t)$. Hence the vector

$$V = (\psi'(t), -\varphi'(t)),$$

which is perpendicular to the tangent vector, since $V \cdot \alpha' = 0$, is normal to the boundary. It remains to convince ourselves that V does point outward, away from B. Let us introduce an orthogonal coordinate system centered at the boundary point $\alpha(t)$. From Fig. 10.13, we see that the unit outer normal vector should play the role of $(1, 0)$, and the unit tangent the role of $(0, 1)$. Since the first coordinate $\psi'(t)$ of V agrees with the second coordinate of $\alpha'(t)$, we conclude that V, just as $(1, 0)$, points outward and is rightly called the outer normal. Consequently, the *unit* outer normal vector $N(t)$ is obtained from the relations (s is arc length)

$$V = (\psi'(t), -\varphi'(t)) = N(t)\frac{ds}{dt}.$$

For a simple example of outer normal, let B be the disk of radius 2 centered at the origin, so that ∂B is the circle of radius 2. See Fig. 10.14. At a point (x, y) on ∂B, the unit outer normal is given by $N = \frac{1}{2}(x, y) = (\cos t, \sin t)$, where ∂B is parametrized by $\alpha(t) = (2 \cos t, 2 \sin t)$. Note here that the unit tangent is $(-\sin t, \cos t) = \frac{1}{2}(-y, x)$.

Let us now state the divergence form of Stokes' Theorem.

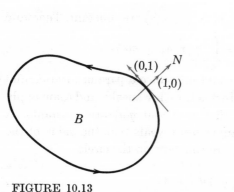

FIGURE 10.13   FIGURE 10.14

THEOREM 4: (Divergence.) If $F = (P, Q)$ is a smooth vectorfield and if $B \subset \mathbb{R}^2$ satisfies the assumptions of Theorem 3, then

$$\iint_B \text{div } F \, dA = \iint_B \left(\frac{\partial P}{\partial x} + \frac{\partial Q}{\partial y} \right) dA = \int_{\partial B} F \cdot N \, ds,$$

where N is the unit outer normal to ∂B and s is arc length on ∂B.

Proof: The first equality is the definition of the divergence:

$$\text{div } F = \frac{\partial P}{\partial x} + \frac{\partial Q}{\partial y}.$$

To prove the second equality, we let $q = P$, $p = -Q$ in Theorem 3. Then

$$\iint_B \left(\frac{\partial P}{\partial x} + \frac{\partial Q}{\partial y} \right) dA = \int_{\partial B} - Q \, dx + P \, dy.$$

Thus, if $\alpha(t) = (\varphi(t), \psi(t))$, $a \leq t \leq b$, is a parametrization of ∂B, we have

$$\iint_B \left(\frac{\partial P}{\partial x} + \frac{\partial Q}{\partial y} \right) dA = \int_a^b (-Q\varphi' + P\psi') \, dt = \int_a^b F \cdot N \, ds. \quad \text{Done.} \ll$$

This theorem has a nice physical interpretation. Think of $F(X)$ as the velocity vector of a fluid—a gas, say—moving in the plane, and assume unit density. Then div $F(X)$ measures how the gas is moving from the point X. Integrating this over the whole region B, we obtain the net change in the amount of fluid in B. But there is another way to measure

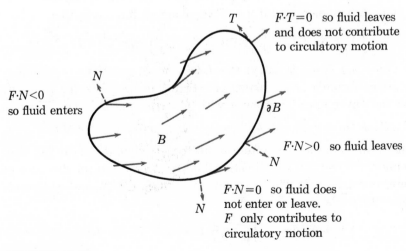

T $F \cdot T = 0$ so fluid leaves and does not contribute to circulatory motion

N

$F \cdot N < 0$ so fluid enters

∂B

B

$F \cdot N > 0$ so fluid leaves

N

N $F \cdot N = 0$ so fluid does not enter or leave. F only contributes to circulatory motion

FIGURE 10.15

the same quantity: merely stand by the walls of B and see how much leaves. See Fig. 10.15. In computing this, we, of course, need only the component of the velocity F perpendicular to the boundary (the component of the velocity tangent to the boundary affects only the circulatory, that is, whirlpool-like, motion of the fluid in B). Theorem 4 is just a statement of the equality of these two ways of measuring the change in the amount of fluid in B. If div $F = 0$, then the net fluid flow across any closed curve is zero. Thus we refer to a vectorfield satisfying div $F = 0$ as having no *sources or sinks* in B.

Perhaps a more familiar statement of the same idea is to let B denote a room in which a party is taking place and $F(X)$ the velocity of the person standing at X. Then one can measure the change in the number of people at the party either by (1) counting the people inside the room B or (2) standing at the doors $(= \partial B)$. The number from (1) is the left side of Theorem 5, and (2) the right side.

One can also see how the one-variable theorem

$$\int_a^b f'(x)\, dx = f(b) - f(a)$$

fits into this pattern. See Fig. 10.16. The minus sign in front of $f(a)$ arises because the outer normal to the interval $a \le x \le b$ is in the negative direction.

FIGURE 10.16

It is easy to prove two identities due to Green. *Green's First Identity* asserts that if B has a smooth boundary and if u and v are smooth functions, then

$$\iint_B v\, \Delta u\, dA = \int_{\partial B} v\, \nabla u \cdot N\, ds - \iint_B \nabla v \cdot \nabla u\, dA,$$

where $\Delta u = u_{xx} + u_{yy}$ is called the *Laplacian* of u and $\nabla v = \operatorname{grad} v = (v_x, v_y)$. This formula is an analog of integration by parts. To prove it, we use the Divergence Theorem 4 with $P = vu_x$ and $Q = vu_y$. Then

$$P_x + Q_y = vu_{xx} + v_x u_x + vu_{yy} + v_y u_y = v\, \Delta u + \nabla v \cdot \nabla u,$$

and the result follows. It is often useful to note that the term $\nabla u \cdot N$ in the boundary integral is the directional derivative of u in the direction of the outer normal to ∂B. This is sometimes written $\partial u/\partial N$, and so Green's first identity reads

$$\iint_B v\, \Delta u\, dA = \int_{\partial B} v\, \frac{\partial u}{\partial N}\, ds - \iint_B \nabla v \cdot \nabla u\, dA.$$

Green's Second Identity asserts that

$$\iint_B (v \, \Delta u - u \, \Delta v) \, dA = \int_{\partial B} \left(v \frac{\partial u}{\partial N} - u \frac{\partial v}{\partial N} \right) ds.$$

This is proved by writing Green's First Identity with the roles of u and v reversed and then subtracting the two equations.

10.2d Applications

Now let us see how Stokes' Theorem may be applied to some issues in mathematical physics.

Let $B \subset \mathbb{R}^2$ be a given region with smooth boundary, and let $u(x, y, t)$ denote the temperature at the point $(x, y) \in B$ at time t. In physics, it is shown that a simple model for heat flow requires that u satisfy the differential equation $u_t = \Delta u$; that is,

$$\frac{\partial u}{\partial t} = \frac{\partial^2 u}{\partial x^2} + \frac{\partial^2 u}{\partial y^2}.$$

To determine the temperature $u(x, y, t)$ for $t \geq 0$, it is physically plausible to require prior knowledge of both the initial temperature (at $t = 0$)

$$u(x, y, 0) = f(x, y), \qquad (x, y) \in B,$$

and the temperature on the boundary for all time $t \geq 0$

$$u(x, y, t) = \varphi(x, y, t), \qquad (x, y) \in \partial B, \qquad t \geq 0.$$

Although we do not prove that a solution does exist, it is easy to show that there is at most one solution—a uniqueness theorem. Thus, once one solution satisfying the initial and boundary conditions has been found, by whatever means, you are guaranteed that there is no other one.

Suppose that v and w are two solutions, and let $u = v - w$. We show that $u(x, y, t) = 0$ for all $(x, y) \in B$. This, of course, proves that $v = w$. Now, since $v_t = \Delta v$ and $w_t = \Delta w$, we see that

$$u_t - \Delta u = v_t - w_t - (\Delta v - \Delta w) = 0,$$

and for $(x, y) \in B$,

$$u(x, y, 0) = v(x, y, 0) - w(x, y, 0) = f(x, y) - f(x, y) = 0,$$

and similarly for $(x, y) \in \partial B$ and $t \geq 0$,

$$u(x, y, t) = v - w = \varphi - \varphi = 0.$$

Now—and here is the trick—we define a kind of "energy" function $E(t)$ by

$$E(t) = \frac{1}{2} \iint_B u^2(x, y, t) \, dx \, dy.$$

Then, differentiating with respect to t under the integral sign (see the theorem in Sec. 9.3), we find that

$$\frac{dE}{dt} = \frac{1}{2} \int\int_B \frac{\partial}{\partial t} u^2(x, y, t) \, dx \, dy = \int\int_B u u_t \, dx \, dy.$$

But $u_t = \Delta u$, and so by Green's First with $v = u$ there,

$$\frac{\partial E}{\partial t} = \int\int_B u \, \Delta u \, dx \, dy = \int_{\partial B} u \, \nabla u \cdot N \, ds - \int\int_B \|\nabla u\|^2 \, dx \, dy.$$

Now we recall that $u = 0$ on ∂B, and so the boundary integral is zero. Thus

$$\frac{dE}{dt} = -\int\int_B \|\nabla u\|^2 \, dx \, dy \leq 0.$$

Consequently $E(t)$ is a decreasing function, $E(t) \leq E(0)$—physically, energy is dissipated. Since $u(x, y, 0) = 0$, however, we see that $E(0) = 0$. Moreover, it is evident from the definition that $E(t) \geq 0$. Hence for all $t \geq 0$

$$0 \leq E(t) \leq E(0) = 0;$$

that is, $E(t) = 0$ for all $t \geq 0$. This implies that $u(x, y, t) = 0$ for all $t \geq 0$, since if not, then $E(t) > 0$ for some $t \geq 0$.

As our second application, we consider vectorfield F satisfying div $F = 0$ in a region $\Omega \subset \mathbb{R}^2$; that is, F has no sources or sinks in Ω. See Fig. 10.17. For instance, we might have

$$F(x, y) = \left(\frac{x}{x^2 + y^2}, \frac{y}{x^2 + y^2}\right)$$

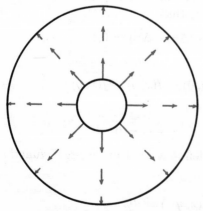

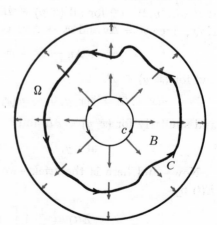

FIGURE 10.17   FIGURE 10.18

with Ω the region $1 \le x^2 + y^2 \le 9$ between two circles of radius 1 and 3, respectively. It is easy to check that, for $(x, y) \ne 0$,

$$\text{div } F = \frac{\partial}{\partial x}\left(\frac{x}{x^2 + y^2}\right) + \frac{\partial}{\partial y}\left(\frac{y}{x^2 + y^2}\right) = 0.$$

Let C be the curve in Fig. 10.18. We claim that

$$\int_C F \cdot N \, ds = \int_c F \cdot N \, ds = 2\pi,$$

where c is the unit circle $\|X\| = 1$ oriented counterclockwise. If F represents the velocity vector of a fluid in Ω, this asserts that the net flow across the curve equals that across the unit circle. This is plausible if one observes that the vectorfield F is radial outward of length 1 at every point on c. It also appears that there is a source at the origin.

The computation is easy. Let B denote the region between C and the unit circle $\|X\| = 1$. Then by the Divergence Theorem 4,

$$0 = \iint_B \text{div } F \, dA = \int_{\partial B} F \cdot N \, ds.$$

But $\partial B = C - c$, where we have $-c$, not c, since its orientation is reversed when considered part of the boundary of B. Therefore

$$0 = \int_{\partial B} F \cdot N \, ds = \int_C F \cdot N \, ds - \int_c F \cdot N \, ds.$$

We now evaluate the integral around c. There are two methods. The easiest is to observe that on c, the radius vector $X = (x, y)$ is an outer normal. Thus $F = N$ is also the unit outer normal, so that $F \cdot N = 1$ on c. Consequently

$$\int_c F \cdot N \, ds = \int_c ds = 2\pi.$$

The second method is to parametrize c as $x = \cos \theta$, $y = \sin \theta$, $0 \le \theta \le 2\pi$. Then $F = (\cos \theta, \sin \theta)$ on c, and, as we saw just before Theorem 4, $N = (\cos \theta, \sin \theta)$. Therefore $F \cdot N = 1$, and

$$\int_c F \cdot N \, ds = \int_0^{2\pi} 1 \, d\theta = 2\pi.$$

This number 2π represents the magnitude of the source at the origin.

A third important application of Stokes' Theorem appears in Sec. 10.3.

10.2e Higher dimensions

Stokes' Theorem generalizes to three and higher dimensions. We content ourselves with an intuitive description of the $\mathbb{R}^3$ cases without precise

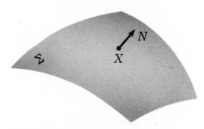

FIGURE 10.19

definitions or proofs. These generalizations require the notion of integration on a surface Σ in $\mathbb{R}^3$. Let f be a real-valued function defined on Σ. See Fig. 10.19. For example, if Σ is a thin sheet of material, then $f(X)$ might be the surface density at the point X in Σ. Then, if dA is thought of as the "element of area on Σ," we think of

$$\iint_{\Sigma} f \, dA$$

as the total mass of Σ. In the important special case where $f = 1$ this becomes

$$\iint_{\Sigma} dA = \text{area of } \Sigma.$$

In addition, we must give the piece of surface Σ an orientation. This is done by arbitrarily designating one "side" of the surface as the "outer" or "positive" side and letting the vector $N(X)$ denote the unit outer normal to Σ at X. (Warning: Not all surfaces can be oriented. The "Mobius strip" is a famous example of this. Ask someone to show you a few of the strange and fascinating properties of this surface. We guarantee that he will be delighted to do so.) Once Σ is oriented, we are automatically led to a positive orientation of the boundary curve $\partial \Sigma$ by thinking of Σ as a piece of a curved plane. See Fig. 10.20, where Σ is a piece of a surface with a hole in it.

As a final preliminary, we extend line integrals to curves C in $\mathbb{R}^3$. Say that $X = \alpha(t)$, $a \leq t \leq b$, is a parametrization of C. Then, just as before, we define

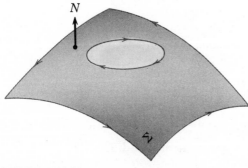

FIGURE 10.20

$$\int_c F(X)\cdot dX = \int_a^b F(\alpha(t))\cdot \alpha'(t)\, dt.$$

The first generalization is often called

STOKES' OR OSTROGRADSKY'S THEOREM 5 : Let $F = (P, Q, R)$ be a smooth vectorfield in $\mathbb{R}^3$ Then for any smooth orientable surface Σ

$$\iint_\Sigma (\operatorname{curl} F)\cdot N\, dA = \int_{\partial\Sigma} F\cdot dX,$$

where $\operatorname{curl} F$ is the vector

$$\operatorname{curl} F = (R_y - Q_z,\, P_z - R_x,\, Q_x - P_y).$$

NOTE: In the special case when Σ is a region in the xy-plane, so that $N = (0, 0, 1)$, $R = 0$, and $F = (P, Q, 0)$, then

$$(\operatorname{curl} F)\cdot N = Q_x - P_y,$$

and the formula simplifies to

$$\iint_\Sigma (Q_x - P_y)\, dA = \int_{\partial\Sigma} F\cdot dX,$$

which is just our Theorem 3.

Since the curl F is a new object, we compute it for the example $F(X) = (xyz, y - 3z, 2y)$. Then $P = xyz$, $Q = y - 3z$, and $R = 2y$, so that $\operatorname{curl} F = (2 + 3, xy - 0, 0 - xz) = (5, xy, -xz)$.

We remark that the complicated formula for curl F is usually remembered by thinking of it as the symbolic cross product of the operator $\nabla = (\partial/\partial x)\mathbf{i} + (\partial/\partial y)\mathbf{j} + (\partial/\partial z)\mathbf{k}$ with the vector F and is written

$$\operatorname{curl} F = \nabla \times F = \begin{vmatrix} \mathbf{i} & \mathbf{j} & \mathbf{k} \\ \dfrac{\partial}{\partial x} & \dfrac{\partial}{\partial y} & \dfrac{\partial}{\partial z} \\ P & Q & R \end{vmatrix}.$$

The second generalization is called the *Divergence Theorem*. Let $F = (P, Q, R)$ be a smooth vectorfield in $\mathbb{R}^3$. Then, by definition,

$$\operatorname{div} F = \frac{\partial P}{\partial x} + \frac{\partial Q}{\partial y} + \frac{\partial R}{\partial z}$$

(recall that div F can be thought of as the symbolic inner product of the operator ∇ with F; that is, div $F = \nabla\cdot F$). If $B \subset \mathbb{R}^3$ is a solid region whose boundary ∂B is a smooth surface with unit outer normal N, then the Divergence Theorem states that

$$\iiint_B \operatorname{div} F \, dV = \iint_{\partial B} F \cdot N \, dA,$$

where $dV = dx \, dy \, dz$ is the element of volume in $\mathbb{R}^3$ and dA is the element of surface area on the surface ∂B. The physical interpretation made before in terms of fluid flow carries over to this three-dimensional case without change.

It is straightforward to prove the three-dimensional versions of Green's identities by imitating the two-dimensional proofs. For example, Green's First becomes

$$\iiint_B v \, \Delta u \, dV = \iint_{\partial B} v \, \nabla u \cdot N \, dA - \iiint_B \nabla v \cdot \nabla u \, dV,$$

where $\Delta u = u_{xx} + u_{yy} + u_{zz}$ and $\nabla v = (v_x, v_y, v_z)$.

Although it might be difficult to believe, these theorems are quite easy to use in most standard applications. Here are a few.

EXAMPLES: 1. Let $F = (x, 3y, -z)$, and let Σ be the hollow sphere of radius a centered at the origin. Then, if B is the solid ball of radius a with $\Sigma = \partial B$, we have

$$\iint_\Sigma F \cdot N \, dA = \iiint_B \operatorname{div} F \, dV = 4\pi a^3.$$

To see this, we compute $\operatorname{div} F = 1 + 3 - 1 = 3$, and so the right side is three times the volume of the ball; that is, $3(4\pi a^3/3) = 4\pi a^3$ as claimed.

2. Let $u(x, y, z)$ be a smooth function that satisfies the *Laplace equation*

$$\frac{\partial^2 u}{\partial x^2} + \frac{\partial^2 u}{\partial y^2} + \frac{\partial^2 u}{\partial z^2} = 0;$$

that is, $\Delta u = 0$, in a domain B in $\mathbb{R}^3$. Such functions are called *harmonic functions*. They arise often in applications. For example, u might be the temperature distribution in a body at thermal equilibrium (this means that the temperature is independent of time). If the temperature is zero on the boundary of the body, $u = 0$ on ∂B, we claim that it must be zero throughout the body. This physically plausible assertion is easy to prove. We apply Green's First Identity, stated above, in the special case when $u = v$. Since $\Delta u = 0$, the integral on the left is zero. Moreover, $u = 0$ on ∂B. Thus the double integral over ∂B is also zero. This leaves

$$\iiint_B \|\nabla u\|^2 \, dV = 0,$$

which implies that $\nabla u = 0$; that is, $u_x = 0$, $u_y = 0$, and $u_z = 0$ throughout B. Therefore $u \equiv$ constant in B. But $u = 0$ on ∂B, and so the value of the constant is zero; that is, $u \equiv 0$ throughout B. The proof is complete.

Exercises

1. Verify Stokes' Theorem 3, by computing both sides separately and showing that they are equal, for each of the following:

   (a) $F = (1 - y, x)$, $\qquad$ B is the disk $x^2 + y^2 \le 4$;

   (b) $F = (x + y, x - 6y)$, $\qquad$ B is the triangle with vertices at $(-4, 1)$, $(2, 1)$, and $(2, 5)$;

   (c) $F = (xy, -xy)$, $\qquad$ B is the disk $x^2 + y^2 \le 9$;

   (d) $F = (y \sin \pi x, y \cos \pi x)$, $\qquad$ B is the rectangle $1 \le x \le 2$, $-1 \le y \le 1$;

   (e) $F = (x^2 + y, e^y - x)$, $\qquad$ B is the region above the parabola $y = x^2$ and below the line $y = 1$;

   (f) $F = (x^2 + y^2, -2xy)$, $\qquad$ B is the half-disk $x^2 + y^2 \le 16$, $y \ge 0$;

   (g) $F = (x, 2x + y)$, $\qquad$ B is the ring region $1 \le x^2 + y^2 \le 4$;

   (h) $F = (y^2, x^2)$, $\qquad$ B is the region outside the circle $x^2 + y^2 = 1$ but inside the rectangle $-2 \le x \le 3$, $-4 \le y \le 5$.

2. (a) Given a domain $B \subset \mathbb{R}^2$, show that

   $$\text{Area } (B) = \frac{1}{2} \int_{\partial B} x \, dy - y \, dx.$$

   (b) Use this to find the area inside the ellipse $x = a \cos \varphi$, $y = b \sin \varphi$, $0 \le \varphi \le 2\pi$.

   (c) Use this to find the area of the region $1 \le x^2 + y^2 \le 25$.

3. If u is a smooth function, prove that

   $$\iint_B \Delta u \, dA = \int_{\partial B} \frac{\partial u}{\partial N} \, ds.$$

4. Let a force field $F = (2 - 5x + y, x)$. Find the work expended in moving a particle of unit mass counterclockwise around the boundary of the triangle with vertices at $(1, 1)$, $(3, 1)$, and $(3, 5)$.

5. Let C_1 denote the circle $x^2 + y^2 = 1$ and C_2 the circle $(x - 2)^2 + (y - 1)^2 = 25$, both oriented counterclockwise, and let B denote the ring region between these curves. If a vectorfield F satisfies div $F = 0$ in B, show that

   $$\int_{C_1} F \cdot N \, ds = \int_{C_2} F \cdot N \, ds.$$

   Note that the proof of this extends immediately to the case where C_1 and C_2 are replaced by more general curves and B is the region between them.

6. For certain kinds of heat flow, the temperature $u(x, y, t)$ in a region $B \subset \mathbb{R}^2$ satisfies

   $$u_t = \Delta u - u.$$

   Prove that there is at most one solution that has a given initial temperature $u(x, y, 0) = f(x, y)$, $(x, y) \in B$, and a given boundary temperature $u(x, y, t)$

$= \varphi(x, y, t)$, $(x, y) \in \partial B$, $t \geq 0$. (The same function $E(t)$ given in Sec. 10.2d works here.)

7. Let $F = (2xy, yz, z)$ be a vectorfield in $\mathbb{R}^3$. Use the divergence form of Stokes' Theorem to evaluate

$$\iint_{\Sigma} F \cdot N \, dA,$$

where Σ is the boundary of the cube $|x| \leq 1$, $|y| \leq 1$, $|z| \leq 1$.

8. Let $F = (x - y, y^2 - z, x^2 + 5z)$ be a vectorfield in $\mathbb{R}^3$. Evaluate

$$\iint_{\Sigma} F \cdot N \, dA,$$

where Σ is the sphere $x^2 + y^2 + z^2 = 9$.

Exercises 9 to 12 contain several simple properties of the curl.

9. Compute curl F for the following vectorfields F:
    (a) $F = (y, z, 2 - z)$,   (c) $F = (x^2 + y^2, 0, 0)$,
    (b) $F = (xy + z, yz + x, zx + y)$,   (d) $F = (y - z, z + x, x + y)$.

10. Let $F = (P, Q, R)$ be a smooth vectorfield in $\mathbb{R}^3$ and u a smooth real-valued function. Prove that:
    (a) div (curl F) = 0;
    (b) curl (grad u) = 0;
    (c) curl $(uF) = u$ curl $F + (\text{grad } u) \times F$;
    (d) if V is a constant vector and $X = (x, y, z)$, then curl $(V \times X) = 2V$.

11. Let F and G be smooth vectorfields in $\mathbb{R}^3$. Show that:
    (a) curl $(F \times G) = (\text{curl } F) \cdot G - F \cdot \text{curl } G$;
    (b) curl $(F + G) = \text{curl } F + \text{curl } G$;
    (c) if curl $F = G$, then div $G = 0$.

12. Is there a vectorfield $F = (P, Q, R)$ such that:
    (a) curl $F = (x, y, z)$? Why?
    (b) curl $F = 2(z - y, x - z, y - x)$? Why?

13. Let Σ be the hemisphere of the sphere $x^2 + y^2 + z^2 = 1$, where $z \geq 0$, oriented with the outer normal to the sphere, and let $F = (x - y, 3z^2, -x)$. Evaluate

$$\iint_{\Sigma} (\text{curl } F) \cdot N \, dA.$$

14. Let Σ be the part of the plane $x + y + z = 1$ in the first octant of $\mathbb{R}^3$, oriented so that the unit outer normal is $N = (1, 1, 1)/\sqrt{3}$. Verify Stokes' Theorem by evaluating both sides of

$$\iint_{\Sigma} (\text{curl } F) \cdot N \, dA = \int_{\partial\Sigma} F \cdot dX,$$

where $F = (z - y, x - z, y - x)$.

15. Let $B \subset \mathbb{R}^3$ be a solid region, say a ball, and F a smooth vectorfield. Show that

$$\iint_{\partial B} (\text{curl } F) \cdot N \, dA = 0.$$

(Suggestion: Let $\Sigma = \partial B$, and observe that $\partial\Sigma$ has no points.)

16. Let $F = (p, q)$ be a vectorfield in $\mathbb{R}^2$, and let $B \subset \mathbb{R}^2$ be a region with a smooth boundary ∂B. If $F \equiv$ constant vector on ∂B, show that

$$\iint_B (q_x - p_y) \, dA = 0.$$

Note that the second application in Sec. 10.2e is a special case.

17. Let $B \subset \mathbb{R}^3$ denote the solid pyramid bounded by the planes $x + y + 2z = 6$, $x = 0$, $y = 0$, and $z = 0$. Use the Divergence Theorem to evaluate

$$\iint_{\partial B} F \cdot N \, dA,$$

where $F = (2x, 2y, 4z)$.

18. Let F and G be two smooth vectorfields in a ball $B \subset \mathbb{R}^3$ such that (1) $F = \nabla\varphi$, $G = \nabla\psi$, and (2) div $F = $ div G in B, while (3) $F \cdot N = G \cdot N$ on the boundary of B, where N is the unit outer normal to ∂B. Prove that $F \equiv G$ in B.

19. (a) Let $\Sigma \subset \mathbb{R}^3$ be the sphere $\|X\| = 1$ and Σ' be some smooth closed surface outside the sphere Σ, so that Σ' might be some larger sphere. If div $F = 0$ in the region between Σ and Σ', prove that

$$\iint_\Sigma F \cdot N \, dA = \iint_{\Sigma'} F \cdot N \, dA.$$

(b) Let $B \subset \mathbb{R}^3$ be the region $1 \leq \|X\| \leq 3$. If $F(X) = X/\|X\|^3$ in B, show by direct computation of both sides that

$$\iiint_B \text{div } F \, dV = \iint_{\partial B} F \cdot N \, dA.$$

20. Prove the Divergence Theorem in $\mathbb{R}^3$ if B is a box: $a_1 \leq x \leq b_1$, $a_2 \leq y \leq b_2$, $a_3 \leq z \leq b_3$.

GRAVITATIONAL FORCE

The purpose of these few pages is to investigate the gravitational force due to a spherically symmetric solid. We prove that in this case the gravitational force is the same as that due to a point mass. Although the problem caused Newton considerable difficulty, we find it quite easy. To use Newton's phrase, however, "We are standing on the shoulders of giants."

Recall (Sec. 7.2) Newton's Law of Gravitation for the force F on a mass m at X due to a mass M at Y:

$$F = -\gamma \frac{mM}{\|X - Y\|^3} (X - Y).$$

It is convenient to introduce the force at X per unit mass, $G = F/m$,

$$G(X) = -\gamma \frac{M}{\|X - Y\|^3}(X - Y).$$

$G(X)$ is called the *gravitational field* at X. By a straightforward computation, one sees that

$$\operatorname{div} G(X) = 0, \qquad \text{for } X \neq Y.$$

Thus if $B \subset \mathbb{R}^3$ is any region not containing Y (see Fig. 10.21, where Y is in a hole in B), then the Divergence Theorem yields

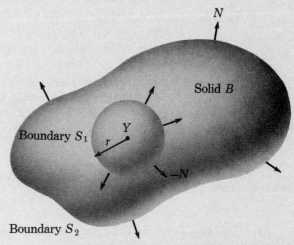

FIGURE 10.21

$$0 = \iiint_B \operatorname{div} G \, dV = \iint_{\partial B} G \cdot N \, dA = \iint_{S_2} G \cdot N \, dA - \iint_{S_1} G \cdot N \, dA.$$

Here we have used $-S_1$, since the orientation given to S_1 in Fig. 10.21 (its outer normal points into B) is the inner normal when regarded as part of the boundary of B (denoted ∂B above). Consequently, for the smooth surfaces S_2, S_1, with $Y \notin B$,

$$\iint_{S_2} G \cdot N \, dA = \iint_{S_1} G \cdot N \, dA.$$

Now we pick a convenient surface for S_1: a sphere of radius r with center at Y. Then for $X \in S_1$, the unit outer normal to S_1 is $N = (X - Y)/$ $(\|X - Y\|)$ and $r = \|X - Y\|$. Therefore, if $X \in S_1$,

$$G \cdot N = -\frac{-\gamma M}{\|X - Y\|^2} = -\frac{\gamma M}{r^2}.$$

Now, since the area of $S_1 = 4\pi r^2$, we have

$$\iint_{S_2} G \cdot N \, dA = \iint_{S_1} G \cdot N \, dA = -\frac{\gamma M}{r^2} \iint_{S_1} dA = -4\pi\gamma M.$$

All this has assumed that the mass M (at Y) lies inside the surface S_1. On the other hand, if M lies outside S_2, then div $G = 0$ throughout B. See Fig. 10.22. Hence

$$\iiint_B \operatorname{div} G \, dV = \iint_{S_2} G \cdot N \, dA = 0.$$

Combining these two cases, we conclude that if S is any closed surface, then

$$\iint_S G \cdot N \, dA = \begin{cases} 0, & M \text{ outside } S, \\ -4\pi\gamma M, & M \text{ inside } S. \end{cases}$$

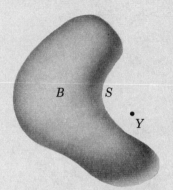

FIGURE 10.22

It is important to observe that this integral, called the total "flux" of the gravitational field through the surface S, depends only on the mass M and whether M is inside or outside S, but not on the particular location Y of the mass nor on the shape of the surface.

Now suppose that there are lots of masses $M_1, M_2, \ldots, M_n$ at the points $Y_1, Y_2, \ldots, Y_n$. Let $G_j(X)$ be the gravitational field at X due to the mass M_j at Y_j. Since force is a vector, the gravitational field G due to all these masses is a sum:

$$G = G_1 + \cdots + G_n.$$

Consequently, if S is any closed surface (see Fig. 10.23), then the result above applied to each mass separately yields

$$\iint_S G \cdot N \, dA = \sum_1^n \iint_S G_j \cdot N \, dA$$
$$= -4\pi\gamma \cdot (\text{total mass inside } S);$$

that is,

$$\iint_S G \cdot N \, dA = -4\pi\gamma \cdot (\text{total mass inside } S).$$

This is called *Gauss' Law*.

More generally yet, if the mass distribution in a region B is given in terms of a density ρ, then

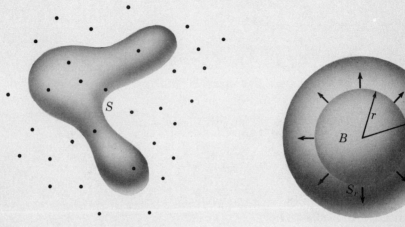

FIGURE 10.23   FIGURE 10.24

$$M = \text{total mass in } B = \iiint_B \rho \, dV.$$

Gauss' Law clearly extends to this case, as one can see by thinking of M as the sum $\sum \Delta M$, where $\Delta M = \rho \, \Delta V$ is the mass in a small element of volume ΔV.

We now apply this to the case of a spherical body B, like the earth, and see how to compute the gravitational field. We assume that the mass distribution is symmetric and place the origin at the center of B. See Fig. 10.24. This implies, by symmetry, that the gravitational force $F(X)$ and hence the gravitational field $G(X)$ are in the radial direction. Consequently, if S_r is any sphere of radius r concentric with ∂B, possibly inside B, we have $G \cdot N = \|G\|$, where N is the unit outer normal, in the radial direction, to S_r. Moreover, again by symmetry, we see that $\|G(X)\|$ is constant for all X on S_r. Therefore

$$\iint_{S_r} G \cdot N \, dA = \|G\| \iint_{S_r} dA = 4\pi r^2 \|G\|.$$

Combined with Gauss' Law, we find that

$$\|G(X)\| = -\frac{\gamma M_r}{r^2} = -\frac{\gamma M_r}{\|X\|^2},$$

for X on S_r. Here M_r is the total mass inside the sphere S_r. Since the direction of $G(X)$ is radial, that is, in the direction of the unit vector $X/\|X\|$, we finally conclude that

$$G(X) = -\frac{\gamma M_r}{\|X\|^3}X,$$

and so

$$F(X) = -\frac{\gamma m M_r}{\|X\|^3}X.$$

Of course, if $r \geq R$, where R is the radius of B, then $M_r = M$, the total mass of B. Since, given any point X, we can construct the sphere S_r with $r = \|X\|$, the last formula proves that the *gravitational force at X due to a spherically symmetric mass distribution is the same as that from a point mass M_r at the center of the body, where M_r is the mass of the body at distance $\leq r = \|X\|$ from the center of the body.*

EXERCISE: Let B be a hollow spherical shell. Show that the gravitational force in the hole in B is zero.

10.3 INDEPENDENCE OF PATH; POTENTIAL FUNCTIONS

10.3a Introduction

Physics immediately leads us to ask a question. Say that X_1 and X_2 are points in the plane and $F = (p, q)$ is a force field in the plane. Then the work done by the force F in moving a unit mass along a curve C from X_1 to X_2, the two endpoints, is

$$\text{Work} = \int_C F(X)\cdot dX.$$

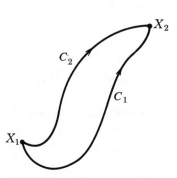

FIGURE 10.25

The question is, under what conditions on F is the work independent of the particular path C joining X_1 to X_2? See Fig. 10.25. In Example 5 in Sec. 10.1b, the line integral *does* depend on the path, and so it would be wrong to believe that one has independence of the path unless further restrictions are imposed on the force F. If the line integral is independent of the path, we say that F is a *conservative force field*. The reason for the name is easy to explain. If F is conservative, then the line integrals along any two paths C_1 and C_2 from

X_1 to X_2 are equal. Therefore, if we go from X_1 to X_2 along C_1 and return from X_2 to X_1 along $-C_2$, the net result is zero; that is, no work is needed to traverse the closed curve $C_1 - C_2$. Consequently, the force F could not have any dissipative aspects, like friction, since dissipation causes an irreversible loss of energy.

The question of a line integral being independent of the path can be raised about any line integral. It does not need the physical conception of work.

10.3b The Main Theorem

We would like some criterion to determine when a line integral is independent of the path. First we must show that some line integrals are indeed independent of the path. Recall the notation $\nabla f = (f_x, f_y)$, where $f(x, y)$ is a smooth real-valued function.

THEOREM 6:   Let $F = \nabla f$, and let X_1 and X_2 be any two points. Then

$$\int_{X_1}^{X_2} F(X) \cdot dX = \int_{X_1}^{X_2} \nabla f(X) \cdot dX = f(X_2) - f(X_1),$$

where the integration is over any piecewise smooth curve C joining X_1 to X_2. Thus, if $F = \nabla f$ the integral is independent of the path.

Proof: First assume that C is smooth, and let $X = \alpha(t)$, $a \leq t \leq b$, be a parametrization of it. Note that $\alpha(a) = X_1$, $\alpha(b) = X_2$. Then

$$\int_C F(X) \cdot dX = \int_a^b \nabla f(\alpha(t)) \cdot \alpha'(t)\, dt.$$

But by the Chain Rule,

$$\frac{d}{dt} f(\alpha(t)) = \nabla f(\alpha(t)) \cdot \alpha'(t)$$

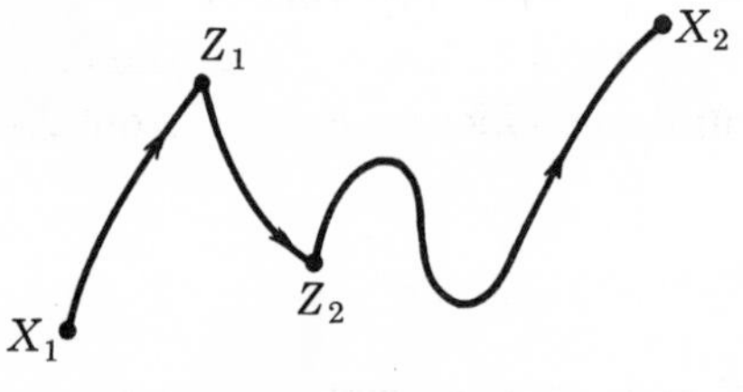

FIGURE 10.26

Therefore, by the Fundamental Theorem of Calculus,

$$\int_C F(X) \cdot dX = \int_a^b \frac{d}{dt} f(\alpha(t))\, dt = f(X_2) - f(X_1).$$

Since the right side is independent of the path C, we conclude that the line integral is independent of the path.

If C is only piecewise smooth, then let its smooth segments be $X_1 Z_1, Z_1 Z_2, \ldots, Z_n X_2$. See Fig. 10.26. We apply the result above to each smooth segment

$$\int_{X_1}^{X_2} \nabla f(X) \cdot dX = \int_{X_1}^{Z_1} + \int_{Z_1}^{Z_2} + \cdots + \int_{Z_n}^{X_2}$$
$$= [f(Z_1) - f(X_1)] + [f(Z_2) - f(Z_1)] + \cdots + [f(X_2) - f(Z_n)].$$

After cancellation, we find that

$$\int_{X_1}^{X_2} \nabla f(X) \cdot dX = f(X_2) - f(X_1),$$

just as desired. This completes the proof. ≪

We remark that this theorem is itself a generalization of the Fundamental Theorem of Calculus, to which it reduces if C is an interval on the x-axis and $f(x, y)$ depends on x only. The next result is our main theorem. In it we collect a variety of conditions for the line integral of a given vectorfield F to be independent of the path of integration. It asserts that essentially the only way this can happen is if there is some function f such that $F = \nabla f$, in which case Theorem 6 gives the independence of path. Recall that a curve C, parametrized by $X = \alpha(t)$, $a \le t \le b$, is *closed* if the initial and final points coincide: $\alpha(a) = \alpha(b)$.

THEOREM 7: Let $F = (p, q)$ be a given smooth vectorfield in a domain $B \subset \mathbb{R}^2$. The following conditions are equivalent:

a. $\int_C F(X) \cdot dX = 0$ for any piecewise smooth closed curve in B.
b. $\int_C F(X) \cdot dX$ is independent of the path for any piecewise smooth curve C joining two points in B.
c. There is a smooth function $f(x, y)$ such that $\nabla f = F$; that is, $f_x = p$, $f_y = q$.

Any of these implies the following condition on F:

d. $p_y = q_x$.

Moreover, if B is a disk, then condition (d) implies (a) to (c). Thus, for a disk conditions (a) to (d) are equivalent.

Proof: We prove the chain of implications (a) ⇒ (b) ⇒ (c) ⇒ (a). This will prove the equivalence of (a), (b), and (c).

(a) ⇒ (b). Let C_1 and C_2 be any two piecewise smooth curves from X_1 to X_2. Then $C = C_1 - C_2$ is a piecewise smooth closed curve in B, which we have pictured having a "hole." See Fig. 10.27. Therefore, by (a),

$$0 = \int_C F(X) \cdot dX = \int_{C_1} + \int_{-C_2} = \int_{C_1} - \int_{C_2},$$

so that

$$\int_{C_1} F(X) \cdot dX = \int_{C_2} F(X) \cdot dX.$$

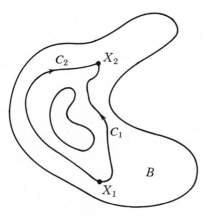

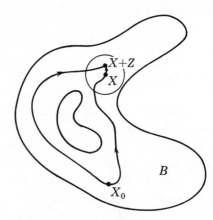

FIGURE 10.27   FIGURE 10.28

(b) ⇒ (c). We must exhibit the function $f(x, y) = f(X)$. Pick a point X_0 in B and fix it throughout the discussion. If X is in B, let

$$f(X) = \int_{X_0}^{X} F(Y) \cdot dY,$$

where we need not specify the path of integration since, by part (b), the result does not depend on the particular path chosen. We must show that $\nabla f = F$.

Since B is an open set and X is in B, some open disk about X is also in B. Therefore, if Z is any sufficiently small vector, the point $X + Z$ is also in B. Let $e = Z/\|Z\|$ be a unit vector in the direction of Z. Then

$$f(X + Z) - f(X) = \int_{X_0}^{X+Z} F(Y) \cdot dY - \int_{X_0}^{X} F(Y) \cdot dY.$$

Since the line integral is independent of the path, we can replace the right side by an integral over the straight line segment from X to $X + Z$ (see Fig. 10.28):

$$f(X + Z) - f(X) = \int_{X}^{X+Z} F(Y) \cdot dY.$$

Parametrize this straight line segment as $Y = X + tZ$, $0 \le t \le 1$. This gives

$$f(X + Z) - f(X) = \int_{0}^{1} F(X + tZ) \cdot Z \, dt.$$

Now divide both sides by $\|Z\|$, and recall from Chap. 5 that

$$\lim_{\|Z\| \to 0} \frac{f(X + Z) - f(X)}{\|Z\|} = f'(X)e = \nabla f(X) \cdot e.$$

Therefore

$$\nabla f(X) \cdot e = \lim_{\|Z\| \to 0} \int_0^1 F(X + tZ) \cdot e \, dt.$$

Because $F(X + tZ) \cdot e$ is a continuous function of $\|Z\|$, we can let $\|Z\|$ $\to 0$ under the integral sign. The result is $\int_0^1 F(X) \cdot e \, dt$. Since the integrand does not depend on t, the integral equals $F(X) \cdot e$; that is,

$$\nabla f(X) \cdot e = F(X) \cdot e$$

for any unit vector e; that is, $(\nabla f(X) - F(X)) \cdot e = 0$ for all unit vectors e. This implies that

$$\nabla f(X) = F(X).$$

(c) $\Rightarrow$ (a). This follows from Theorem 6, since for a closed curve C in that theorem, we have $X_2 = X_1$.

Part (d) is a consequence of part (c) and the fact that $f_{xy} = f_{yx}$ for smooth functions $f(x, y)$—compare Sec. 5.2. To see this, we observe that, by part (c), $f_x = p$, and $f_y = q$. Therefore

$$p_y = f_{xy} = f_{yx} = q_x.$$

It remains to prove that for a disk, if $p_y = q_x$, then there is a function f such that $\nabla f = F$. Let X_0 be the center of the disk, and define

$$f(X) = \int_{c_1} F(Y) \cdot dY,$$

where c_1 is the path from X_0 to X in Fig. 10.29, composed of a horizontal and vertical segment. Then, as in the proof of (b) $\Rightarrow$ (c), we see that $f_y = q$. To show that $f_x = p$, we observe that if R is the rectangle in Fig. 10.29, then by Stokes' Theorem, since $\partial R = c_1 - c_2$,

$$\int_{c_1} - \int_{c_2} = \int_{\partial R} F(Y) \cdot dY = \iint_R (q_x - p_y) \, dA = 0,$$

and thus

$$f(X) = \int_{c_2} F(Y) \cdot dY.$$

Just as before, we find that $f_x = p$. Thus, for a disk (d) $\Rightarrow$ (c). This proves that for any disk conditions (a) to (d) are equivalent. Done. $\ll$

REMARK: If the line integral of a force field F is independent of the path, we say that the force is *conservative* and call the function f such that ∇f $= F$ the *potential function* of F.

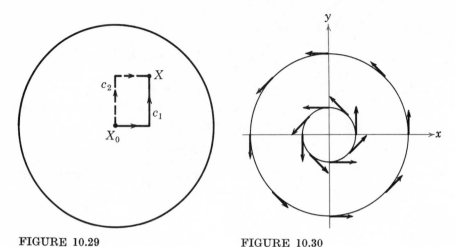

FIGURE 10.29   FIGURE 10.30

10.3c Examples:

One wonders if the proof that (d) $\Rightarrow$ (c) can be extended to more general regions than disks. The following example shows that it cannot always be generalized.

1. Let $B \subset \mathbb{R}^2$ be the ring domain $1 \leq x^2 + y^2 \leq 9$, and let

$$p(x, y) = -\frac{y}{x^2 + y^2}, \qquad q(x, y) = \frac{x}{x^2 + y^2}.$$

Then a computation shows that $p_y = q_x$. See Fig. 10.30. If we integrate $F = (p, q)$ around the circle $x = 2 \cos \theta$, $y = 2 \sin \theta$, $0 \leq \theta \leq 2\pi$, then

$$\int_C F(X) \cdot dX = \int_0^{2\pi} (\sin^2 \theta + \cos^2 \theta)\, d\theta = 2\pi.$$

In other words, the line integral of F around some closed curve C is not zero. Thus (d) $\not\Rightarrow$ (a) in this case.

It turns out that (d) implies (a) only if the region B has no "holes" in it. In the example just done, B does have a "hole," the disk $x^2 + y^2 < 1$, and at the origin the given vectorfield F "blows up." This is typical of the problems that arise if the region has one or more holes.

2. Let C be a smooth curve from $X_1 = (2, 3)$ to $X_2 = (3, -1)$. We evaluate

$$\int_C (3x^2 + y)\, dx + (e^y + x)\, dy.$$

There are two methods for evaluating this. Both rely on the observation that the integral is independent of the path, since $p_y = 1 = q_x$, for all $(x, y) \in \mathbb{R}^2$ (here B is a big disk, or even all of $\mathbb{R}^2$).

Method 1. We pick a convenient path—the straight line C:

$$x = 2 + t, \qquad y = 3 - 4t, \qquad 0 \le t \le 1.$$

Then

$$\int_C = \int_0^1 [3(2 + t)^2 + (3 - 4t) - 4e^{3-4t} - 4(2 + t)]\, dt$$

$$= 10 + e^{-1} - e^3.$$

Method 2. Since $p_y = q_x$ and B is a disk containing X_1 and X_2, by the theorem, there is a function f such that $\nabla f = F$,

$$\nabla f = (f_x, f_y) = (3x^2 + y, e^y + x).$$

We find f. Since $f_x = 3x^2 + y$, by integrating with respect to x and holding y constant, we find that

$$f(x, y) = \int (3x^2 + y)\, dx = x^3 + xy + K(y),$$

where $K(y)$ is a function (the "constant" of integration) that depends only on y. Now we compute f_y from this expression and compare it with what f_y is known to be, that is, $f_y = e^y + x$:

$$e^y + x = f_y = x + \frac{dK}{dy}.$$

This yields $K' = e^y$, and so $K(y) = e^y$; that is,

$$f(x, y) = x^3 + xy + e^y.$$

It is now elementary to compute the desired line integral, using Theorem 6:

$$\int_{X_1}^{X_2} (3x^2 + y)\, dx + (e^y + x)\, dy = \int_{X_1}^{X_2} \nabla f(X)\cdot dX = f(X_2) - f(X_1)$$

$$= x^3 + xy + e^y \Big|_{(2,\, 3)}^{(3,\, -1)}$$

$$= (3^3 - 3\cdot 1 + e^{-1}) - (2^3 + 2\cdot 3 + e^3)$$

$$= 10 + e^{-1} - e^3.$$

3. Let $F(x, y) = (3x^2 + y,\, e^y + x)$ be a given force field. We show that F is a conservative force field and find the potential function. Since F is smooth and satisfies $p_y = q_x$ for all $(x, y) \in \mathbb{R}^2$, by the previous theorem, the force F is conservative. The problem of finding the potential function was solved in the previous example (method 2), where we found that

$$f(x, y) = x^3 + xy + e^y.$$

You might wonder why one prefers to work with the potential function instead of the force. The answer is that the force is a vector, but the potential is a scalar. It is easier to work with scalar-valued functions, like $f(x, y) = x^3 + xy + e^y$, than with vector-valued functions.

Exercises

1. Evaluate the line integrals $\int_C F(X) \cdot dX$ by finding a potential function f, as in Example 2, method 2, such that $\nabla f = F$ and then, applying Theorems 6 and 7:

    (a) $F(X) = (x, y)$, C is $x = \cos t$, $y = 2 \sin t$, $0 \le t \le \pi/2$;
    (b) $F(X) = (y, x + 3)$, C is $x = t^2$, $y = t + 1$, $-1 \le t \le 2$;
    (c) $F(X) = (e^y, xe^y)$, C is the boundary of the rectangle $-1 \le x \le 2$, $5 \le y \le 9$, oriented counterclockwise;
    (d) $F(X) = (y^2, 2xy + 2y)$, C is the boundary of the half-disk $x^2 + y^2 \le 9$, $x \ge 0$, oriented counterclockwise;
    (e) $F(X) = (x + y, x - 7y)$, C is a pentagon with successive vertices at $(2, 0)$, $(1, 1)$, $(0, 1)$, $(-1, 0)$, and $(0, -1)$.

2. (a) Show that the force field $F(X) = (x + e^x \sin y, e^x \cos y)$ is conservative. and find a potential function for it.
    (b) Let C be a smooth curve from $(1, \pi)$ to $(0, \pi/2)$. Evaluate the work needed to move a particle of unit mass along C:

    $$\text{Work} = \int_C F(X) \cdot dX.$$

3. Let C be a smooth path in the disk $x^2 + (y - 3)^2 \le 6$ joining the two points $(-2, 2)$ and $(2, 2)$. For which of the following vectorfields is the line integral on C independent of the path C joining the two points?

    (a) $F(X) = (x - 1, e^y)$,
    (b) $F(X) = (3x - y, 2y + x)$,
    (c) $F(X) = \left(-\dfrac{y}{x^2 + y^2}, \dfrac{x}{x^2 + y^2} \right)$,
    (d) $F(X) = (e^y \cos x, e^y \sin x)$,
    (e) $F(X) = (e^y, e^x)$,
    (f) $F(X) = \left(\dfrac{x}{1 + x^2 + y^2}, \dfrac{y}{1 + x^2 + y^2} \right)$.

4. Let $F(X) = \left(-\dfrac{y}{x^2 + y^2}, \dfrac{x}{x^2 + y^2} \right)$. Evaluate $\int_C F(X) \cdot dX$, where:

    (a) C is the shorter arc of the circle $x^2 + y^2 = 8$ from $(-2, 2)$ to $(2, 2)$.
    (b) C is the longer arc of the circle $x^2 + y^2 = 8$ from $(-2, 2)$ to $(2, 2)$.
    (c) Is this line integral independent of the path C in the disk $x^2 + y^2 \le 100$ from $(-2, 2)$ to $(2, 2)$? Does this agree with the "moreover" part of Theorem 7?

5. The gravitational force between a mass M at the origin in $\mathbb{R}^2$ and a mass m at X in $\mathbb{R}^2$ is

$$F(X) = \frac{-\gamma m M X}{\|X\|^3},$$

where γ is a constant. Show that F is conservative by finding a potential $\varphi(X)$ such that $F = \operatorname{grad} \varphi$ at all points $X \neq 0$.

6. Let C be a smooth closed curve, and let $p(x)$ and $q(y)$ be smooth functions. Show that

$$\int_C p(x)\,dx + q(y)\,dy = 0.$$

7. Show that $\int_C (x - y)\,dx + 2y\,dy$ is not independent of the path C by showing that *each* of the statements a, b, d in Theorem 7 do not hold in this case.

8. If F is a conservative smooth vectorfield and G is any smooth vectorfield, then for any smooth closed curve C

$$\int_C (F(X) + G(X)) \cdot dX = \int_C G(X) \cdot dX.$$

Proof or counterexample.

9. Let F and G be two smooth vectorfields in $\mathbb{R}^2$. If F is conservative and if the vectors $F(X)$ and $G(X)$ have the same direction at every point $X \in \mathbb{R}^2$, then G is conservative too. Proof or counterexample.

10. Let $\varphi(r)$ be a smooth scalar-valued function of the one-variable r. If a vectorfield F is given by $F(X) = \varphi(\|X\|)X$, prove that F is conservative by verifying that $F(X) = \nabla \psi(\|X\|)$, where

$$\psi(r) = \int_0^r \rho \varphi(\rho)\,d\rho.$$

In particular, since any central force field F is of the form above, this shows that *central force fields are conservative*.

Answers to Selected Exercises

CHAPTER 1

Section 1.1

1.   (a) $(0, 5)$,   (c) $(0, 9)$,   (e) $(3, -9)$,   (g) $(-1, 4)$

2.   (a) $(-1, -\frac{10}{3})$,   (c) $(-\frac{9}{2}, -\frac{19}{2})$

4.   length of $Z = \sqrt{5}$

8. $\alpha = -5$, $\beta = -4$

Section 1.2

1.   (a) $(1, 4, -1)$,   (c) $(2, 0, 0)$,   (e) $(2, 12, 0)$,   (g) $(7, 0, -21)$

2.   (a) $(1, 4, -2, 1)$,   (c) $(1, -2, 1, 1)$,   (e) $(-1, -4, 2, -1)$

4.   (b) $X = e_1 - 2e_2 + 3e_3$

Section 1.3

1.   (a) subspace,   (c) subspace,   (e) not subspace,   (g) not subspace

2.   (a) not subspace,   (c) subspace,   (e) subspace,   (g) subspace

4.   (a) false,   (c) true,   (e) false,   (g) false,   (i) false,   (k) false,   (m) false

7.   (a) All straight lines through the origin in $\mathbb{R}^2$.

Section 1.4

1. (a) no,   (c) no,   (e) $Y = 0X$,   (g) no

2. (a) in,   (c) not in,   (e) not in,   (g) in,   (i) in

3. (a) $2X_1$,   (g) $-X_2$,   (i) $X_1 - 2X_2$

4. (a) $(0, 0, 0) = 0(1, -1, 2)$, $(-2, 2, -4) = -2(1, -1, 2)$,
    (c) $(2, 0, 8) = 2(1, 0, 4)$, $(-4, 0, -16) = -4(1, 0, 4)$

5. (a) $(3, 1, 2) = (1, 1, 0) + 2(1, 0, 1)$,   (c) $(3, 3, 1) = 2(1, 1, 1) + (1, 1, -1)$

6. (a) $\mathbb{R}^3$,   (c) plane,   (e) $\mathbb{R}^3$

7. (many correct answers)   (a) $X_1 = (1, 0, 0)$, $X_2 = (0, 1, 0)$,
    (c) $X_1 = (1, 1, 0)$, $X_2 = (0, 1, 1)$

8. (a) line, $X_1 = (0, 1, 0)$,   (c) line, $X_1 = (2, 1, 0)$,
    (e) plane, $X_1 = (1, 1, 0)$, $X_2 = (0, 0, 1)$

9. (a) $\alpha_1 = -2c$, $\alpha_2 = c$, $\alpha_3 = 3c$ for any $c \neq 0$

10. (a) $-2x_1 + x_2 + 3x_3 = 0$

11. (many correct answers)   (a) $(2, -1, 0)$, $(1, 0, -1)$

12. (a) $2x_1 - x_2 = 0$, $x_1 - x_3 = 0$

Section 1.5

1. (a) $x_1 + 2x_2 = 0$,   (c) yes

2. (a) $x_1 - x_2 + x_3 = 0$,   (c) yes

3. (a) $x_2 = 0$

4. (a) $2x_1 + 2x_2 + x_3 = 0$,   (b) $2x_1 + 2x_2 + x_3 = 2$

5. (a) $x_2 = 0$

CHAPTER 2

Section 2.1

2. (a) 4,   (c) $\sqrt{2}$,   (e) 4,   (g) $\sqrt{3}$,   (i) 4,   (k) 2

4. (a) 4,   (c) $\sqrt{2}$

6. (e) $(0, 0, -1)$,   (g) $(1, 1, 1)/\sqrt{3}$

Section 2.2

1. (a) 0,   (c) 3

3. $\cos \theta = 1/\sqrt{2}$

4. (a) $Y = c(5, -1)$ for any $c \neq 0$,   (b) two, $Y = \pm(5, -1)/\sqrt{26}$

Section 2.3

1. (a) $2x_1 - x_2 + 3x_3 = 0$

2. (a) $\mathscr{S}'$ is a straight line through the origin. It is a subspace containing Z. $\mathscr{S}' \perp \mathscr{S}$.

3. (a) $(2, -12, 11)$, (c) $(1, -1, 1)$

5. (a) $Z = c(-4, 1, 2)$ for any c

6. (a) false, (c) true, (e) true, (g) false, (i) true

Section 2.4

1. (a) $-3\mathbf{i} - 3\mathbf{j} - 3\mathbf{k}$, (c) $-4\mathbf{i} + 2\mathbf{j}$, (e) 0, (g) $-9\mathbf{i} + 6\mathbf{j} + \mathbf{k}$,
   (i) $9\mathbf{i} - 6\mathbf{j} - \mathbf{k}$, (k) $-10\mathbf{i} - 4\mathbf{j} - 6\mathbf{k}$, (m) $6\mathbf{i} + 6\mathbf{j} + 6\mathbf{k}$

5. $\pm(5\mathbf{i} - 2\mathbf{j} + 3\mathbf{k})/\sqrt{38}$

7. $\pm(4\mathbf{i} - 3\mathbf{j})/5$

12. $\sqrt{77}$

13. (a) $X = 0$

14. $c(2\mathbf{j} - \mathbf{k})$ for any c

CHAPTER 3

Section 3.1

1. (a) linear, (c) not linear, (e) linear, (g) not linear,
   (i) not linear, (k) not linear, (m) linear

2. Image of L is the line $y_2 = 2$.

3. (a) $LX = (3x_1 + x_2, 4x_1 - x_2)$, $L(13, -2) = (37, 54)$

4. (a) No, $L(x_1, x_2) = (x_1 + 4x_2)(2, 3)$ so every $Y \in \mathscr{I}(L)$ has the form $(2c, 3c)$.
   $(1, 1)$ is not of this form.
   (c) The straight line $3y_1 = 2y_2$.

5. (a) $Le_1 = (1, 2)$, (c) $LX = (3, 20)$

6. $[L] = \begin{bmatrix} 1 & -1 \\ 2 & 0 \end{bmatrix}$

7. (a) $Le_1 = 2$, (c) $L(3e_1 + 2e_2) = 0$

8. (a) $[L] = [4 \quad 3]$

9. (a) $Le_1 = \bar{e}_2$, (c) $Le_3 = \bar{e}_1 + \bar{e}_2$

11. For all of these $\mathscr{I}(L)$ is the line $y_1 = 2y_2$.

13. (a) $\{0\}$, (c) line $x_1 = 0$, (e) line $x_1 = x_2$

14.   (a) No. $0 \subset \mathcal{N}(L)$ always, (c) $X = 0$

15.   (a) false, (c) true, (e) false, (g) true, (i) true, (k) true, (m) true,
       (o) false

Section 3.2

1.   (a) linear, (c) linear, (e) not linear, (g) not linear

2. $\mathcal{I}(L)$ is all of $\mathbb{R}$

3.   (a) $Z = (2, -5)$, (c) $[L] = [2 \quad -5]$

4.   (a) the line $2x_1 = 5x_2$, (c) No, $LX = \langle Z, X \rangle$ so if $X \perp Z$ then $X \in \mathcal{N}(L)$.

6.   The sketch is the plane $x_1 + x_2 + z = 0$. Its intersection with $z = 0$ is the
       line $x_1 + x_2 = 0$.

Section 3.3

1.   (a) $3e_1 + e_2$, (c) $8e_1 + 4e_2$, (e) $2e_1 + 4e_2$, (g) $e_1 - 3e_2$,
       (i) $2e_1 + 2e_2$, (k) $2e_1$

2.   (a) $[L] = \begin{bmatrix} 1 & 1 \\ 1 & -1 \end{bmatrix}$, $[M] = \begin{bmatrix} 2 & 1 \\ 0 & 3 \end{bmatrix}$

4.   (a) true, (c) true, (e) true, (g) true

5.   (a) true, $X = L^{-1}(LX) = L^{-1}Y$, (c) true, (e) true, (g) true,
       (i) false, for example $-I$ or M in Ex. 3

6.   (a) $L^{-1}e_1 = -e_2$, $L^{-1}e_2 = e_1$

7.   (a) $X = L^{-1}Y = 1728L^{-1}e_1 - 1984L^{-1}e_2 = -1984e_1 - 1728e_2$

8.   (a) nonsense, (c) sense, (e) nonsense, (g) nonsense

12.   (a) False; $L(x_1, x_2) = (x_2, 0)$ has $L^2 = 0$.
       (c) False; $L(x_1, x_2) = (-x_1, x_2)$ has $L^2 = I$.

Section 3.4

1. $A = \begin{bmatrix} 2 & 0 & 4 \\ -3 & -1 & 144 \end{bmatrix}$

2.   (a) $\begin{bmatrix} 3 & 1 & 6 \\ -3 & 0 & 146 \end{bmatrix}$, (c) $\begin{bmatrix} 1 \\ 0 \\ 0 \\ 1 \end{bmatrix}$

3.   (a) $\begin{bmatrix} 1 & -2 \\ 6 & 5 \end{bmatrix}$, (c) $\begin{bmatrix} 21 & 28 \\ 7 & -7 \\ 7 & 0 \end{bmatrix}$

4.   (a) $\begin{bmatrix} -3 & 5 \\ 3 & 9 \end{bmatrix}$, (c) $\begin{bmatrix} 2 \\ 0 \end{bmatrix}$, (e) $\begin{bmatrix} 1 \\ 3 \end{bmatrix}$

$$\text{(g)} \begin{bmatrix} 1 & -1 & 3 \\ 0 & 0 & 2 \\ 3 & -3 & -3 \end{bmatrix}, \quad \text{(i)} \begin{bmatrix} 5 & 10 & 0 \\ 15 & 5 & 10 \\ -5 & 0 & 10 \end{bmatrix}$$

5. (a) $\alpha = \frac{1}{2}, \beta = -\frac{1}{2}, \gamma = 0, \delta = 1$

6. (a) $\begin{bmatrix} -2 & 1 \\ 1 & 3 \end{bmatrix}$, (c) $[Le_1] = $ first column of $[L]$, (e) $\begin{bmatrix} 4 \\ 5 \end{bmatrix}$

7. (a) $[M] = \begin{bmatrix} 2 & 1 \\ 0 & 3 \end{bmatrix}$

8. (a) $Le_1 = -2e_1 + e_2, \quad Le_2 = e_1 + 3e_2,$
   (b) $L(x_1e_1 + x_2e_2) = x_1Le_1 + x_2Le_2,$
   (c) $y_1 = -2x_1 + x_2, \quad y_2 = x_1 + 3x_2$

9. (a) $[5]^{-1} = [\frac{1}{5}]$, (b) $x = (\frac{1}{5})12 = \frac{12}{5}$,
   (c) all except $[0]$, (d) $A^{-1} = \begin{bmatrix} \frac{1}{5} & 0 \\ 0 & \frac{1}{6} \end{bmatrix}$,
   (e) & (f) $X = A^{-1}(AX) = A^{-1}Y = \begin{bmatrix} \frac{12}{5} \\ -\frac{13}{6} \end{bmatrix}$, (g) $X = 0$

10. (a) $A^{-1} = \begin{bmatrix} -\frac{1}{2} & \frac{1}{2} \\ 0 & 1 \end{bmatrix}$, $C^{-1} = \begin{bmatrix} -\frac{1}{2} & 0 \\ 0 & 1 \end{bmatrix}$,
    (b) $x_1 = (-y_1 + y_2)/3, x_2 = (y_1 + 2y_2)/3$

11. (a) true, (c) true, (e) true, (g) true, (i) false

13. (a) $A^{-1} = \begin{bmatrix} 1 & -2 & -2 \\ 0 & 1 & 1 \\ 0 & 0 & 1 \end{bmatrix}$, (c) $A^{-1} = \begin{bmatrix} 1 & -1 & 1 & -1 \\ 0 & 1 & -1 & 1 \\ 0 & 0 & 1 & -1 \\ 0 & 0 & 0 & 1 \end{bmatrix}$

14. $A^2 = A, B^2 = 0, AB = 0, BA = B, ABA = BAB = (AB)^2 = (BA)^2 = 0,$
    $A^5 = A, (A + 2B)^2 = A + 2B, (3A - B)^2 = 9A - 3B$

18. (a) $C = A^{-1}B = \begin{bmatrix} -1 & -2 \\ 1 & 1 \end{bmatrix}$, (c) $C = 0,$
    (e) $C = A^{-2} = \begin{bmatrix} 7 & -12 \\ -4 & 7 \end{bmatrix}$

19. $A = \begin{bmatrix} 0 & -1 \\ 1 & 0 \end{bmatrix}$, which is a rotation by $\pi/2$ counterclockwise.

Section 3.5

1. (a) not affine, (c) affine, (e) affine, (g) affine (i) affine,
   (k) not affine, (m) affine

2. (a) 6, (c) 7, (e) $L(x_1, x_2) = (-x_1 + 4x_2)$

3. (a) $T(X) = (7 - 3x_1 + x_2)/6$

4. (a) $T(X) = 7 - 8x_1 - 3x_2$, unique; (c) $T(X) = 7 - x_1 + 4x_2$

5.   (a) false,   (c) false,   (e) false,   (g) true,   (i) true,   (k) false

7.   (a) $T(X) = (1 - x_1, -2 + 3x_1 + x_2, 3 - 2x_1 - 3x_2)$, $T(1, 2) = (0, 3, -5)$

CHAPTER 4

Section 4.1

1.   (a) neither,   (c) neither,   (e) affine,   (g) linear (and hence affine too),
           (i) neither,   (k) linear (affine too)

2.   (a) $Z = h(X)$ is a downward opening paraboloid.
           (b) $h(X) = 1$ at $X = (0, 0)$, $h(X) = 0$ on the circle $\| X \| = 1$, $h(X) = -3$
           on the circle $\| X \| = 2$.

4. $f(X) = 0$ on the quarter circle $\| X \| = 1$, $x_1 \geq 0$, $x_2 \geq 0$; hottest at $X = 0$
           where $f(0, 0) = 1$.

Section 4.2

1.   (a) Theorem 1, part 2,   (c) 1

2. $\delta = \frac{1}{70}$

3.   (a) Theorem 1, part 2

4.   (a) 5,   (c) (1, 1, 1)

Section 4.3

1.   (a) $T(x) = 1$,   (c) $T(x) = -3 + 4(x + 2)$

2.   (d) $T(X) = 1$,   (f) $f(X_1) = -1$

3.   (a) $y = 2 + 3x$,   (c) $f(X) = T(X)$

4.   (a) $T(X) = 9 + 2(x_1 - 5) - (x_2 - 3)$,
           (c) $f(X) = \frac{15}{8} + (x_1 + 1)^2 + 2(x_2 - \frac{1}{4})^2$

5.   (a) $y_0 = 9$, $[L] = [2 \quad -1]$,   (c) $T(X) = 2 + 2x_1 - x_2$, $y_1 = 2$,   (e) yes

6.   (a) $T(x) = 1 + 2(x - 1)$,   (d) $T(X) = 1 + 2(x_2 - 1)$

CHAPTER 5

Section 5.1

1.   (a) $\frac{6}{5}$,   (c) $-\frac{1}{18}$,   (e) -1,   (g) $\frac{6}{7}$

2.   (b) $2, 2\sqrt{2}, 2, 0, -2, -2\sqrt{2}, -2, 0$,   (c) increase most rapidly
           $(1, 1)/\sqrt{2}$, decrease most rapidly $-(1, 1)/\sqrt{2}$, not change
           $(-1, 1)/\sqrt{2}, -(-1, 1)/\sqrt{2}$

6.   (a) $\nabla_{-e}f(X) = -\nabla_e f(X)$

Section 5.2

1. (a) $f_1 = x_2^2 + 2$, $f_2 = 2x_1x_2$,

(c) $h_x = (1 + x)e^{x+2y}$, $h_y = 2xe^{x+2y} + 2y$,

(e) $g_\theta = -\sin(\theta - 3\varphi)$, $g_\varphi = 3\sin(\theta - 3\varphi)$,

(g) $g_1 = x_2x_3$, $g_2 = x_1x_3$, $g_3 = x_1x_2$,

(i) $v_x = 2x/(x^2 + y^2 + z^2 + 3)$, $v_y = 2y/(\ldots)$, $v_z = 2z/(\ldots)$

2. (a) $f_{11} = 0$, $f_{21} = f_{12} = 2x_2$, $f_{22} = 2x_1$,

(c) $h_{xx} = (2 + x)e^{x+2y}$, $h_{xy} = h_{yx} = 2(1 + x)e^{x+2y}$, $h_{yy} = 4xe^{x+2y} + 2$,

(e) $g_{\theta\theta} = -\cos(\theta - 3\varphi)$, $g_{\theta\varphi} = g_{\varphi\theta} = 3\cos(\theta - 3\varphi)$, $g_{\varphi\varphi} = -9\cos(\theta - 3\varphi)$,

(g) $g_{11} = g_{22} = g_{33} = 0$, $g_{12} = g_{21} = x_3$, $g_{13} = g_{31} = x_2$, $g_{23} = g_{32} = x_1$,

(i) $v_{xx} = (-4x^2 + 2g)/g^2$, $v_{yy} = (-4y^2 + 2g)/g^2$, $v_{zz} = (-4z^2 + 2g)/g^2$,

$v_{xy} = v_{yx} = -4xy/g^2$, $v_{xz} = v_{zx} = -4xz/g^2$, $v_{yz} = v_{zy} = -4yz/g^2$,

where $g = x^2 + y^2 + z^2 + 3$

3. (a) All are zero except $f_{122} = f_{212} = f_{221} = 2$.

(g) All are zero except those having exactly one x_1, one x_2, and one x_3 derivative, as g_{123}, which all equal 1.

4. $f(x, y) = -x + 3y - 3$

13. (a) $u_t(x, t)$ is the velocity, $u_{tt}(x, t)$ the acceleration, and $u_x(x, t)$ the slope of the point x at time t.

(b) $u(0, t) = u(\pi, t) = 0$, so the end points are fixed. The initial position is $u(x, 0) = 3\sin 2x$, initial velocity $u_t(x, 0) = 0$, while $u_t(\pi/4, 3) = 0$. The slope there is $u_x(\pi/4, 3) = 0$.

Section 5.3

1. (a) $T(x) = 1 + 2(x - 1)$, (c) $T(x) = 1 + (x + 1)$

2. (a) $[f'(X_0)] = [0 \quad 0]$; $f(X) - T(X) = 2(x_1 - 1)^2 + 2x_2^2 = 2\|X - X_0\|^2$

so $\lim_{X \to X_0} \dfrac{f(X) - T(X)}{\|X - X_0\|} = \lim_{X \to X_0} \dfrac{2\|X - X_0\|^2}{\|X - X_0\|} = \lim_{X \to X_0} 2\|X - X_0\| = 0$

(c) $[f'(X_0)] = [1 \quad 0]$; $f(X) - T(X) = (x - 1)^2$

so $\lim_{X \to X_0} \dfrac{f(X) - T(X)}{\|X - X_0\|} = \lim_{X \to X_0} \dfrac{(x - 1)^2}{\|X - X_0\|} = 0$

(note $|x - 1| \le \|X - X_0\|$).

(e) $[f'(X_0)] = [2 \quad -1]$; for the limit, note $2|xy| \le x^2 + y^2$.

Section 5.4

1. (a) $[f'(2, -1)] = [-2 \quad 3]$, (c) $[h'(1, 2)] = [-\frac{1}{18} \quad -\frac{1}{9}]$

(e) $[k'(1, \pi)] = [0 \quad -1]$, (g) $[u'(1, 0, 3)] = [0 \quad 3 \quad 0]$

2. (a) $\frac{6}{5}$, (c) $-\frac{1}{18}$, (e) -1, (g) $\frac{6}{7}$

3. (a) $T(X) = -6 - 2(x_1 - 2) + 3(x_2 + 1)$,

(c) $T(X) = (3 - (x - 1) - 2(y - 2))/18$, (e) $T(X) = \pi - y$,

(g) $T(X) = 3y$

4. (b) $T(X) = 2 - 2(x - 1) - 2(y - 1)$

6. (b) $T(X) = -2(x - 1) + 2(y - 1)$

9. (a) $f(x, y) = \sqrt{x^3 + y^3}$, $X_0 = (1, 2)$, $a \approx 3.01$,
  (c) $f(x, y) = \sqrt{(y/x)}$, $X_0 = (4, 4)$, $c \approx 1.03$,
  (e) $f(X) = \sqrt{x^2 + y^2 + z^2}$, $X_0 = (2, 6, 3)$, $e \approx 6.98$

10. (a) increase: $e = (-2, 5)/\sqrt{29}$; decrease: $e = -(-2, 5)/\sqrt{29}$
  (c) increase: $e = (4, 7, -10)/\sqrt{165}$; decrease: $e = -(4, 7, -10)/\sqrt{165}$
  (e) standard methods fail—but observe $p(x, y) = (x + 1)(y - 1) - 1$
  so by inspection, increase: $e = \pm(1, 1)/\sqrt{2}$;
  decrease: $e = \pm(1, -1)/\sqrt{2}$

CHAPTER 6

Section 6.0

1. true

5. true

7. false, $f(x) = x^4$ is a counterexample

11. true

15. true

19. false

23. true

3. true

9. true

13. true

17. false

21. true

25. true

Section 6.1

1. (a) $(3, 0)$, (c) $(0, n\pi)$, $n = 0, \pm1, \pm2, \ldots$, (e) $(0, 0), \pm(1, 1), \pm(1, -1)$,
  (g) all points on the ξ and η axes, (i) $(-1, 0, 2)$

2. (a) $\varphi(h, k) = h^2 + 2k^2 \geq 0$, local minimum,
  (e) At $(0, 0)$: $\varphi(h, k) = -h^2 - k^2 + h^2k^2 \leq -(h^2 + k^2)/2 \leq 0$ so p has a
  local maximum at $(0, 0)$; at $(1, 1)$: $\varphi(h, k) = hk(h + 2)(k + 2)$ which
  can be either positive or negative so p has a saddle at $(1, 1)$. Similarly,
  p has a saddle at $(-1, -1)$ and $\pm(1, -1)$.
  (g) $\varphi(h, k) = h^2k^2 \geq 0$ so u has local minima on the lines $\xi = 0$ or $\eta = 0$.
  (i) $\varphi(h, k, l) = h^2 + 3k^2 + l^2 \geq 0$ so f has a local minimum at $(-1, 0, 2)$.

3. (a) 2, (c) $(0, 0, \frac{1}{4})$ where distance is $\frac{1}{4}$

Section 6.2

1. (a) $\begin{bmatrix} 2 & -4 \\ -4 & 2 \end{bmatrix}$, (c) $\begin{bmatrix} 9e^{3x-2y} & -6e^{3x-2y} \\ -6e^{3x-2y} & 4e^{3x-2y} \end{bmatrix}$

  (e) $\begin{bmatrix} 0 & x_3 & x_2 \\ x_3 & 0 & x_1 \\ x_2 & x_1 & 0 \end{bmatrix}$

2.   (a) By the Mean Value Theorem, $f(X) = f(X_0)$ for any X, X_0.

        (b) Let $h = f - g$, observe $h' = 0$ and apply (a).

        (c) $f(X) = 3x - 2y - 1$. Unique by (b).

4.   Let $T(X)$ be the best affine approximation to f at X_0, then $z = T(X)$ is the tangent plane at $(X_0, f(X_0))$. We must show $f(X) - T(X) \geq 0$ for all X. But, by Taylor's Theorem,
$$f(X) - T(X) = \tfrac{1}{2}\langle f''(X^*)(X - X_0), (X - X_0)\rangle$$
$$= 2(x - x_0)^2 + 3(y - y_0)^2 \geq 0.$$

Section 6.3

1.   (a) positive definite,   (c) indefinite,   (e) positive non-definite,

        (g) positive non-definite,   (i) indefinite,   (k) positive definite

2.   (a) $(2, 0)$ minimum,   (c) $(-1, 1)$ saddle point,

        (e) $(0, \tfrac{1}{3})$ maximum, $(0, 1)$ saddle,   (g) $(3, 3)$ minimum,

        (i) $(0, 0)$ saddle,   (k) $(0, 1)$ minimum, $(0, -1)$ maximum,

        (m) $(0, 0, -1)$ and $(-1, -1, -1)$ saddles, $(1, -1, -1)$ minimum,

        (o) $(1, 0, 0)$ minimum, $(-1, 0, 0)$ saddle

3. $\varphi(x) \equiv f(x, b)$ has a local minimum at $x = a$. Hence $f_{xx}(a, b) = \varphi''(a) \geq 0$. Similarly, $f_{yy}(a, b) \geq 0$.

5.   See Exercise 4, Section 6.2.

7.   (a) Let $\varphi(t) = f(X + tY)$, then $\varphi'(0) = \langle f'(X), Y \rangle$ by Lemma 1 in Section 6.2. But, $\varphi(t) = \langle X, AX \rangle + 2t\langle AX - Z, Y \rangle + t^2\langle Y, AY \rangle - 2\langle X, Z \rangle$. Thus $\varphi'(0) = 2\langle AX - Z, Y \rangle$ for any Y, so $\langle f'(X) - 2(AX - Z), Y \rangle = 0$ for any Y, which implies $f'(X) = 2(AX - Z)$. At a critical point of f, $AX = Z$.

        (b) *Method 1:* For any Y, $f(X_0 + Y) - f(X_0) = 2\langle Y, AX_0 - Z \rangle + \langle Y, AY \rangle = \langle Y, AY \rangle \geq 0$.

        Method 2: By a computation $f''(X) = A$. Therefore by Taylor's Theorem (about X_0), $f(X) \geq f(X_0)$.

11.   (a) counterexample: $\begin{bmatrix} 2 & -1 \\ -1 & 2 \end{bmatrix}$

Section 6.4

1.   (a) maximum $(3, 4)/5$,   minimum $-(3, 4)/5$,

        (c) maximum $\pm(1, 1)$ and $\pm(1, -1)$,   minimum $(0, 0)$,

        (e) maximum $(2, 3)$,   minimum $(-1, 5)$

2.   (a) $(3, -3)/2$,   (c) $(2, \pm 2, \pm 2)$ and $(-2, \mp 2, \pm 2)$

4. $2' \times 1' \times 1'$

6. $2' \times 2' \times 2'$

8. $x = 12, \ y = 7$

10. $C + F + G = 100$ so   Profit $= 10C + 15F + 12G = 1200 + 3F - 2C$.
C and F must lie in the region $10 \le C \le 20$, $F \le 60$, $C \le F$. To maximize
Profit in this region, we maximize F and minimize C, so $F = 60, C = 10$,
and $G = 30$.

CHAPTER 7

Section 7.0

1.   (a) a circle, radius a, traversed twice counterclockwise beginning at $(a, 0)$.

3. $F(t) = (\cos t, -\sin t, t)$ is one example.

4. $F(t) = (2t^{3/2}, t^3)$. The parabola is $y = x^2/4$. $F(4) = (16, 64)$

6.   many possible answers (a) $F(t) = (t, t)$, $0 \le t \le 1$;
(c) $F(t) = (1 + t, 1 + 3t, 1 - 7t)$, $0 \le t \le 1$

10.   (a) $F(t) = (-2 + t, 1, -1 + t)$

Section 7.1

1.   (a) $F'(t) = (-7, 9)$,   (c) $H'(\theta) = (e^\theta, 3 - 14\theta)$,   (e) $\Phi'(t) = (-1, 4, -9)$,
(g) $Q'(r) = (2r \sin r + r^2 \cos r, 2re^{3r} + 3r^2 e^{3r})$

2.   (a) $T(t) = (1 - 7t, 5 + 9t)$ for $t_0 = 0, 1, -2$,
(c) $T(\theta) = (3, 0) + (1, 3)\theta$,
(e) $T(t) = (3 - t, 1 + 4t, -9t)$ for $t_0 = 0, -1, 3$,   (g) $T(r) = 0$

3.   At $t_0 = 1, G(t) - T(t) = (0, t^2 - 2t + 1)$ so
$$\frac{G(t) - T(t)}{t - 1} = (0, t - 1) \longrightarrow 0 \text{ as } t \longrightarrow 1.$$

4.   (a) $F(0) = (1, 5)$, $F'(0) = (-7, 9) = F'(1)$,
$\| F'(0) \| = \sqrt{130} = \| F'(1) \|$, $F(1) = (-6, 14)$;
(e) $\Phi(0) = (3, 1, 0)$, $\Phi(1) = (2, 5, -9)$, $\Phi'(0) = (-1, 4, -9) = \Phi'(1)$,
$\| \Phi'(0) \| = \sqrt{98} = \| \Phi'(1) \|$

5.   (a) $F''(t) = 0$,   (c) $H''(\theta) = (e^\theta, -14)$,   (e) $\Phi''(t) = 0$,
(g) $Q''(r) = ((2 - r^2) \sin r + 4r \cos r, (2 + 12r + 3r^2)e^{3r})$

6.   (a) $F(t) = (1 + t + t^2, 3 + 3t)$,
(c) $F(t) = (2 + t - 2t^2, 3t + t^2, -2 + t + t^3)$

7.   (a) $(3, 8, 3)$

8.   (a) $F(t) = (6t + 3t^2 + 2t^3, 18t + 9t^2)/6$,
(c) $F(t) = \left(-\frac{11}{6} + 2t + \frac{t^2}{2} - \frac{2}{3}t^2, -\frac{11}{6} + \frac{3}{2}t^2 + \frac{t^3}{3}, \frac{9}{4} - 2t + \frac{t^2}{2} + \frac{t^4}{4} \right)$

9.   (a) $F(t) = \left(2 - 2t^2, t^3, t - t^2 + \frac{t^4}{3} \right)$, $F(3) = (-16, 27, 21)$

10.   (a) $\varphi'(t) = 27t^2 - 32t - 60$

14. (b) $t = 1$: $T(t) = (2, 2)t$ so $x = 2t, y = 2t$; i.e., $x = y$
$t = -1$: $T(t) = (2, -2)t$ so $x = 2t, y = -2t$; i.e., $x = -y$

15. (a) $F(t) =$ constant vector since $f_j'(t) = 0, j = 1, 2, 3$ implies
$f_j(t) =$ constant.
(b) $F(t) = A + Bt$, where A, B are constant vectors, since
$f_j''(t) = 0, j = 1, 2, 3$ implies $f_j(t) = a_j + b_j t$.

18. $G(4) = (5, 12, -8)$

19. Write out everything using components of the vectors.

20. Let $\varphi(t) = \langle Z, F(t) \rangle$ and observe $\varphi(0) = 0, \varphi'(t) = 0$.

22. (a) counterexample: $F(t) = (\cos 2\pi t, \sin 2\pi t)$. Think of a particle moving around a closed path with speed greater than one.
(b) Think of speed less than one. Let $\varphi(t) = \langle F(t), Z \rangle$, where
$Z = F(1) - F(0)$. Then by Schwarz, $|\varphi'(t)| \leq \|Z\|$. Thus by the Mean
Value Theorem, for some t^* between 0 and 1,
$|\langle F(1) - F(0), Z \rangle| = |\varphi(1) - \varphi(0)| = |\varphi'(t^*)| \leq \|Z\|$; that is,
$\|F(1) - F(0)\|^2 \leq \|F(1) - F(0)\|$ so $\|F(1) - F(0)\| \leq 1$.

27. time $= \frac{1}{2} + 7\pi/4$.

Section 7.2

1. $R_{\text{moon}} = (gR_e^2 T^2/4\pi^2)^{1/3} \approx 238{,}000$ miles

6. $v^2 \geq gR_e^2/10^4$, $v \approx 11{,}400$ miles/hour

Section 7.3

1. (a) 15, (c) 12, (e) $\frac{14}{3}$, (g) 36

2. (a) 72

3. $\|P - Q\|$

6. (a) Let $Z = \int_a^b F(t)\, dt$. Since the result is correct if $Z = 0$, we assume
$Z \neq 0$. Then, by Schwarz's inequality,
$$\|Z\|^2 = \langle Z, \int_a^b F(t)\, dt \rangle = \int_a^b \langle Z, F(t) \rangle\, dt \leq \int_a^b \|Z\| \|F(t)\|\, dt$$
$$= \|Z\| \int_a^b \|F(t)\|\, dt.$$
Now cancel $\|Z\|$.

(b) Use part (a) and the integral for arc length.

CHAPTER 8

Section 8.1

1. (a) $\begin{bmatrix} 1 & 1 \\ 1 & -1 \end{bmatrix}$, (c) $\begin{bmatrix} 1 & 0 \\ 0 & 2\pi \end{bmatrix}$, (e) $\begin{bmatrix} -\frac{1}{2} & \frac{1}{2} \\ \frac{1}{2} & \frac{1}{2} \\ 0 & -1/\sqrt{2} \end{bmatrix}$

2.   (a) $T(x_1, x_2) = \begin{bmatrix} 1 + x_1 + x_2 \\ -2 + x_1 - x_2 \end{bmatrix}$,   (c) $T(y_1, y_2) = \begin{bmatrix} y_1 \\ 2\pi y_2 \end{bmatrix}$,

(e) $T(\theta, \varphi) = \dfrac{1}{2} \begin{bmatrix} 1 - \theta + \varphi \\ 1 - \pi/2 + \theta + \varphi \\ \sqrt{2}\,(1 + \pi/4 - \varphi) \end{bmatrix}$

3.   (b) $T(\theta, \varphi) = \begin{bmatrix} 1 \\ \theta \\ \pi/2 - \varphi \end{bmatrix}$,   (c) the plane $x_1 = 2$,

(d) $T(\theta, \varphi) = \dfrac{1}{\sqrt{2}} \begin{bmatrix} \pi/2 - \theta \\ 1 - \pi/4 + \varphi \\ 1 + \pi/4 - \varphi \end{bmatrix}$, the plane is $y + z = \sqrt{2}$

Section 8.2

1.   The following make sense: (a) $G \circ F$,   (c) $F \circ G$,   (e) both

2.   (a) $[F'(X)] = \begin{bmatrix} e^{x+y^2} & 2ye^{x+y^2} \\ 2xe^{x^2+y} & e^{x^2+y} \end{bmatrix}$ $[G'(r, s, t)] = \begin{bmatrix} 1 & 2s & 2t \\ 3r^2 & 1 & 2t \end{bmatrix}$

(b) $F \circ G$ makes sense. $G(-1, 0, 0) = (-1, -1)$,
   $(F \circ G)'(-1, 0, 0) = F'(-1, -1)G'(-1, 0, 0)$

$$= \begin{bmatrix} 1 & -2 \\ -2 & 1 \end{bmatrix} \begin{bmatrix} 1 & 0 & 0 \\ 3 & 1 & 0 \end{bmatrix} = \begin{bmatrix} -5 & -2 & 0 \\ 1 & 1 & 0 \end{bmatrix}$$

3.   (a) At $X = (\pi, 1)$, $u = 0$, $v = -1$. If $h(x, y) = f(u(x, y), v(x, y))$,

$$h'(\pi, 1) = [f_u(0, -1) \quad f_v(0, -1)] \begin{bmatrix} u_x(\pi, 1) & u_y(\pi, 1) \\ u_v(\pi, 1) & v_y(\pi, 1) \end{bmatrix}$$

$$= [1 \quad 0] \begin{bmatrix} -1 & 0 \\ 0 & -1 \end{bmatrix} = [-1 \quad 0]$$

so $h_x(\pi, 1) = -1$, $h_y(\pi, 1) = 0$.

5.   (a) $\varphi'(1) = [f_x(0, -1) \quad f_y(0, -1)] \begin{bmatrix} -2 \\ 2 \end{bmatrix} = 4$

6. $h'(1) = -1$

9.   (b) $u_x = a\varphi'(ax + by)$, $u_y = b\varphi'(ax + by)$ so $bu_x - au_y = 0$.

10.   At $t = 1$, where $X(1) = (0, -1, 1)$ and temperature $f(X(1)) = 2$.

12. $g_s(1, 1) = -2$, $g_t(1, 1) = -2$

17.   (a) $r^2 = x^2 + y^2$ so $r_x = x/r$, $u_x = \varphi'(r)r_x = \varphi'x/r$,
         $u_{xx} = (r^2 - x^2)\varphi'/r^3 + x^2\varphi''/r^2$. u_{yy} is similar.
         Thus, $0 = u_{xx} + u_{yy} = \varphi'' + \varphi'/r$.
         (b) Let $\psi = \varphi'$. Then $r\psi' + \psi = 0$ so $\psi = a/r$, i.e., $\varphi' = a/r$. Thus
         $\varphi(r) = a \log r + b$.

Section 8.3

2.   (a) 0,   (c) 2,   (e) 0,   (g) 0,   (i) $2x$

3.   (a) $(2x, 2y) = \text{grad}\,(x^2 + y^2)$

      (b) If $F = \text{grad}\,\varphi = (\varphi_x, \varphi_y)$, then $p = \varphi_x, q = \varphi_y$ so $p_y = \varphi_{xy} = q_x$.

      (c) 1-a, d, e, i are not irrotational since $p_y \neq q_x$

          (1-b) $(1, 2) = \text{grad}\,(x + 2y)$

          (1-c) $(x, y) = \text{grad}\,(x^2 + y^2)/2$

          (1-f) $(y, -x)/(x^2 + y^2) = \text{grad}\,(\text{arc tan}\,(x/y))$

          (1-g) $(x, -y) = \text{grad}\,(x^2 - y^2)/2$

          (1-h) $(2x, y) = \text{grad}\,(x^2 + y^2/2)$

CHAPTER 9

Section 9.1

14.   (c) (i) 704,   (iii) $\frac{104}{3}$

15.   (b) (i) 64,   (iii) -120

Section 9.2

1.   (a) 9,   (c) $\frac{64}{5}$

2.   (a) $\int_1^3 \left(\int_1^5 xy \, dy \right) dx = \int_1^5 \left(\int_1^3 xy \, dx \right) dy = 48$

      (c) $\int_1^2 \left(\int_{x^3}^8 xy \, dy \right) dx = \int_1^8 \left(\int_1^{y^{1/3}} xy \, dx \right) dy = \frac{513}{16}$

      (e) $\int_{-1}^0 \left(\int_{-x}^1 xy \, dy \right) dx + \int_0^2 \left(\int_{x/2}^1 xy \, dy \right) dx = \int_0^1 \left(\int_{-y}^{2y} xy \, dx \right) dy = \frac{3}{8}$

      (g) $\int_0^1 \left(\int_{x/2}^{2x} xy \, dy \right) dx + \int_1^2 \left(\int_{x/2}^{(x+3)/2} xy \, dy \right) dx + \int_2^3 \left(\int_{2x-3}^{(x+3)/2} xy \, dy \right) dx$

         $= \int_0^1 \left(\int_{y/2}^{2y} xy \, dx \right) dy + \int_1^2 \left(\int_{y/2}^{(y+3)/2} xy \, dx \right) dy + \int_2^3 \left(\int_{2y-3}^{(y+3)/2} xy \, dx \right) dy$

3.   0

5.   (a) 6,   (c) $\frac{4}{15}$

6.   (a) 36,   (c) 0,

      (e) 6π (the $dx \, dy$ integral is most easily evaluated using polar coordinates)

7.   (a) π,   (c) 0,   (e) $\pi(e^{50} - e^8)/16$

8. $\log\left(\frac{3}{2}\right)$, evaluate the x integral first

10. $125\pi/3$, use polar coordinates

11.   (a) $(e - 1)/6$

13. $4\pi abc/3$

16.   (b) use $xy = (x^2y^2/4)_{xy}$

17.   volume $= \pi^2 a^4/2$

19.   (a) 6

Section 9.3

1. $J = 3 \iint_B dA = 3 \text{ (area } B) = 36$

4.   (a) $f'(t) = \int_1^4 x^2 e^{tx^2}\, dx$

   (c) $u'(x) = 2\int_0^x (x - t)\sin t\, dt, \quad u''(x) = 2 - 2\cos x$

5. $v_x(x, t) = 2g(2x + t) - 2g(2x - t),\ v_t(x, t) = g(2x + t) + g(2x - t),$
   $v_{xx}(x, t) = 4g'(2x + t) - 4g'(2x - t) = 4v_{tt}(x, t)$

CHAPTER 10

Section 10.1

1.   (a) 1,   (c) $-12\pi^2$,   (e) $4 - 3\pi/2$,   (g) -15,   (i) 4,   (k) 4

2.   All equal 2. See Theorem 6 in Section 10.3, observing that
   $(6x - y^2,\ -2xy) = \nabla(3x^2 - xy^2)$

4. $C = C_1 - C_{10}$ so $\int_C = \int_{C_1} - \int_{C_{10}} = A - B = 3$

6. $\int_{C_1} = -\int_{C_2} = 2\sin 1$

8.   (a) $-\frac{1}{3}$,   (h) 3,   (j) $-\frac{1}{2}$

9.   work $= 0$

10.   (a) True. Let $x = a,\, y = t$ be a parametrization of C.

   (c) False. Consider $\int_C dx$, where C is $x = -t,\ \ y = 0, \qquad 0 \le t \le 1.$

13. $\frac{25}{6} - 2/\pi$

14.   (a) $\frac{5}{3}$,   (c) $\frac{8}{3}$

Section 10.2

1.   (a) 8π,   (c) 0,   (e) $-\frac{8}{3}$,   (g) 6π

2.   (b) πab,   (c) 24π

4.   0

5.   Apply the Divergence Theorem to B, observing $\partial B = C_2 - C_1$.

8. 216π

9.   (a) $(-1, 0, -1)$,   (c) $(0, 0, -2y)$

13. $-\pi$. Use Stokes' Theorem and integrate over $\partial\Sigma = \{x^2 + y^2 = 1, z = 0\}$, counterclockwise.

17.   144

Section 10.3

1.   (a) $(x^2 + y^2)/2$, $\frac{3}{2}$,   (c) xe^y, 0,   (e) $(x^2 - 7y^2)/2 + xy$, 0

2.   (a) $F(X) = \nabla(x^2/2 + e^x \sin y)$,   (b) $\frac{1}{2}$

3.   (a) independent,   (c) independent,   (e) dependent

5. $F(X) = \text{grad}\,(mM/\|X\|)$

Index